# Biometeorology

Volume 4

**Series Editor**

Glenn McGregor, Department of Geography, University Durham Department of Geography, Durham, UK

*For Instructions to Authors please contact the Publishing Editor: margaret.deignan@springer.com*

Biometeorology has its linguistic origins in Greek; bio refers to life while meteoros makes reference to the study of phenomena near and above the earth's surface. These are the essential elements of biometeorology, such that it is concerned with the interaction between living organisms and variations in atmospheric processes at a range of time and space scales.

Since its emergence as a discipline in the mid 1950s, Biometeorology has grown to have great societal relevance. Much of its subject matter focuses on how biophysical, human, animal and plant systems are influenced by change and variability in the atmospheric environment. Understanding such relationships provides insights into how the provisioning, supporting and regulating services provided by the biosphere may be affected by atmospheric variations and ultimately the societal consequences of atmosphere related changes in biophysical and ecosystem processes.

The purpose of this series in biometeorology is to:

- complement the material published in the International Society of Biometeorology's International Journal of Biometeorology,
- to communicate the interdisciplinary philosophy and science of biometeorology to as wide an audience as possible,
- introduce scientists and policy makers to the societal relevance of and recent developments in its subfields.
- demonstrate how a biometeorological approach can provide entry points to the understanding and possible solution of cross-cutting environmental issues such as climate change, desertification, deforestation, biodiversity loss, land degradation, emerging diseases and the availability of clean water and air.

The series in biometeorology comprises theme-based volumes organised around subject matter within the subfields of biometeorology, which broadly include human, animal, plant, ecosystems, tourism, artificial or indoor climates, air quality and urban. In addition to volumes that synthesise the state of knowledge in the subfields, a primary focus will be on important, new and emerging topics in biometeorology that have a bearing on the health of the biosphere and thus human society. An editorial board, the broad role of which will be to elicit and review suggestions for volume themes, will oversee the series.

More information about this series at http://www.springer.com/series/8101

Eduardo L. Krüger
Editor

# Applications of the Universal Thermal Climate Index UTCI in Biometeorology

## Latest Developments and Case Studies

Springer

*Editor*
Eduardo L. Krüger
Department of Civil Construction
Federal University of Technology
Curitiba - PR, Brazil

Biometeorology
ISBN 978-3-030-76715-0 ISBN 978-3-030-76716-7 (eBook)
https://doi.org/10.1007/978-3-030-76716-7



This Springer imprint is published by the registered company Springer Nature Switzerland AG
The registered company address is: Gewerbestrasse 11, 6330 Cham, Switzerland

*To my beloved father Helmuth, who always instigated me for carrying on with my research, an indefatigable adviser and a true inspiration in all my struggles. In memoriam.*

***Epigraph***

*"Wer nie sein Brot mit Tränen aß,*

*Wer nie die kummervollen Nächte*

*Auf seinem Bette weinend saß,*

*Der kennt euch nicht, ihr himmlischen Mächte."*

*—Johann Wolfgang von Goethe (1749–1832)*

# Preface

This book marks the 10-year anniversary of the outdoor thermal comfort index called Universal Thermal Climate Index 'UTCI'. The planning for putting through a book proposal to Springer's Biometeorological Series started in 2017 and from then on, several actions were taken in order to carry out this mission. In 2019, a meeting was held in Warsaw, Poland, in the shape of an International Conference dedicated to UTCI applications. The conference was organized by the Polish Academy of Sciences, the International Society of Biometeorology, the National Research Institute and the University of Warsaw, with a great deal of effort from Prof. Krzysztof Błażejczyk. The aim of this conference was similar to that of the current book, namely, discussing current achievements in human bioclimatology based on the UTCI. The event gave me the opportunity to have an overview of the diverse applications of the index so far and also to identify potential collaborators for book chapter contributions. As a follow-up of this conference, special issues found place at the International Journal of Biometeorology, Geographica Polonica and Miscellanea Geographica, with papers presented at the Warsaw Conference.

The year 2020, however, was severely struck by the COVID-19 pandemic, which affected in many ways the anticipated start for preparing this book. Throughout the several months living with the pandemic (declared as such by the World Health Organization on March 11, 2020), several articles have been published that draw parallels between the current COVID crisis with the challenges we face ahead with the progress of the climate change crisis. To the already established climate resilience discourse, we will need now to add pandemic resilience as an essential requirement for human settlements. At the moment, there is a debate on which steps need to be taken for humanity to achieve herd immunity, i.e. the benchmark at which a sufficient number of people are immune to the virus to stop its spread. This debate needs also to take into account the efficacy of existing vaccines, the pace of the many vaccination rollouts taking place all over the world versus the rise of new virus variants. It is very likely that COVID-19 will become endemic and stay for a longer time. Therefore, we will need to live with and constantly struggle against both crises (actually, as duly put by E. Felsenthal, in his editorial for the Time Magazine's 1–8 February 2021 issue, the crises we face are manifold and we can talk about a trust crisis, an inequality crisis, an economic crisis and a democracy crisis as well). In the case of the COVID crisis, the multiple applications of the UTCI (as described in this book as well as in the literature) are well aligned with the fields currently addressed by studies on the pandemic (e.g. urban areas, outdoor spaces).

Applications of thermal biometeorological indices such as the UTCI can be strongly associated to research initiatives aimed at climate change mitigation. Included here are urban planning initiatives, health and safety measures for outdoor spaces, improvements in environmental quality and well-being and biometeorology in general, the latter having a precursor role as a discipline that goes beyond single disciplines to safeguard human health.

The book series on Biometeorology launched by Prof. Glenn McGregor in 2009 had among its aims, as well put by Prof. McGregor "to demonstrate how a biometeorological approach can provide insights to the understanding and possible solution of cross-cutting environmental issues", including adaptation to climate change. This book pursues this aim and presents a current snapshot of applications of the UTCI starting with an in-depth literature review on the subject. It would fulfill a broader purpose if applications, approaches and methods presented here would catalyze progress across multiple areas in human biometeorology not only those covered by the book.

The book also showcases research initiatives across a wide range of disciplines and highlights the multidisciplinary nature of what is termed 'Outdoor Thermal Comfort'. The UTCI was the successful result of a collaboration, which was carried out following an interdisciplinary cross-country teamwork approach, though headed and pushed energetically forward by Prof. Gerd Jendritzky under the auspices of the International Society of Biometeorology. The application of the UTCI in different fields of knowledge reinforces the importance of connecting ideas, methods and people beyond single disciplinary boundaries.

Editor

Curitiba, Brazil  Eduardo L. Krüger

# Acknowledgements

I first wish to thank the academic Civil Construction department, the post-graduate department and the directorate of post-graduate studies of the Federal University of Technology of Paraná (Universidade Tecnológica Federal do Paraná—UTFPR) for the concession of the annual leave, without which I would not have been able to carry out this mission. In this regard, I am thankful to my students who were patient enough to cope with a reduced and sporadic communication from my side during the months of preparation of this book. I am indebted to colleagues from my academic department who kindly replaced me in my academic duties during my leave, especially José Alberto Cerri and our visiting professor and my dear friend Eduardo González, who have both taken responsibility for teaching and research during my absence.

I wish to express my gratitude to the chapter contributors that agreed to submit their valuable work to this book in order to cover as much as possible a wide range of applications of the UTCI. Among contributors, I wish to thank sincerely my colleague Peter Bröde from IfADo, for his encouragement and support since the first thoughts and ideas toward starting this project. Peter was also the person who introduced me to the UTCI, bringing to my attention its usefulness and novelty, during the Thermal Comfort Conference held in Windsor, UK, in 2010, at the initial phase of divulging the index to the scientific community. We soon started a very proficuous collaboration that ultimately leads to the elaboration of this book.

I am also greatly indebted to Krzysztof Błażejczyk who not only was enthusiastic about the book project but also kindly invited me to take part of the International Conference UTCI—10 years of applications, which was held in Warsaw, Poland, in May 2019. During that meeting I was able to get an overview of the very diverse applications of the UTCI taking place worldwide and to meet personally Jan Geletič, Michal Lehnert and Claudia Di Napoli, who also contributed to this book with their interesting mapping approach to UTCI applications.

I wish to thank Andreas Wagner and Marcel Schweiker from the Karlsruher Institut für Technologie for their collaboration and willingness to 'lend' us their state-of-the-art climate chamber (the LOBSTER) during the field campaigns carried out by me and my former Ph.D. student Cintia Tamura in Karlsruhe back in 2014/2015.

I would like to thank Charlie Lam, who agreed to share a chapter with me, for the many discussions we had on which methods to take in our analyses, and later during the organization and the many revisions of the chapter.

I am thankful to Katerina Pantavou, who agreed to prepare a chapter for this book. As a follower of her research output on outdoor thermal comfort, duly cited in several instances of this book, I was honored to receive her chapter contribution.

I am indebted to Kevin Lau, who was also an attendee at the Warsaw Conference and who I had the chance of welcoming in Brazil for starting a collaborative work on the global database of outdoor comfort studies, presented in one of the chapters of the book.

To my colleagues from Brazil who agreed to share their data from outdoor comfort surveys, my sincere thanks. Ivan Callejas was an indefatigable collaborator, revising parts of the text and providing useful advice. Solange Leder, Simone Hirashima, Patricia Drach and students provided useful data for analysis and kindly disclosed details on their databases. I wish to thank Eduardo Grala da Cunha and Luisa Alcantara for their willingness to conduct additional surveys, for their trust in my distant guidance during that process and for the careful revision of their data.

I would like to express my gratitude to the publishing team at Springer for the constant technical support I got from them during the elaboration of this book. Particularly, I wish to thank Anthony Doyle, from Springer UK, who contacted me back in 2017 with a first invitation to submit a book to Springer.

Last but not least, I need to thank my family for the many hours devoted to this project, while staying right next to them during the very long and everlasting home-office period amidst the COVID-19 pandemic.

This work was supported by the Brazilian research funding agency CNPQ (Conselho Nacional de Desenvolvimento Científico e Tecnológico), who sponsored my 6-month stay in Karlsruhe for elaborating most parts of the book.

# Contents

# Chapter 1
# Introduction

**Eduardo L. Krüger**

**Abstract** The outdoor thermal comfort index termed 'Universal Thermal Climate Index' or 'UTCI' was launched in 2009 as successful output of interdisciplinary work performed by a group of experts that composed the European COST Action 730, Commission 6, within the framework of the International Society of Biometeorology (ISB). COST Action 730 was centered on the development of a physiological assessment model of the thermal environment in order to significantly enhance applications related to health and well-being in the fields of public weather service, public health systems, precautionary planning and climate impact research. After a full decade from the finalization and official launch of the UTCI, it is timely relevant to check on the progress and current applications of the index so far. UTCI is to be included in the revision of one of the guidelines of the Association of German Engineers (Verein Deutscher Ingenieure 'VDI'), VDI Guideline 3787-2, which is aimed at providing "evaluation methods of human biometeorology as a standard for taking into account climate and air quality (bioclimate) in relation to man in overall physical planning". The chapter describes motivations that lead to the organization and writing of the book, putting this in the context of the COVID-19 pandemic.

**Keywords** UTCI · Outdoor thermal comfort · Pandemic · Climate change

## 1 Context

The organization and writing of this book were done during the Coronavirus disease (COVID-19) pandemic. One of the side-effects of the spread and the speed with which this disease propagated around the globe was the sheer uncertainty it created in many aspects of life. The pandemic installed an 'Age of Uncertainty' that affected economies worldwide.

E. L. Krüger (✉)
Departamento de Construção Civil, Universidade Tecnológica Federal do Paraná - UTFPR, Campus Curitiba - Sede Ecoville, Rua Deputado Heitor Alencar Furtado, 4900, 81280-340 Curitiba, Brazil
e-mail: ekruger@utfpr.edu.br



E. L. Krüger (ed.), *Applications of the Universal Thermal Climate Index UTCI in Biometeorology*, Biometeorology 4,
https://doi.org/10.1007/978-3-030-76716-7_1

Uncertainties regarding economic development increased alongside uncertainties in the political realm, including the effectiveness of implemented policies such as the duration and extent of social distancing. Altig et al. (2020) analyzed a variety of economic uncertainty measures during the pandemic in 2020. They all showed that the COVID-19 economic crisis exhibits an unprecedented nature and the uncertainty generated by it has skyrocketed in its wake. Uncertainties were also observed in the case of the controversial policy of the closing of schools and the associated risk of transmission in classroom settings vis-à-vis the adverse impacts on academic progress from distance learning. So far, little evidence has been found to corroborate that schools have contributed meaningfully to increased community transmission (Honein et al. 2021; European Centre for Disease Prevention and Control ECDC 2020). The ECDC December 23 Report recommends that school policies should instead aim at non-pharmaceutical interventions such as physical distancing that prevent crowding as well as hygiene and safety measures for preventing transmission.

Uncertainties also arise with respect to our own personal development: impacts on work and academic productivity, well-being and mental state during a lockdown, among others.

Intrinsic uncertainties naturally exist regarding the virus itself, its lethality, infectiousness and resilience over time and more recently regarding its progress, with more harmful and contagious variants found in different parts of the world. Add to that, the uncertain containment of the disease amidst the many vaccination rollouts, taking place at the current time in various countries.

With respect to the subject matter of this book, many of the motivations and priorities we had set before the pandemic have changed with the advance of the disease. It may at first seem that the current global COVID-19 crisis has dwarfed the common goal of promoting climate change resilience in cities. Indeed, the rhythm of research carried out on topics related to the book contents got severely disrupted. Yet, during the pandemic, summer periods present enhanced and concomitant risk factors amidst the pandemic and heat stress poses additional challenges to the public health system (Martinez et al. 2020). This multi-hazard crisis requires governments to plan policies for mitigating the risk of both heat exposure and COVID-19 transmission. In particular, government initiative and community-level measures (Shumake-Guillemot et al. 2020) are warranted to minimize the risk from COVID-19 transmission and heat exposure among vulnerable populations, such as the elderly and children. Traditional heat intervention strategies (e.g. cooling shelters) can become infeasible during COVID-19. Therefore, it is vital to alleviate thermal discomfort at an early stage to prevent future severe heat-related illnesses.

In the aftermath of the pandemic, it might well be that new, unforeseen applications of the outdoor thermal comfort indices like the Universal Thermal Climate Index 'UTCI' will arise. Possible future applications could then be related to improvements of hygiene and safety conditions in urban outdoor spaces and how such spaces should be designed to curb future outbreaks. Strategies such as the provision of natural ventilation within building arrays as well as avoiding unnecessarily constrained spaces and encroachments in urban areas might become in vogue requiring thermal comfort analyses for such structural modifications of the urban realm.

Developments such as the current rate of urbanization and trends toward more compact cities might undergo a rethinking process. The urban character of the pandemic and the need to reduce agglomerations in urban areas might attract more people to small towns and mid-sized cities. Locally, outdoor recreational spaces (public squares and parks) might go through a revival. Such rediscoveries of rural life and outdoor areas can have an impact on microclimate and will therefore require thermal comfort assessments associated with those changes. Tourism in National Parks and nature reserves, one of the application areas of the UTCI identified in the literature review, will most likely call for dedicated thermal comfort assessments in order to guide mobility and to identify risk areas in connection with heat waves and global warming. As correctly put by Więckowski (2020), from every crisis an opportunity arises for reflection and for change and there are noticeable signs that the pandemic has already performed substantial changes in the Anthropocene, with nature benefiting rather tangibly and rapidly with improvements in air and water in cities and in wildlife in National Parks.

The area termed human biometeorology will also be affected by the "new normal" after the pandemic. Despite its remote origins during the times of Hippokrates in ancient Greece (Höppe 1997), the science of biometeorology formally started as an interdisciplinary area in the 1960s and was later subdivided into three specialized sub-fields (animal, plant and human biometeorology) and in more recent decades into further branches (Burton et al. 2009). Defined as the science of the influences of the atmospheric environment on man (Höppe 1997), at least for the human sub-field biometeorology will need to adapt to new potential usages after the pandemic, opening new avenues of interdisciplinary research that deal with thermal comfort, boundary-crossing with hygiene and safety measures.

Parting from the basic question posed to the field of biometeorology in the ISB website "How does weather and climate impact the well-being of all living creatures?" and looking at the study areas proposed for that field of investigation, the years to come will witness an increase in research focusing in well-being, morbidity, mortality of humans that face extreme weather and inadvertent microclimatic changes resulting from urbanization (albeit now additionally aimed at diminishing the risk of new contagions) next to changes in land use in non-urban areas (e.g. by deforestation, irrigation schemes etc.).

In urban planning, climate considerations will be required alongside post-COVID planning measures. In a recent review paper of research published during the pandemic, the environmental quality of cities was found to be a dominant subject matter (Sharifi and Khavarian-Garmsir 2020). Overall, an optimistic future for research and actions toward environmentally sound city planning was envisaged by the authors as the current COVID crisis brought to life the issue of urban vulnerability to pandemics. While partial and total lockdowns pointed to improvements in air quality in cities due to reductions in traffic-related pollution, there is an apparent direct relationship between pollution and the spread of the pandemic. Regarding climatic parameters, the reviewed studies showed conflicting outcomes for the relationship between air temperature and the COVID-19 confirmed cases. An evident advantage of promoting higher wind speeds in urban areas is noted, specifically in reducing air pollutants and consequently diminishing the spread. The renewed interest in these topics adds thus pandemic resilience to the climate change discourse towards adaptation measures in cities. As pointed out by Allan and Jones (2020), it is important to "enable resilience strategies to not only in respect of addressing climate change, but also regarding pandemic preparedness and mitigation because these have also proven to be disastrous to our current and rapidly urbanized world".

An attempt towards better planning of urban areas in view of the 2003 SARS mini pandemic (Cherry 2004) was the launch of an expert task team to propose urban design guidelines for Hong Kong (Ng 2009). Due to the air-borne spread of SARS (Severe Acute Respiratory Syndrome) and to the congested urban conditions of hot-humid Hong Kong, guidelines focused primarily on ventilation strategies with the development of a technical methodology of an air ventilation assessment system. In the context of the current pandemic, in dealing with an airborne transmission of a deadly virus, the mission of that expert group "how to design and plan the city fabric for better natural air ventilation?" might grow in importance in many other urban centers worldwide.

Still, one of the main challenges facing humanity today is climate change. During the present COVID-19 crisis, similarities have been found between both challenges as they both affect people globally, such as time-lagged response to crises, irreversible changes they cause with unforeseeable consequences, social and spatial inequality, weakening of international solidarity and the fact that it is less costly to prevent than to cure (Manzanedo and Manning 2020). In terms of the latter, climate-change mitigation measures seem to be far more economically sound than adaptation, although both will certainly be required. Associated costs, however, depend on how we quantify the cost of adaptation, and the types of adaptation we consider. Many adaptation studies use global RCP (Representative Concentration Pathways) or SSP (Shared Socio-economic Pathways) scenarios. This type of adaptation is known as planned adaptation, which include government policies and heat-health warning systems. However, adaptation might also occur through autonomous adaptation, such as acclimatization and behavioral adjustment such as changes in clothing choices or spending less time outdoors during hot weather. For example, Gosling et al. (2017) combined both planned and autonomous adaptation and examined their impact of future heat-related mortality. Without considering autonomous adaptation in the equation, an overestimation of the future effects of heat on mortality is likely to occur (Huang et al. 2011).

As adequately put by Manzanedo and Manning (2020), in both the pandemic and the climate crises it becomes evident the need for early action and prevention. Other than considering that climate change matters are less important than pressing concerns related to the pandemic, the current crisis offers insights for more adequately managing climate change. Similar comparisons between both crises are presented by Klenert et al. (2020), again reinforcing the need for consistent policies actions in fighting them and reminding that delay in dealing with them is usually costly. Yet, in drawing such parallels between the two crises, we need to recognize that climate change is much more difficult to confront than the current pandemic. Those authors conclude that "drawing the right lessons from this crisis [the pandemic] will prepare policy-makers and citizens for the long-term challenges presented by climate change." That line of thought seems to be accepted by the scientific community and several other authors (e.g. Watts et al. 2020) agree that relevant lessons can be drawn from the actual crisis, which are applicable to climate change management.

## 2 Climate Change and Human Thermal Comfort

Climate change is a wicked problem, that is, one that is difficult or even impossible to be solved. It would be helpful to highlight the co-benefits in addressing climate change and pandemics such as COVID-19, in order to facilitate international climate change mitigation efforts. One example of the literature on relationships between human health and climate change mitigation strategies is given by Smith et al. (2014). Those authors evaluate direct, ecosystem-mediated and human-mediated impacts of climate change on health outcomes, suggesting that climate change will likely and with very high confidence exacerbate health problems that already exist. Working or spending more time outdoors, for example, will be severely compromised with rising temperatures and high humidity. At least in the case of air pollution, there might be co-benefits with "health gains from strategies that are directed primarily at climate change, and mitigation of climate change from well-chosen policies for health advancement" (Smith et al. 2014).

Public awareness and media coverage have increased substantially with climate-related events becoming more and more frequent over the last decades. A few months before the outbreak, in September 2019, TIME magazine devoted an entire issue to climate change with the hope that this was an occasion we could all rise to. Despite the disruption in the climate change discourse caused by the pandemic, it is still a major issue in human biometeorology and the contents of this book are in some way related to it. According to the last Intergovernmental Panel on Climate Change (IPCC) Report, published in 2015 (the next one is due 2022), human influence on climate is proven, clear and growing with extended impacts on human and natural systems. The 2015 Report states that it is very likely that the number of cold days and nights has decreased and that of warm days and nights has increased globally and there is "medium confidence that the observed warming has increased heat-related human mortality and decreased cold-related human mortality in some regions" (IPCC 2014). Regarding human biometeorology, as stated in the IPCC Report, deleterious impacts of such climate extremes include human morbidity and mortality with consequences for mental health and human well-being, which can be accentuated at the lowest levels of socio-economic development. At present, the climate change discourse broadly justifies any struggle to study urban climate and outdoor thermal comfort, which can be supported by indices such as the UTCI.

It is unlikely that mankind will be able to keep global warming lower than 2 °C relative to pre-industrial levels as proposed by the Paris Agreement and many authors such as Hanna and Tait (2015) rely on human thermoregulation and acclimatization to cope with rising temperatures, particularly in cities. However, physiology alone will not be sufficient for ensuring comfort conditions outdoors, so that behavioral and technological adaptation measures (Auliciems 2009) will also be required. Matters related to outdoor thermal comfort are thus of multi and interdisciplinary nature. The area covers a wide range of disciplines, including but not limited to physiology, health and behavioral sciences, occupational health, urban planning and urban design, landscape architecture, meteorology and urban climatology. Also lying within the research area called 'outdoor thermal comfort' or OTC research, in connection to climate change and urbanization, health aspects deserve full attention and are gaining momentum amidst health professionals and policy makers toward heat adaptation and adaptation to extreme events such as heat waves (McGregor and Vanos 2018). Particularly vulnerable populations and professionals who are frequently exposed to the outdoor environment are more prone to heat-induced illnesses and combined risks of multi-focal environmental aspects (thermal stress, urban noise, pollutants, e.g. Candas and Dufour 2005).

In this context, thermal comfort indices are aimed at diverse applications in outdoor spaces, providing relevant information to planners who seek more adequate conditions for humans as well as an integrated assessment of perceived thermal comfort for several issues related to health and safety and biometeorology (as in the case of heat waves or future climate projections, for example). The rationale for developing integrated energy-balance approaches for existing meteorological variables instead of using single variables was already introduced by Büttner in the 1930s and the advantages of such indices became evident (Höppe 1997). Such indices combine relevant meteorological data like air temperature and humidity, wind velocity and solar radiation into a single value, which can be then translated to categorical scales expressing the predicted degree of human thermal stress and serve as important indicators for weather forecasting systems. Advanced applications include urban climatology, where index results can be used as important indicators of advantages or disadvantages arising from the implementation of changes in urban morphology, vegetation and water bodies.

As mentioned above, the subject matter is also relevant to human health research. As relevant support to studies involving different adaptive opportunities (physiological, psychological, social and behavioral) of humans in outdoor spaces, comfort/stress indices can also gauge the extent of such adaptations, such as assessing the impact of a continuous usage of air-conditioned indoor spaces on tolerance of pedestrians to heat or cold stress. Another important application is the possibility of better assessing thermal stress levels in particular locations accounting for target age and gender groups, metabolic rate and clothing insulation. Again, hygienic needs should go hand in hand with the thermal component in decision making processes of planners (Höppe 1997).

# 3 UTCI

The Universal Thermal Climate Index 'UTCI' was developed within the framework of the International Society of Biometeorology (ISB), who in 2000 established the UTCI ISB Commission 6. Prior to that, during the International Congress of Biometeorology in Sydney, Australia, 1999, first discussions took place toward the elaboration of such index. In 2004, the idea of a new COST Action on the UTCI was launched, which was subsequently approved as COST Action 730.

COST Actions (European Cooperation in Scientific and Technical Research) are a networking instrument for researchers, engineers and scholars to cooperate and coordinate nationally funded research activities. They allow European researchers to jointly develop their own ideas in any science and technology field.

The COST Action 730 was centered on the development of a physiologically relevant model of the thermal environment in order to significantly enhance applications related to health and well-being in the fields of public weather service, public health systems, precautionary planning and climate impact research. Started in 2005 and finalized in 2009, it involved researchers from 19 European countries and also from Israel, Canada, New Zealand and Australia. During its elaboration, both the UTCI expert group and the ISB defined a methodological framework for using the UTCI, which involved the preparation of input meteorological variables, calculation procedure and assessment scale referring to thermal stress and physiological reactions of the human organism in actual weather conditions under given assumptions, i.e. walking at 4 km/h (with a metabolic rate of 135 W/m$^2$) and wearing clothing adjusted to prevailing thermal conditions. After its conclusion, the index was presented at several conferences and in two special issues, respectively, of the International Journal of Biometeorology (vol. 56, 2012) and Geographia Polonica (vol. 86(1), 2013).

What is the UTCI? In their executive summary, the expert group of COST Action 730 defines the index as the air temperature of the reference condition causing the same thermophysiological model response as the actual condition. The equivalence between the actual and the reference thermal environment is based on the dynamic physiological response between both. A computed UTCI value or data point is an equivalent temperature obtained from a set of air temperature, radiation, wind and humidity data that corresponds to the air temperature in the reference condition of radiation, humidity and wind speed, both producing the same thermal strain condition. The associated assessment scale was developed from the simulated physiological responses comprising 10 different categories ranging from extreme cold stress to extreme heat stress (Table 1).

In addition, the UTCI band between 18 and 26 °C, that range closely corresponds to the definition of the "thermal comfort zone" defined in the Glossary of Terms for Thermal Physiology (2003) as: "the range of ambient temperatures, associated with specified mean radiant temperature, humidity, and air movement, within which a human in specified clothing expresses indifference to the thermal environment for an indefinite period".

**Table 1** UTCI assessment scale—UTCI categorized in terms of thermal stress (adapted from Bröde et al. 2012)

| UTCI (°C) range | Stress category |
|---|---|
| Above +46 | Extreme heat stress |
| +38 to +46 | Very strong heat stress |
| +32 to +38 | Strong heat stress |
| +26 to +32 | Moderate heat stress |
| +9 to +26 | No thermal stress |
| +9 to 0 | Slight cold stress |
| 0 to −13 | Moderate cold stress |
| −13 to −27 | Strong cold stress |
| −27 to −40 | Very strong cold stress |
| Below −40 | Extreme cold stress |

The UTCI is based on the advanced multi-node 'Fiala' thermoregulation model coupled with a state-of-the-art clothing model that accounts for the behavioural adaptation of clothing insulation to ambient temperature. Fiala's multi-node thermoregulation model simulates human heat transfer processes within the body and at its surface, and was perfected after several validation studies took place for a wide range of thermal exposure conditions (Fiala et al. 2003).

From the three interdisciplinary working groups (WGs) formed during the development of the index and described in the executive summary, one of them was devoted to applications of the UTCI. The basic task of WG3 (Applications) was to bridge the gap between science and application. At that point, applications of existing thermal comfort and thermal sensation indices, particularly PT, SET* and PET in diverse field were identified, but not of the UTCI. After the project's completion, dissemination of results took place in short-term scientific missions of young researchers, presentations at several conferences and in a training school for young scientists. The final outcome was presented in a symposium with the World Meteorological Organization (WMO).

Due to the time-consuming calculation of this equivalent temperature, faster options were considered for the calculation of UTCI data, among them look-up tables with pre-calculated UTCI values for various combinations of climatic parameters and a polynomial regression function, which predicts the equivalent temperature values for those same exposure conditions (Bröde et al. 2012).

Several studies published on the UTCI pointed to a number of methodological issues that still needed to be clarified. Bröde (2019) showed an account of caveats of the UTCI, which are related to calculation procedures, meteorological input data and the proposed thermal comfort assessment scale for the UTCI. One of the limitations refers to the procedure adopted for assessing UTCI values, whether using interpolations from look-up tables or the polynomial regression function, when compared to UTCI-Fiala model calculations. The range of meteorological parameters for obtaining valid UTCI data is well presented by Bröde et al. (2012) as well as the procedures to be taken for using consistent input data. Within those boundaries, UTCI calculations should be reliable within acceptable error margins.

The assessment scale proposed for the UTCI (Table 1) is asymmetrical as the diverse categories are based on limits for thermoregulatory variables and effector functions (human physiology). Thus, when defining the thermal comfort zone for a given population, the limits can deviate from the described 'no thermal stress' range, in order to account for regional adaptation to local climatic conditions. The associated index Dynamic Thermal Sensation (DTS) (Fiala et al. 2003) is, in this case, a more adequate predictor for the thermal comfort sensation.

In his presentation, Bröde (2019) listed the main recommendations for a correct application of the UTCI, some of which have been mentioned here. Developments of the index have also been proposed such as adjustments in calculated UTCI data to account for modified clothing, activity level and exposure duration. Presently, UTCI is about to be included in the revision of one of the guidelines of the Association of German Engineers (Verein Deutscher Ingenieure 'VDI') VDI Guideline 3787-2 and an R package is to be launched with the polynomial function, the look-up table, and a DTS calculator.

## *3.1 The International Conference UTCI—10 Years of Applications*

In May 2019, a conference was held in Warsaw, Poland, "UTCI—Assessment Measure in Human Bioclimatology—10 Years of Applications" alongside the "1st European Biometeorologist Regional Meeting". The conference had the aim to celebrate the 10th anniversary of the UTCI, showcasing diverse areas of application of the index in order to discuss achievements in human bioclimatology based on the UTCI and to review the progress in biometeorological research in Europe. The conference was organized by the Institute of Geography and Spatial Organization, Polish Academy of Sciences, by the International Society of Biometeorology, by the Institute of Meteorology and Water Management National Research Institute and by the University of Warsaw, Faculty of Geography and Regional Studies. During that conference, a plethora of presentations covered diverse application areas of the UTCI, such as occupational thermal stress, physiology, forecasting models, health-related aspects of climate change, urban climate, bioclimate assessment, weather extremes and heat waves.

As a follow-up to that conference, two special issues have been launched, by the International Journal of Biometeorology (IJBM) and Miscellanea Geographica, respectively and another special issue is currently in progress at Geographia Polonica. Those special issues are a complement to the literature review shown in the next chapter (Chap. 3: Literature Review on UTCI Applications).

## 4 Chapters Distribution

This book is structured as follows.

This chapter gives an Introduction on the subject, relating the current pandemic to the subject matter, summarizing the UTCI and new developments and updates to it.

In Chap. 2, Peter Bröde summarizes calculation procedures for obtaining UTCI data, limitations noticed and modifications performed, presenting the way forward for further improvements for the assessment of UTCI output.

Chap. 3 presents a comprehensive literature review of peer-reviewed papers on UTCI with special focus on its applications. The chapter is a mix between a basic bibliometric analysis and a standard literature review on relevant papers, with the aim of categorizing applications of the index.

In Chap. 4, Peter Bröde and colleagues compare predictions of thermal sensation, expressed as the Dynamic Thermal Sensation (DTS) from the UTCI-Fiala model of human thermoregulation, to thermal sensation votes as recorded on the 7-unit ASHRAE scale for two Brazilian cities. Field data from outdoor comfort surveys are used for that purpose. Aim is to evaluate the influence of biometrical data (age, sex, body composition), site morphology (open space, street canyon), climatic state (comfort/discomfort) and clothing choice on the model's predictive power.

In Chap. 5, Charlie Lam and colleagues look at human acclimatization in the short term and in the long term, which potentially affect reported thermal sensation. Instead of testing the sensitivity of the index as in the preceding chapter, the UTCI and, more precisely, the DTS is used here as a reliable reference for comparisons between subgroups of respondents using field data from a longitudinal study and from a cross-sectional study.

In Chap. 6, we investigate an important component of outdoor thermal perception, namely, the adaptation of local populations to prevailing climatic conditions at a given location. This factor has been noticed in several studies, as pointed out in the literature review, and many of them attempted to calibrate comfort and thermal stress ranges of the UTCI in order to account for regional adaptation. Field data refer to locations in Brazil and encompass a broad latitude range and various climate types.

In Chap. 7, Katerina Pantavou analyzes adverse health effects arising from heat-related thermal discomfort thereby examining the susceptibility of pedestrians to outdoor thermal environment, assessed using the UTCI.

The subsequent three chapters refer to UTCI mapping with special focus on urban planning. In Chap. 8, Krys Błażejczyk and Anna Błażejczyk present an overview of mapping approaches for the UTCI at different spatial scales. In Chap. 9, Jan Geletič and colleagues present potential applications of the UTCI in urban planning for Prague, in the Czech Republic. Authors evaluate a selected neighborhood in terms of heat stress and outdoor thermal comfort in streets and courtyards, using a city-scale model (PALM).

In Chap. 10, Claudia Di Napoli and colleagues describe the UTCI-based forecasting systems developed in Czech Republic, Italy, Poland, Portugal and at the pan-European scale. The chapter describes their characteristics and discusses their potential as warning systems for thermal hazards.

Finally, in Chap. 11, Kevin Lau presents the rationale and current stand of the proposed repository for a worldwide outdoor thermal comfort database, which can facilitate diverse comparisons between locations in terms of outdoor comfort data and thereby disseminate the usage of the UTCI.

## References

Allan Z, Jones DS (2020) Pandemic stricken cities on lockdown. Where are our planning and design professionals [now, then and into the future]?. Land Use Policy 97:104805

Altig D, Baker S, Barrero JM, Bloom N, Bunn P, Chen S, Thwaites G (2020) Economic uncertainty before and during the COVID-19 pandemic. J Public Econ 191:

Auliciems A (2009) Human adaptation within a paradigm of climatic determinism and change. In: Biometeorology for adaptation to climate variability and change. Springer, Dordrecht, pp 235–267

Bröde P, Fiala D, Błażejczyk K, Holmér I, Jendritzky G, Kampmann B, Havenith G (2012) Deriving the operational procedure for the Universal Thermal Climate Index (UTCI). Int J Biometeorol 56(3):481–494

Bröde P (2019) UTCI application—basics, caveats, new developments. In: International Conference UTCI—assessment measure in human bioclimatology—10 years of application. Warsaw, 22–23 May 2019

Burton I, Ebi KL, McGregor G (2009) Biometeorology for adaptation to climate variability and change. In: Biometeorology for adaptation to climate variability and change. Springer, Dordrecht, pp 1–5

Candas V, Dufour A (2005) Thermal comfort: multisensory interactions? J Physiol Anthropol Appl Human Sci 24(1):33–36

Cherry JD (2004) The chronology of the 2002–2003 SARS mini pandemic. Paediatr Respir Rev 5(4):262–269

European Centre for Disease Prevention and Control. COVID-19 in children and the role of school settings in transmission–first update. 23 Dec 2020. Accessed 3 Feb 2021. https://www.ecdc.europa.eu/en/publications-data/children-andschool-settings-covid-19-transmission

Fiala D, Lomas KJ, Stohrer M (2003) First principles modeling of thermal sensation responses in steady-state and transient conditions. ASHRAE Trans 109:179

Glossary of terms for thermal physiology (2003) Third Edition revised by the commission for thermal physiology of the international union of physiological sciences (IUPS Thermal Commission). J Therm Biol 28:75–106

Gosling SN, Hondula DM, Bunker A, Ibarreta D, Liu J, Zhang X, Sauerborn R (2017) Adaptation to climate change: a comparative analysis of modeling methods for heat-related mortality. Environ Health Perspect 125(8):

Hanna EG, Tait PW (2015) Limitations to thermoregulation and acclimatization challenge human adaptation to global warming. Int J Environ Res Public Health 12(7):8034–8074

Honein MA, Barrios LC, Brooks JT (2021) Data and policy to guide opening schools safely to limit the spread of SARS-CoV-2 infection. JAMA

Höppe P (1997) Aspects of human biometerology in past, present and future. Int J Biometeorol 40(1):19–23

Huang C, Barnett AG, Wang X, Vaneckova P, FitzGerald G, Tong S (2011) Projecting future heat-related mortality under climate change scenarios: a systematic review. Environ Health Perspect 119(12):1681–1690

IPCC (2014) Climate change 2014: synthesis report. contribution of working groups I, II and III to the fifth assessment report of the intergovernmental panel on climate change. In: Pachauri RK, Meyer LA (eds) Core writing team. IPCC, Geneva, Switzerland, p 151

Klenert D, Funke F, Mattauch L, O'Callaghan B (2020) Five lessons from COVID-19 for advancing climate change mitigation. Environ Resource Econ 76(4):751–778

Manzanedo RD, Manning P (2020) COVID-19: lessons for the climate change emergency. Sci Total Environ 742:

Martinez GS, Linares C, De'Donato F, Diaz J (2020) Protect the vulnerable from extreme heat during the COVID-19 pandemic. Environ Res 187:109684

McGregor GR, Vanos JK (2018) Heat: a primer for public health researchers. Public Health 161:138–146

Ng E (2009) Policies and technical guidelines for urban planning of high-density cities–air ventilation assessment (AVA) of Hong Kong. Build Environ 44(7):1478–1488

Sharifi A, Khavarian-Garmsir AR (2020) The COVID-19 pandemic: impacts on cities and major lessons for urban planning, design, and management. Sci Total Environ 142391

Shumake-Guillemot J, Amir S, Anwar N, Arrighi J, Böse-O'Reilly S, Brearley M, Zaitchik B (2020) Protecting health from hot weather during the COVID-19 pandemic. http://www.ghhin.org/assets/technicalbrief-COVID-and-Heat-final.pdf

Smith K, Woodward A, Campbell-Lendrum D, Chadee D, Honda Y, Liu Q, Butler C (2014) Human health: impacts, adaptation, and co-benefits. In: Climate change 2014: impacts, adaptation, and vulnerability. Part A: global and sectoral aspects. Contribution of working group II to the fifth assessment report of the intergovernmental panel on climate change. Cambridge University Press, pp 709–754

Watts N, Amann M, Arnell N, Ayeb-Karlsson S, Beagley J, Belesova K, Costello A (2020) The 2020 report of The Lancet Countdown on health and climate change: responding to converging crises. The Lancet

Więckowski M (2020) Can a pandemic stop or slow the Anthropocene? Geographia Polonica 93(4):473–492

# Chapter 2
# Issues in UTCI Calculation from a Decade's Experience

**Peter Bröde**

**Abstract** Computing values of the Universal Thermal Climate Index (UTCI) from meteorological input of air temperature, wind speed, humidity and mean radiant temperature is straightforward thanks to the publicly available simple calculation algorithms provided by the operational procedure. These have triggered numerous UTCI implementations, especially relying on the regression polynomial following its publication. Nevertheless, several issues that will require consideration for facilitating successful UTCI application have emerged from a decade's experience in counselling UTCI users. These comprise the huge errors introduced when extrapolating the regression function beyond its domain of definition; the question on how to upscale wind speed input to the required height of 10 m above ground; and the interpretation of the asymmetrical UTCI assessment scale in terms of physiology and thermal comfort. This chapter provides hints and guidelines on how to handle these issues, and especially encourages the application of the hardly used look-up table approach, which will help avoiding many, if not all concerns related to UTCI calculation via the regression polynomial.

**Keywords** UTCI calculation · Algorithms · Data table · Wind speed · Assessment scale · Thermal comfort

## 1 Introduction

A decade after its first release, the Universal Thermal Climate Index (UTCI) has become an established tool for assessing the outdoor thermal environment in the major fields of human biometeorology (Jendritzky et al. 2012; Jendritzky and Höppe 2017). The UTCI summarises the interaction of ambient temperature, wind, humidity

P. Bröde (✉)
Leibniz Research Centre for Working Environment and Human Factors at TU Dortmund (IfADo), Ardeystrasse 67, 44139 Dortmund, Germany
e-mail: broede@ifado.de

 
E. L. Krüger (ed.), *Applications of the Universal Thermal Climate Index UTCI in Biometeorology*, Biometeorology 4,
https://doi.org/10.1007/978-3-030-76716-7_2

and radiant fluxes on human physiology. The dynamic physiological responses underlying the UTCI were simulated by an advanced multi-node model of human thermoregulation (Fiala et al. 2012), which had been extensively validated (Psikuta et al. 2012; Kampmann et al. 2012) and coupled with a model of adaptive clothing choice in urban populations (Havenith et al. 2012). UTCI values are expressed on an equivalent temperature scale. This involved the definition of a reference environment with 50% relative humidity (but vapour pressure not exceeding 2 kPa), with still air and radiant temperature equalling air temperature, to which all other climatic conditions are compared. The UTCI is supplemented by an assessment scale classifying UTCI values into ten categories of thermal stress from extreme cold to extreme heat.

As illustrated by Fig. 1, the major, or even only concern from the users' perspective is the provision of the meteorological input, as the operational procedure (Bröde et al. 2012a) offers simplified algorithms to compute UTCI values from air temperature ($T_a$), wind speed ($v_a$), mean radiant temperature ($T_{mrt}$) and water vapour pressure ($p_a$). These algorithms rely on the database of the offsets to air temperature reflecting the deviation in humidity, wind and radiation of the actual compared to the reference condition. Extensive simulations with the advanced UTCI-models (Fiala et al. 2012; Havenith et al. 2012) yielded these offsets once for all relevant ($T_a$, $v_a$, $T_{mrt}$, $p_a$)-combinations.

It is crucial to understand that this database, which is available as supplemental information (Fig. 1) to the published operational procedure (Bröde et al. 2012a), is the pivotal source for calculating the UTCI according to Eq. 1:

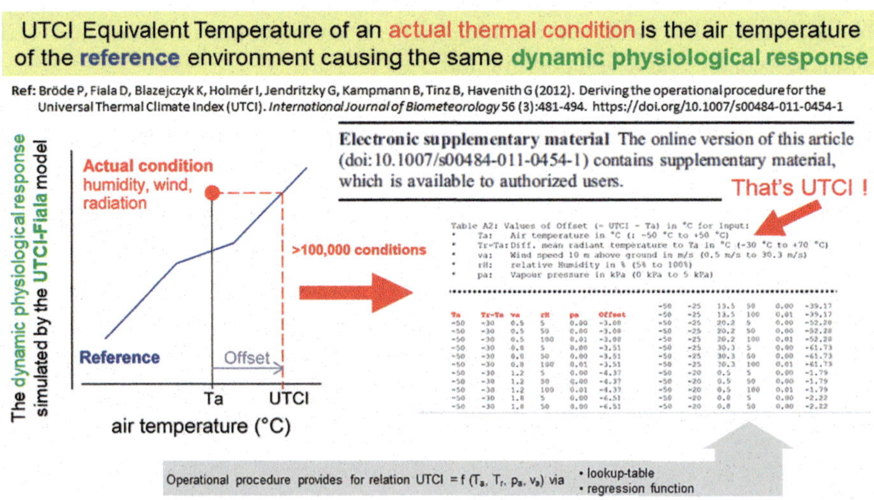


**Fig. 1** UTCI application via simplified calculation procedures provided by the operational procedure (Bröde et al. 2012a), which were based on the tabulated offsets to air temperature reflecting the deviation in humidity, wind and radiation of the actual condition compared to the reference environment. These offsets result from extensive simulations with the advanced UTCI-models (Fiala et al. 2012; Havenith et al. 2012)

$$UTCI = T_a + Offset\,(T_a, v_a, T_{mrt}, p_a) \tag{1}$$

The operational procedure (Bröde et al. 2012a) suggests a look-up table approach and a regression polynomial for approximating the tabulated offsets (Fig. 1), and thus the UTCI according to Eq. 1, however, the overwhelming part of UTCI applications curently rely on the regression polynomial.

The UTCI website (UTCI—Universal Thermal Climate Index 2009) offers an online calculation tool with access to the FORTRAN code for computing the UTCI (Bröde 2009) using the regression polynomial. The publicly available code and the online supplements to the operational procedure (Bröde et al. 2012a) have triggered numerous implementations of UTCI calculation within application programming interfaces, e.g. the Human Heat Exchange API (Ravanelli and Kingma 2019), or using Python (Tartarini and Schiavon 2020) or R (Schweiker 2016). Furthermore, the UTCI was integrated into application software like Bioklima (Błażejczyk and Błażejczyk 2010), PALM (Fröhlich and Matzarakis 2020), Grasshopper and Ladybug (Perini et al. 2017; Milosevic et al. 2017; Bajšanski et al. 2015), or RayMan (Matzarakis et al. 2007), to name a few.

In addition, UTCI calculation has been implemented within many individual research projects. As the author of this chapter was the corresponding author of the paper describing the operational procedure (Bröde et al. 2012a) and also acts as contact person of the UTCI website (UTCI—Universal Thermal Climate Index 2009), we received frequent requests for help and/or clarification on UTCI application.

Thus, based on a decade's experience, this chapter will summarize the major issues that occurred in UTCI application, which are related to

- the features of the regression polynomial compared to the look-up table approach to UTCI approximation;
- the errors due to ignoring the domain of definition of UTCI input parameters;
- the input of wind speed at the height of 10 m above ground;
- the features of the UTCI assessment scale in terms of its asymmetry and interpretation regarding physiology and thermal comfort.

## 2 UTCI Approximation: Regression Polynomial Versus Look-up Table

The tabulated offsets (Fig. 1) in connection with the algorithms offered by the operational procedure (Bröde et al. 2012a) provide for a straightforward execution of UTCI calculation without the need to implement and operate the advanced and complex models of human thermoregulation and clothing (Fiala et al. 2012; Havenith et al. 2012).

The look-up table algorithm will easily find the corresponding tabulated offset, if the ($T_a$, $v_a$, $T_{mrt}$, $p_a$)-combination represents a point on the 4-dimensional grid of input parameters. Values for intermediate points will be interpolated as average of the $2^4 = 16$ neighbouring points on the grid. The storage of the tabulated offsets and the search for the 16 neighbours pose challenges to the implementation of this algorithm, which will result in longer execution time compared to the simpler regression approach.

On the other hand, the look-up table algorithm is almost as accurate as the direct UTCI computation via the physiological model, whereas the easier-to-implement regression polynomial works with acceptable accuracy for most conditions, with the approximation error increasing for high wind speeds above 17 m/s (Bröde et al. 2012a).

Another advantage of the look-up table algorithm is that it can only be applied over the grid of valid input values, for which tabulated offsets exist. This means, it possesses an intrinsic robustness against non-observance of the domain of definition of UTCI input parameters. Inappropriate implementations of the regression polynomial, however, are highly susceptible to this issue, which is detailed in the following section.

## 3 Domain of Definition of UTCI Input Parameters

The domain of definition with valid UTCI values stretches over the range of ($T_a$, $v_a$, $T_{mrt}$, $p_a$)-combinations with tabulated offsets (Fig. 1). This results in the following intervals representing valid input to the UTCI: $-50\ °C \leq T_a \leq +50\ °C$, $0.5\ m/s \leq v_a \leq 30.3\ m/s$, $-30\ °C \leq T_{mrt}\text{-}T_a \leq +70\ °C$, $5\% \leq RH \leq 100\%$ (with $p_a < 5$ kPa and RH denoting relative humidity). These boundaries do not restrict routine application as indicated by comparisons to the control run of the general circulation model ECHAM4 (Stendel and Roeckner 1998). For wind speeds and relative humidity values below 0.5 m/s or 5%, respectively, the oprational procedure (Bröde et al. 2012a) advises to perform the offset calculations with the lower bounds. Similarly, for $T_a$ below −50 °C, the offset should be calculated for the lower bound and then applied to compute the UTCI according to Eq. 1. A corresponding procedure can be applied for $T_a$ exceeding +50 °C by computing the offset for the upper bound of $T_a$, however, this extrapolation will likely underestimate heat stress for air temperatures above 55 °C.

It is important to note that the approach using a look-up table intrinsicly respects the above mentioned boundaries simply because the data table does not contain any offset values outside the domain of definition.

With the regression polynomial, however, the input requires properly checking for validity as exemplified in the FORTRAN code available from the UTCI website (Bröde 2009), otherwise as a consequence of extrapolation results will become unreliable; e.g. UTCI values below zero Kelvin (!) were reported for wind speeds above 34 m/s, i.e. outside the domain of definition (Novák 2011).

As another example, the chair of the COST Action 730 on the UTCI, Gerd Jendritzky (personal communication, June 23, 2014), was confronted by a colleague

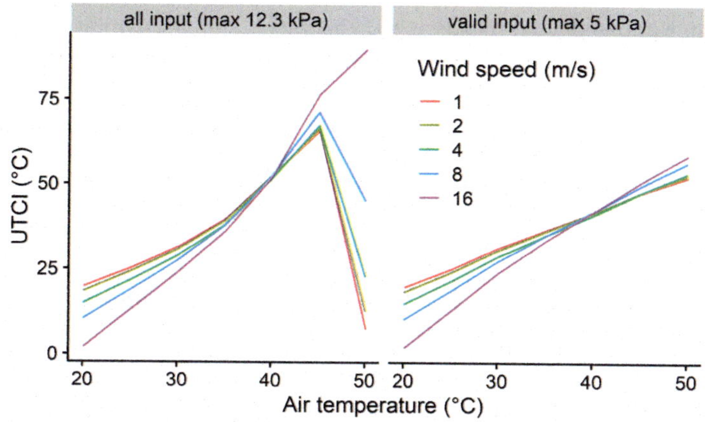


**Fig. 2** Wind effect on the UTCI calculated by the regression polynomial in relation to air temperature for simulated data averaged over relative humidity ranging from 10 to 100%. The left panel shows the curves for all input data, the right panel displays the results for input adhering to the upper bound of 5 kPa for water vapour pressure input

with an alleged 'strange response' of the UTCI to wind in the heat, which is shown in the left panel of Fig. 2. This study using the regression polynomial was based on simulated data averaged over relative humidity ranging from 10 to 100%, thus with water vapour pressure up to 12.3 kPa, far above the upper limit of 5 kPa valid for UTCI input, and also far above any water vapour pressure values observed in natural environments relevant to real world applications (Stendel and Roeckner 1998). Restricting the input to valid (and reasonable) vapour pressure values yielded the curves in the right panel of Fig. 2, which were well in line with previously reported results (Bröde et al. 2012a, 2013a).

## 4 Input of Wind Speed at 10 m Above Ground

Following meteorological convention, the input of wind speed ($v_a$ in m/s) is taken as value measured 10 m above ground (Bröde et al. 2012a). If wind speed in an actual application represents a value $v_{a,x}$ at a height of $x$ m, the following equation provided by the operational procedure (Bröde et al. 2012a) will transform the wind speed $v_{a,x}$ to the required $v_a$ at reference height of 10 m:

$$v_a = v_{a,x} \times \log(10/0.01) / \log(x/0.01) \quad (2)$$

The value 0.01 in the above equation refers to the roughness length (z0 in m), which was set to 0.01 m representing open outdoor terrain (grassland) conditions (Oke 1987) for the purpose of the UTCI (Fiala et al. 2012). UTCI users, who have measured or estimated wind speeds at body height (1, 2 m or a similar height) in non-open terrains potentially advocating for different values of z0, e.g. around buildings or in urban areas, are advised still to use Eq. 2 for upscaling wind speed to 10 m above ground. This will ensure that the physiological response as calculated by the UTCI-Fiala model (Fiala et al. 2012) considers the actually measured (or estimated) wind speed at body height.

## 5 UTCI Assessment Scale—Asymmetry, Physiology and Thermal Comfort

UTCI includes an assessment scale establishing UTCI threshold values that define different categories of thermal stress from extreme cold to extreme heat. This scale was developed by comparing the thermoregulatory variables and effector functions simulated by the UTCI-Fiala model (Fiala et al. 2012) to criteria from thermo-physiology and ergonomics (Bröde et al. 2012a, 2013a; Błażejczyk et al. 2013).

As shown by Fig. 3, the scale is asymmetric with more categories covering wider intervals in the cold compared to the heat, which reflects differences in the capacity to adapt clothing insulation to ambient temperature as assumed by the UTCI-clothing model (Havenith et al. 2012).

Especially, thermoregulatory responses simulated for UTCI values from 18 to 26 °C, e.g. the absence of pronounced sweating or shivering, comply with the definition of the thermal comfort zone TCZ as "range of ambient temperatures, associated

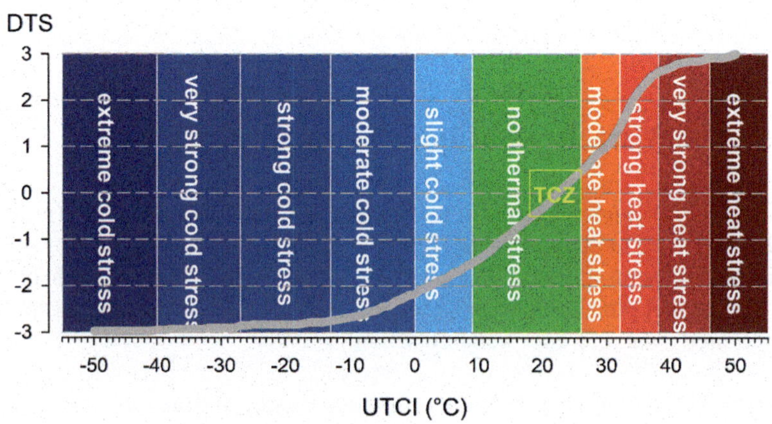


**Fig. 3** Dynamic Thermal Sensation (DTS) calculated by the UTCI-Fiala model (Fiala et al. 2012) mapping thermal sensation votes ranging from −3:'cold' over 0:'neutral' to +3:'hot' according the 7-unit-ASHRAE scale (ASHRAE 2004) to UTCI. Vertical reference lines indicate the thermal stress categories from the UTCI assessment scale with the sub-interval 18 °C ≤ UTCI ≤ 26 °C indicating the thermal comfort zone (TCZ)

with specified mean radiant temperature, humidity, and air movement, within which a human in specified clothing expresses indifference to the thermal environment for an indefinite period" (IUPS Thermal Commission 2003).

Notably, the dynamic thermal sensation (DTS) shown in Fig. 3 as calculated by the UTCI-Fiala model (Fiala et al. 2012, 2003) for UTCI reference conditions and averaged over the 2 h exposure time, yielded values between $-0.5$ and $+0.5$ for the TCZ. This suggests considering DTS for thermal sensation prediction by UTCI in field studies or surveys on outdoor thermal comfort. This approach, which is facilitated by the published UTCI-to-DTS relation (Bröde 2019), had been applied previously (Bröde et al. 2013b, 2012b) and is again showcased in the chapter on the sensitivity of UTCI thermal comfort prediction of this book (Chap. 4).

## 6 Conclusions and Recommendations

UTCI application is straightforward thanks to the easy to implement calculation algorithms offered by the operational procedure that are publicly available (Bröde et al. 2012a; Bröde 2009; UTCI—Universal Thermal Climate Index 2009). Nevertheless, the following hints and recommendations will facilitate UTCI application and will ensure better understanding of the procedures and the reliability of the results.

- Do not mistake the UTCI-Fiala model or the regression polynomial for representing UTCI; rather the pre-calculated **grid of offsets** to air temperature reflecting the deviation in humidity, wind and radiation of the actual condition compared to the reference environment **are key to establishing UTCI** application.
- Any UTCI application must respect the domain of definition, i.e. the allowed **range for UTCI input parameters** ($T_a$, $v_a$, $T_{mrt}$, $p_a$). Note that using the **look-up table approach completely prevents** any such issue, which can only occur with the regression polynomial.
- **Upscaling wind speed** measured (or estimated) **at person level**, e.g. in an urban environment, by **Eq.** 2 ensures that human responses are calculated by UTCI with the **most appropriate** information on **wind speed input**.
- The thermal stress categories of the **assessment scale** reflect limits for thermoregulatory variables and effector functions, i.e. they are **based on physiology**, but not on (symmetric) thermal sensation. The scale's **asymmetry** reflects **different adaptive capacities** under heat and cold stress in donning and doffing clothing.
- Do not confuse the thermal stress categories with thermal sensation. **Outdoor thermal comfort** studies comparing thermal sensation votes should **rely on** the published **UTCI-to-DTS** relation (Bröde 2019) with the UTCI subrange of 18–26 °C marking the thermal comfort zone (**TCZ**).

As an outlook, we noticed that the look-up table approach to UTCI application is currently hardly implemented, though it is more accurate and more robust compared to the regression polynomial. Thus, we plan to make a corresponding calculation procedure available in due course at the UTCI website (UTCI—Universal Thermal Climate Index 2009), hoping that it will still broaden the range of UTCI applications.

**Acknowledgements** UTCI was developed within COST Action 730, the COST office is supported by the EU framework program Horizon 2020. Parts of this chapter were presented as a talk during the 'International Conference UTCI—assessment measure in human bioclimatology—10 years of application' held at the Institute of Meteorology and Water Management in Warsaw from 22nd to 23rd May 2019 with support from the International Society of Biometeorology and the S.W. Tromp Foundation, which we gratefully acknowledge.

## References

ASHRAE (2004) Standard 55—thermal environmental conditions for human occupancy. ASHRAE Inc., Atlanta

Bajšanski IV, Milošević DD, Savić SM (2015) Evaluation and improvement of outdoor thermal comfort in urban areas on extreme temperature days: applications of automatic algorithms. Build Environ 94:632–643

Błażejczyk K, Błażejczyk M (2010) BioKlima© 2.6, Program komputerowy, http://www.igipz.pan.pl/Bioklima-zgik.html

Błażejczyk K, Jendritzky G, Bröde P, Fiala D, Havenith G, Epstein Y, Kampmann B (2013) An introduction to the universal thermal climate index (UTCI). Geogr Pol 86(1):5–10

Bröde P (2009) UTCI–Version a 0.002–October 2009. http://utci.org/public/UTCI%20Program%20Code/UTCI_a002.f90. Accessed 15 Mar 2021

Bröde P, Fiala D, Błażejczyk K, Holmér I, Jendritzky G, Kampmann B, Havenith G (2012a) Deriving the operational procedure for the universal thermal climate index (UTCI). Int J Biometeorol 56(3):481–494

Bröde P, Krüger EL, Rossi FA, Fiala D (2012b) Predicting urban outdoor thermal comfort by the universal thermal climate index UTCI—a case study in Southern Brazil. Int J Biometeorol 56(3):471–480

Bröde P, Blazejczyk K, Fiala D, Havenith G, Holmer I, Jendritzky G, Kampmann B (2013a) The universal thermal climate index UTCI compared to ergonomics standards for assessing the thermal environment. Ind Health 51(1):16–24

Bröde P, Krüger EL, Fiala D (2013b) UTCI: validation and practical application to the assessment of urban outdoor thermal comfort. Geogr Pol 86(1):11–20

Bröde P (2019) UTCI-to-DTS. 1.0 edn., Zenodo. https://doi.org/10.5281/zenodo.3372231

Fiala D, Lomas KJ, Stohrer M (2003) First principles modeling of thermal sensation responses in steady-state and transient conditions. ASHRAE Trans 109:179

Fiala D, Havenith G, Bröde P, Kampmann B, Jendritzky G (2012) UTCI-Fiala multi-node model of human heat transfer and temperature regulation. Int J Biometeorol 56(3):429–441

Fröhlich D, Matzarakis A (2020) Calculating human thermal comfort and thermal stress in the PALM model system 6.0. Geosci Model Dev 13(7):3055–3065

Havenith G, Fiala D, Błazejczyk K, Richards M, Bröde P, Holmér I, Jendritzky G (2012) The UTCI-clothing model. Int J Biometeorol 56(3):461–470

IUPS Thermal Commission (2003) Glossary of terms for thermal physiology. J Therm Biol 28:75–106

Jendritzky G, de Dear R, Havenith G (2012) UTCI—why another thermal index? Int J Biometeorol 56(3):421–428

Jendritzky G, Höppe P (2017) The UTCI and the ISB. Int J Biometeorol 61(1):23–27

Kampmann B, Bröde P, Fiala D (2012) Physiological responses to temperature and humidity compared to the assessment by UTCI, WGBT and PHS. Int J Biometeorol 56(3):505–513
Matzarakis A, Rutz F, Mayer H (2007) Modelling radiation fluxes in simple and complex environments—application of the RayMan model. Int J Biometeorol 51(4):323–334
Milošević DD, Bajšanski IV, Savić SM (2017) Influence of changing trees locations on thermal comfort on street parking lot and footways. Urban for Urban Green 23:113–124
Novák M (2011) UTCI–first test in the Czech republic. Prace i Stud Geogr 47:319–325
Oke TR (1987) Boundary layer climates, 2nd edn. Routledge, London
Perini K, Chokhachian A, Dong S, Auer T (2017) Modeling and simulating urban outdoor comfort: coupling ENVI-Met and TRNSYS by grasshopper. Energy Build 152:373–384
Psikuta A, Fiala D, Laschewski G, Jendritzky G, Richards M, Błażejczyk K, Havenith G (2012) Validation of the Fiala multi-node thermophysiological model for UTCI application. Int J Biometeorol 56(3):443–460
Ravanelli N, Kingma B (2019) Human heat exchange API. https://www.humanheat.exchange/. Accessed 15 Mar 2021
Schweiker M (2016) COMF: an R package for thermal comfort studies. R J 8(2):341–351
Stendel M, Roeckner E (1998) Impacts of horizontal resolution on simulated climate statistics in ECHAM4. MPI Report 253. Max-Planck-Institute for Meteorology, Hamburg
Tartarini F, Schiavon S (2020) Pythermalcomfort: a python package for thermal comfort research. SoftwareX 12:100578
UTCI—Universal Thermal Climate Index (2009). http://www.utci.org/. Accessed 15 Mar 2021

# Chapter 3
# Literature Review on UTCI Applications

**Eduardo L. Krüger**

**Abstract** This chapter presents a comprehensive literature review of peer-reviewed papers on the UTCI with special focus on its applications. A search in Scopus and Web of Science has been conducted in February 2021 yielding 320 and 304 documents, respectively, for the time frame from 2000 to March 2021. Results have been classified according to 8 different categories, which roughly define the areas of application of the index: (1) Outdoor Thermal Comfort (OTC) and thermal stress; (2) Urban Climate and Planning studies; (3) Climate-related impacts on human health; (4) Bioclimate; (5) Comparisons with other thermal comfort indices; (6) Meteorological analyses; (7) Climate change research; (8) Tourism. The bulk of research carried out on the UTCI is primarily concentrated on the first two topics, reaching about 60% of papers output. Clusters identified in VOSviewer from co-occurrences of author keywords closely match the main areas of application of the index. Research output shows an intrinsic multidisciplinary nature but it is still concentrated in a few countries. Areas of application such as public weather service and climate-related impacts on the health sector still need to become aware of the potentialities and practicalities of the UTCI.

**Keywords** UTCI · Literature review · VOSviewer · Bibliometric analysis

## 1 Introduction

This chapter presents a comprehensive literature review of peer-reviewed papers on UTCI with special focus on its applications. An Article title/Abstract/Keywords search in Scopus and Web of Science 'WoS' (search string "UTCI" AND "THERMAL" OR "UTCI" AND "CLIMATE", to account for thermal aspects and climate-related issues) yielded 320 and 304 documents, respectively, for the time frame from

E. L. Krüger (✉)
Departamento de Construção Civil, Universidade Tecnológica Federal do Paraná - UTFPR, Campus Curitiba - Sede Ecoville, Rua Deputado Heitor Alencar Furtado, 4900-81280-340 Curitiba, Brazil
e-mail: ekruger@utfpr.edu.br

 
E. L. Krüger (ed.), *Applications of the Universal Thermal Climate Index UTCI in Biometeorology*, Biometeorology 4,
https://doi.org/10.1007/978-3-030-76716-7_3

2000 to 2021, as of Mid-March 2021, about 85% of them published as articles. Both searches (Scopus and WoS) were combined in our literature review, complementing each other. Figure 1 presents search results until the year 2020 in order to account only for complete yearly output. Even though WoS has a slightly lower amount of papers, the overall trend and growth are similar in both databases.

## *1.1 Initial Papers (2000–2010)*

The publication output on the UTCI over the time frame 2000–2020 shows that only one article was published even before the index was under development. This particular paper came out in a Special Issue of Energy and Buildings edited by Fergus Nicol and Ken Parsons with contributions from the Conference *Moving Thermal Comfort Standards into the 21st Century*, held in Windsor, England, in 2020. Höppe (2002) announced the effort at that time to introduce models other than steady-state approaches for outdoor comfort research. Höppe's paper became a classic, pointing to the risk of existing models to overestimate thermal discomfort in outdoor settings. Particularly in cold climates, it would take several hours for a person to reach steady-state conditions, therefore Höppe proposed the non-steady state UTCI, which was then under development by a working group of the International Society for Biometeorology. The sentiment was that thermal comfort models developed for indoors were not fully applicable to outdoors. The paper finishes pointing out possible applications of the index including weather reports and forecasts, climate change scenarios and tourism.

Nevertheless, it was just after the index was launched (in 2009, with the finalization of Cost Action 730) that further publications began to appear (Fig. 1). The very first papers were published by or in co-authorship with former members of the ISB expert group. Some of these publications consisted of a formal presentation of the index with regard to its thermophysiological principles, modeling of human thermoregulation and clothing insulation (Błażejczyk et al. 2010a, b; Bröde et al. 2010). In those papers, potential applications of the index were introduced such as in weather forecasts, bioclimatological assessments, bioclimatic mapping in all scales (from microscale to macroscale), urban design, planning of outdoor spaces, measures towards urban quality of life, outdoor recreation and health, epidemiology and climate research. At the same time, the operational procedure with source code were made available at the UTCI website (www.utci.org) and a tool was created to perform calculations of the UTCI and other indices, the BioKlima software (Błażejczyk and Błażejczyk 2010). From 2014 (Bruse 2014), the ENVI-met V4 package, that allows microclimate simulations of urban environments for assessing the effects of atmosphere, vegetation, architecture and materials, included BioMet as a post-processing tool for calculating Human Thermal Comfort Indices, among them the UTCI.

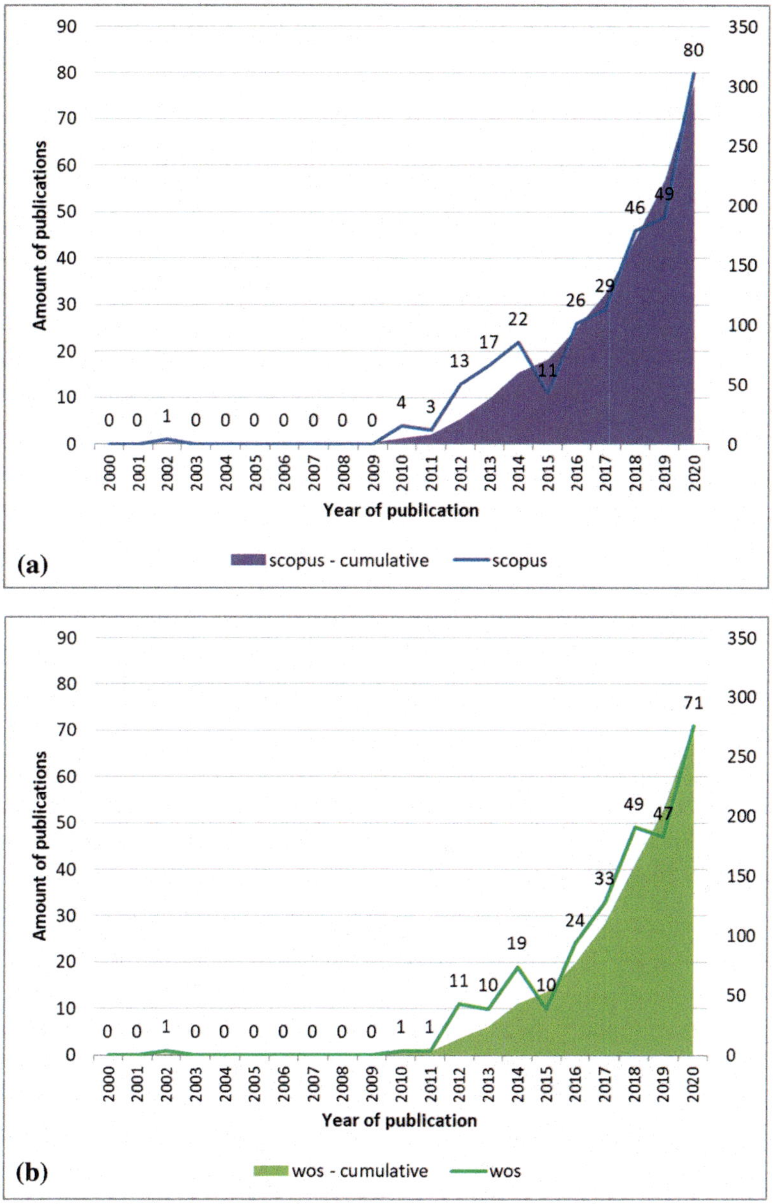


**Fig. 1** **a** Amount of publications in absolute numbers per annum and cumulative—Scopus database (2000–2020). **b** Amount of publications in absolute numbers per annum and cumulative—web of science 'WoS' database (2000–2020)

## 1.2 Early Applications of the UTCI and the IJBM Special Issue

Early applications of the index can be found in local bioclimatic analyses (Idzikowska 2010; Novák 2011; Chabior 2011; Mateeva 2011). The very first paper on such analyses focused on the use of the UTCI for Paris, Rome and Budapest (Idzikowska 2010), conducted by a researcher from the University of Warsaw/Faculty of Geography and Regional Studies. In that particular study, data from a European Project for the Assessment and Prevention of Acute Health Effects of Weather Conditions were used, which already hints at the interdisciplinary nature of the UTCI. The same author (Idzikowska 2011) subsequently performed an analysis of the UTCI with respect to mortality in chosen European cities, situated in different climate zones, thereby starting a new avenue of research with the UTCI as relevant indicator for matters involving human health in outdoor spaces. Noteworthy at this point is also an initial paper dealing with calibration procedures of the thermal comfort assessment scale, developed by Mateeva (2011) for Bulgaria.

In 2012, a special issue of the International Journal of Biometeorology (IJBM) had the purpose of disseminating the UTCI. In the foreword, the interdisciplinary and multidisciplinary character of the research team responsible for developing the index was stressed: "UTCI not only represents a scientific advance but is an exemplar of the co-production of knowledge" (McGregor 2012). The contents of this special issued ranged from describing the rationale behind the index, uncertainties and limitations to a general application of the index in the assessment of thermal climate in outdoor environments. One of the papers in this special issue (Bröde et al. 2012a) resulted from a collaborative work between authors of this book, started just after the Windsor Conference 2010, using relevant field data from outdoor comfort campaigns initiated in 2009 in Curitiba, Brazil.

In that special edition, Jendritzky et al. (2012) presented the reasons that lead to its development pointing out that the use of the UTCI will potentially "standardize applications in the major fields of human biometeorology, thus making research results comparable and physiologically relevant". Areas expected to profit from the newly introduced index were: public weather service, public health system, precautionary planning and climate impact research in the health sector. The thermophysiological model embedded in the UTCI, based on Fiala's multi-node human physiology and thermal comfort model, was one of the most advanced (multi-node) thermo-physiological models identified by ISB Commission 6 at the time.

Also in the special issue, a rigorous validation of the Fiala's model was presented by Psikuta et al. (2012), working on previous validation procedures within the framework of a multi-institutional research initiative conducted by Psikuta et al. (2006, 2007a, b) against a wide range of climatic conditions, metabolic rates and clothing levels. The validity of the model in terms of physiological strain for combined variations of air temperature and humidity was verified by Kampmann et al. (2012). Weihs et al. (2012) focused on the accuracy of the UTCI due to imprecisions in measuring radiant fluxes, showing that uncertainties in measurements of meteorological variables could be critical in the extreme (hot or cold) conditions of thermal stress.

Extensions of the model were also introduced during the development of the index (Fiala et al. 2012). An important feature of the UTCI, its clothing model (Havenith et al. 2012), predicts clothing insulation based on ambient temperature and considering seasonal clothing adaptation habits of Europeans (Błażejczyk et al. 2012). Subsequent studies have shown some inconsistencies of the clothing model for estimating garments actually worn by different populations (European versus South-American) within a narrow band of ambient temperatures albeit the model was in good agreement with field data at higher and lower temperatures (Bröde et al. 2014). Pantavou et al. (2013) noticed in their questionnaire survey that estimated clothing by the UTCI clothing model had reached negative values with high air temperatures (>31.6 °C), a feature that has been adjusted later. Those authors also pointed to the need for calibrating the UTCI thermal comfort/stress assessment scale in order to account for local climatic and population needs, though recognizing at that point that seasonal acclimatization and adaptation effects might have played a role in the mismatch found between observed thermal neutrality and the neutrality range proposed for the UTCI. Noteworthy at this point is also an initial paper dealing with calibration procedures of the thermal comfort assessment scale, developed by Mateeva (2011) for Bulgaria.

Initial shortcomings of the UTCI have been identified in the first papers published on the index. Nastos and Matzarakis (2012), upon performing an analysis of the influence of ambient temperature and calculated values for the thermal indices PET and UTCI for Athens, Greece, noticed that in the inter-comparison between thermal indices, "UTCI does not interpret the mortality risk of strong/extreme heat stress, which is better approached by PET analysis". Despite that, an initial comparison between both indices by Błażejczyk et al. (2012), including a number of other thermal indices, ranging from simple (WBGT, Humidex) to complex, heat-budget models such as the PET, pointed to a strong correlation between both with an r-squared of 0.97 for a large variation of meteorological parameters (65,500 random samples worldwide). Such initial comparison suggested that the UTCI was better able to represent thermal sensation due to its clothing model whereas PET lagged behind as calculations took into account fixed clothing level. Indeed, there were asymmetrical differences between UTCI and PET particularly for cold conditions which were verified in the respective assessment scales for a long series of meteorological data for Freiburg, Germany.

Another shortcoming referred to wind speed. The problem was identified by Novák (2010) for extreme winter conditions and for extremely windy conditions (>30 mps). In a subsequent paper, Novák (2011) mentioned that the problem had been solved by then. The inter-comparison between the UTCI and commonly used indices in biometeorological forecast performed by that author for a warm summer day showed that index results had greater variability, as it accounted for radiant heat exchanges, which lead the author to conclude that the UTCI expresses "the real load of a human body better, including the short-time fluctuations of this load" with similar or better response when compared to other indices.

Bröde et al. (2013a) listed a few limitations of the UTCI, in particular the limited number of physical variables used for calculations whereas metabolic rate is assumed as fixed and clothing is estimated as a function of ambient temperatures, which might limit the applicability of the index in working conditions with protective clothing and with varying workloads. Modifications of the index for accounting for issues related to occupational thermal stress (metabolic rate, clothing levels and exposure duration) have been explored more recently by Bröde et al. (2018) and future developments should further address such issues and potential limitations.

## 2 Applications of the UTCI

With the aim of categorizing potential applications for the UTCI, an attempt was made to group publications found into different categories. Basically, they can be split into the following ones:

- Comparisons with other existing thermal comfort indices—such comparisons were particularly necessary for testing the suitability of the UTCI in human biometeorological evaluations;
- Outdoor Thermal Comfort (OTC)—with studies on thermal perception, thermal stress, related or not to outdoor activities, walkability and the identification of walkable routes in outdoor urban spaces, observational and behavioral studies, calibration of the thermal assessment scale; such studies are centered on the user of outdoor spaces and can have implications for urban planning;
- Climate-related impacts on human health—UTCI is in this case mostly used for assessing health risks in outdoor environments, generally associated to excess heat and heatwave episodes, by means of localized studies or in health-related UTCI mapping;
- Bioclimate—such application generally involves spatial and temporal distributions of the index over a given area, sometimes accompanied by mapping of the UTCI;
- Meteorological analyses—here we include the usage of the UTCI as a complement to atmospheric studies that deal with particular aspects related to Climatology, such as the interpretation of historical data in terms of comfort parameters;
- Climate change research—weather extremes resulting from climate change and also long-term projections of the UTCI for given areas;
- Urban Climate and Planning studies—the application of the UTCI for understanding patterns related to urban climate aiming at urban planning;
- Tourism—one of the expected areas of application of the index; tourism-related studies investigate the relationship between bio climate and OTC conditions on tourism attractiveness, spatiotemporal distribution of the UTCI in tourism areas and tourism management strategies.

## 2.1 Comparisons with Other Thermal Comfort Indices

The applicability of the UTCI in diverse fields has been assessed by several studies, particularly as regards its performance when compared to existing indices. Comparisons between different thermal indices to the UTCI have been carried out by a number of authors: Nastos and Matzarakis (2012), Błażejczyk et al. (2012), Novák (2013), Matzarakis et al. (2014), Lai et al. (2014), Pantavou et al. (2014), Tumini and Perez Fargallo (2015), Provençal et al. (2016), Coccolo et al. (2016), Golasi et al. (2018), Potchter et al. (2018), Zare et al. (2018), Charalampopoulos (2019), Fang et al. (2019a), Staiger et al. (2019), Charalampopoulos and Nouri (2019), Zare et al. (2019), Katavoutas and Founda (2019), Asghari et al. (2019), Santurtún et al. (2020), Lian et al. (2020). Such studies employed diverse methods for the comparisons, such as correlations to meteorological variables, statistical tests with respect to particular information (e.g. mortality data, hospital admissions, data from field measurements, including thermal perception votes), suitability for human biometeorological evaluations in urban and regional planning and sensitivity analyses.

## 2.2 Outdoor Thermal Comfort (OTC)

In this subcategory, the focus of interest is the user of the outdoor space, for leisure activities, for transiting or for using it as a working environment. The reviewed studies can be here further subdivided into the following classifications: thermal perception; thermal stress; walkability; observational and behavioral studies; confounding and interfering factors on OTC; calibration of the UTCI's thermal comfort/stress assessment scale; applications in indoor and transitional spaces.

THERMAL PERCEPTION: Generally, OTC studies are based on measured meteorological variables, which are post-processed as UTCI data for comparisons with reported thermal perception. For that, questionnaire-based surveys are conducted with local population. In the case of the UTCI, such procedure, starting with the afore-mentioned study by Bröde et al. (2012a) for Curitiba, Brazil, was adopted in diverse climates and locations: arid climatic conditions (Abdel-Ghany et al. 2013, 2014; Hadianpour et al. 2018), mediterranean climate (Pantavou et al. 2013), temperate conditions (Mateeva 2011; Nidzgorska-Lencewicz and Mąkosza 2013; Maras et al. 2016; Xu et al. 2018; Zhang et al. 2020; Zhu et al. 2020), in comparisons between subtropical versus maritime temperate climate (Krüger et al. 2012; Bröde et al. 2013b), warm temperate (Hamanaka and Bueno-Bartholomei 2017), humid subtropical (Watanabe et al. 2014; Li et al. 2018; Silva and Hirashima 2020), in a monsoon-oriented subtropical humid climate (Manavvi and Rajasekar 2020; Das and Das 2020), among others. Chen et al. (2020) and An et al. (2021) compared thermal preferences of pedestrians in cold cities in China, yearlong and for winter, respectively.

OUTDOOR ACTIVITIES/THERMAL STRESS: Under such classification, we could identify studies conducted for particular exposure conditions. For outdoor working conditions, the UTCI was used in diverse studies that focused on occupational heat stress: in a spatial and temporal analysis of heat stress conditions and potential risks of heat hazard of outdoor workers in Warsaw, Poland (Błażejczyk et al. 2014a); for predicting heat stress conditions when compared to measured physiological parameters in occupational heat stress assessments in the brick industry in Iran (Vatani et al. 2015); in a physiological and perceptive evaluation of open-pit mines in Iran (Nassir et al. 2017); in a cross-sectional study conducted on physiological responses of male farmers exposed to the hottest conditions in Boukan, West Azerbaijan, Iran (Zamanian et al. 2017); in a cross-sectional study on the human susceptibility of farmworkers in India exposed to outdoor heat during paddy and potato cropping activities (Sen and Nag 2019); in a study of thermal stress levels young soldiers of the Portuguese Army are exposed to during training or field activities under hot conditions (Galan and Guedes 2019); and, for evaluating the influence of OTC conditions on people working in scaffoldings in Poland (Szer et al. 2019). Worth mentioning in this context is the EU's Horizon 2020 HEAT-SHIELD project, particularly a study that evaluated heat stress among workers in two agricultural companies and one construction company in Europe, with questionnaire surveys and concurrent monitoring of meteorological variables (Messeri et al. 2019).

Climate change impacts on workability in warm outdoor environments were studied by Bröde et al. (2018) with respect to different heat stress assessment metrics, including the UTCI. On a similar topic, another study reviewed several heat stress indices, including the UTCI for assessing occupational heat stress risks in hot climate areas, discussing advantages and disadvantages in relation to meteorological data, local workplace environments, body heat production and the use of protective clothing (Gao et al. 2018).

Another subset under this classification involves sport activities. Honjo et al. (2018) predicted heat stress levels with the UTCI along the marathon course in Tokyo, the venue of the Olympic Games in 2020 if it were not for the COVID-19 Pandemic, using high-resolution meteorological data combined with short-term analyses of shading along the course. Fang et al. (2019b) focused on thermal perception of athletes during their break throughout outdoor training sessions at the Guangzhou University campus in Guangzhou, China, using post-processed UTCI values from field measurements and questionnaire-based surveys. Thorsson et al. (2020) compared different thermal indices, among them the UTCI, in terms of their predictive power for ambulance-required assistances and collapses during a city half marathon in Gothenburg, Sweden. Gasparetto and Nesseler (2020) looked at the relationship between athletes' performance and thermal exposure conditions using historical data for marathon runners in the last 12 New York City Marathons. Konefał et al. (2020) analyzed soccer players' performance during the 2018 FIFA Worldcup in Russia with respect to thermal stress conditions at training centers and during tournament matches. Niu et al. (2020) studied the relationship between physiological and subjective thermal responses for various activity levels with the UTCI.

Unrelated to performing activities in the outdoors, some studies focused on the analysis of spatial variability of thermal stress conditions within a given urban area with GIS tools (Dobek et al. 2013; Milewski 2013; Błażejczyk et al. 2014b). Also for Poland, following studies were conducted on this topic: a recent multi-city study focused on hazardous thermal stress conditions by means of spatial and temporal analysis of diverse cities (Kuchcik et al. 2021; Miszuk 2020, 2021) and also Głogowski et al. (2020a) evaluated heat stress intensity in selected meteorological stations in Lower Silesia and in the Polish-Saxon region and Krzyżewska et al. (2020) presented UTCI variability in Poland during summer in connection to tourism. Pecelj et al. (2020) evaluated heat stress conditions during heatwaves in Serbia. Dong et al. (2020) proposed a new heatwave-induced health risk framework that could be used as a urban planning tool, from data gathered in a spatial analysis for Wuhan City, China.

WALKABILITY: The search for walkable routes in outdoor urban spaces that are able to provide physical, psychological and community benefits (Lee et al. 2020), with respect to microclimatic and pedestrian thermal stress conditions has been explored with the UTCI. Basically, three different approaches could be identified in our review: pedestrian walking behavior from observational analysis; dynamic thermal monitoring along specific routes alongside reported thermal responses; and, thermal mapping from fixed locations in a given area. Lee et al. (2020) carried out a walkability study focusing on thermal and OTC factors, using the UTCI as one of the analysis parameters from measurements and onsite observations of pedestrians in a walkway turned into an experimental site in Hong Kong, China. Talhi et al. (2020) conducted micrometeorological walks along different paths in Algiers, Algeria, a location with a south Mediterranean climate. Predefined routes were taken by subjects that were asked to respond a questionnaire at different 12 urban configurations. Objective data were then used for UTCI calculations. As for the third approach mentioned above, a comfort-based navigational study, based on meteorological data recorded at different sites and post-processed as predicted outdoor comfort in UTCI units, was proposed by Liu and Li (2018), facilitating the choice of the most thermally adequate path among different alternatives over a given area in the Fuzhou University Campus, Fuzhou, China.

CONFOUNDING AND INTERFERING FACTORS ON OTC: As pointed out by Höppe (2002), thermal comfort is comprised of three different components: a psychological, a thermophysiological and one based on energy balance relationships. In investigating human responses in OTC surveys, potential interfering factors have been raised in a number of studies, some of them looking at physical components of the thermal environment, such as the interaction between acoustic and thermal aspects (Bjerre et al. 2017), the influence of wind patterns (Hanipah et al. 2016, 2017; Ohashi et al. 2018; Abadla et al. 2019; Haidanpour et al. 2019), shading and solar exposure (Lam and Hang 2017; Xu et al. 2019a) and an ensemble of variables, as analyzed through a sensitivity analysis by Li et al. (2020). Some studies focused on potential physiological components of thermal perception related to thermal history and acclimatization, biometrics, metabolism and ethnicity such as: thermal history and air-conditioning usage (Krüger et al. 2015; Lam et al. 2020, 2021a, b), short- and long-term acclimatization effects (Krüger et al. 2017a; Lam and Lau 2018), heatwave episodes and their implications on perceived thermal comfort (Lam et al. 2018, 2019), exposure time length on OTC (Cheung and Jim 2019), activity and metabolic rates (Liu et al. 2020), anthropometric effects (Krüger and Drach 2017; Jin et al. 2020), personal factors and other physical parameters of the thermal environment (Fang et al. 2018; Xue et al. 2020; Lam et al. 2020; Lehnert et al. 2021), ethnicity (He et al. 2020). And, finally, the psychological component was also examined: psychological effects from site-related context (Krüger et al. 2017b; Krüger 2017).

OBSERVATIONAL AND BEHAVIORAL STUDIES: Perhaps following the tradition of observing people's activities in outdoor spaces started by Jan Gehl in the 1970's for evaluating the livability of urban areas (Gehl 2011), some OTC studies look at influences of microclimate, expressed as the UTCI, on behavior and preferences of users for shade or sun in outdoor spaces. Watanabe and Ishii (2016) analyzed the effect of microclimatic conditions in the choice of a place in sun/shade next to traffic lights, using for that unobtrusive observations of passers-by. A novel approach to observational studies of urban dwellers in outdoor environments was proposed by Reinhart et al. (2017). Such approach is based on Wi-Fi data informing the frequency and extent of usage of a public courtyard in Cambridge, United States alongside UTCI predictions from measured microclimatic conditions onsite. A subsequent paper expands the same approach for the hot-arid conditions of the United Arab Emirates (UAE) to evaluate the effect of outdoor evaporative coolers on frequency of usage of a public courtyard in Abu Dabi (Dhariwal et al. 2019). The spatial distribution of UTCI estimates for the periods of observation with available Wi-Fi data were obtained by means of numerical simulations in ENVI-met. Outdoor activity observations and microclimate measurements were carried out in public outdoor spaces in Australian cities, presenting thresholds for the UTCI due to observed declines in outdoor living, as a function of activities and adaptation measures (clothing, activity level) and meteorological data (Sharifi and Boland 2018). A further study by the authors focused on adaptation measures of people during the passive activity observation periods (Sharifi and Boland 2020). A mix between an observational and a questionnaire-based study with concurrent thermal monitoring was conducted by Huang et al. (2016a) in subtropical Wuhan, China. The study also involved urban interventions (renovations of a playground) which lead authors to conclude on implications for urban design from output obtained in term of UTCI data.

CALIBRATION: The thermal assessment scale originally proposed for the UTCI is one way to interpret results in terms of thermal stress. The scale was defined from strain reactions in lab tests as has a pronounced human-physiological component (Bröde et al. 2012b). A correspondence to thermal comfort categories as defined for other scales is shown in Pantavou et al. (2018). Using survey data, some studies attempted to locally calibrate the UTCI's thermal comfort/stress assessment scale: Bröde et al. (2013b), Pantavou et al. (2013), Lai et al. (2014), Pantavou et al. (2014), Jin et al. (2019), Krüger et al. (2020), Borges et al. (2020), Chen et al. (2020). A more thorough calibration of the UTCI was performed by Pantavou et al. (2018) using the extensive OTC database from the RUROS research initiative (Rediscovering the Urban Realm and Open Spaces) that encompassed seven cities in three European countries. A recent review of studies focusing on the calibration of OTC indices was carried out by Binarti et al. (2020) with 31 reviewed papers on the calibration of diverse indices in hot-humid cities, four of which proposing ranges for the UTCI thermal comfort/stress assessment scale.

INDOOR AND TRANSITIONAL SPACES Even though the UTCI was meant for outdoor environments, some papers have used the UTCI in indoor environments (Knott and Evins 2013; Langner et al. 2013; Mazon 2014; Adekunle 2018; Walikewitz et al. 2018; Adekunle 2019; Adekunle and Nikolopoulou 2019; Adekunle 2020) and in transitional spaces: Huang et al. (2017) used the UTCI for evaluating pedestrian thermal perception in areas underneath elevated buildings and Zhang, Y. et al. (2020) evaluated thermal perception in semi-open transition spaces with overhead layer connecting university buildings with surveys, controlled actions and measurement of physiological parameters. In regard to this matter, Gamero-Salinas et al. (2021) list the UTCI among othe relevant thermal comfort indices but cautiously do not use it as their study involved semi-outdoor spaces and the index is primarily meant for outdoor spaces. Wu et al. (2021) drew comparisons of index performance for predicting thermal sensation in outdoor, transitional and indoor spaces, proposing adaptation of diverse indices for the evaluation of different spaces.

## 2.3 Climate-Related Impacts on Human Health

Climate-related impacts on human health, expressed as changes in the UTCI, are explored by some authors. An initial paper by Idzikowska (2011) performed an analysis of the UTCI with respect to mortality in chosen European cities, situated in different climate zones, thereby starting a new avenue of research with the UTCI as relevant indicator in terms of human health.

In some of these studies, the relationship between UTCI data and medical records has been sought. An example of this is a series of papers conducted for the Polish city of Olsztyn, characterized by cold climate type. For that location, the relationship between meteorological parameters and the calculated UTCI versus consultations for Respiratory Tract Infections (RTI) was analyzed by Romaszko et al. (2019), concluding that the UTCI may be considered a valuable predictive parameter for forecasting seasonal increases in RTI cases. A similar work, focusing on consultations for hypertension and their relationship to the UTCI was presented by Skutecki et al. (2019a). A further paper studied the relationship between lipid parameters and meteorological and OTC conditions (including the UTCI) for Olsztyn (Skutecki et al. 2019b). More recently, a paper by Lindner-Cendrowska and Bröde (2021) analyzed biometeorological data, expressed as the UTCI and influenza-like incidence in Warsaw, from medical data provided by the Voivodship Unit of the State Sanitary Inspection in Warsaw.

Based on Emergency Room (ER)-visit records from three major hospitals in Beijing, China, Ma et al. (2018a) used the UTCI for assessing the climatic thermal environment, correlating the frequency of ER visits and morbidity with thermal stress data. Also for Beijing, Ma et al. (2018b) used ER-visit records for correlating with UTCI data for respiratory tract diseases.

Diverse aspects related to human health in connection to UTCI variations can thus be found in the literature: heat-related morbidity (Hartz et al. 2013); heat- or cold-related mortality (Idzikowska 2011; Morabito et al. 2014a, Kuchcik 2020), or both, from epidemiological information (Błażejczyk et al. 2014c); heat- and cold-related effects on cardiovascular mortality (Urban and Kysely 2014); connections between heat stress and overexposure to unperceivable ultraviolet radiation (Morabito et al. 2014b); incidence of asthma exacerbation (Romaszko-Wojtowicz et al. 2020); and, analysis of health impacts originated from changes in daylighting conditions in melatonin secretion in human subjects (Wieczorek et al. 2013).

Cross-disciplinary studies involving atmospheric sciences, OTC surveys and health-related aspects could also be identified. Romaszko-Wojtowicz et al. (2020) analyzed the relationship between atmospheric conditions and meteorological variables using medical data and the UTCI for Olsztyn, Poland. Changes in mortality due to weather patterns were also researched by Ghalhari and Mayvaneh (2016), using air temperature and UTCI data to correlate with respiratory disease mortality in Iran. Pappenberger et al. (2015) analyzed the application potential of the UTCI for forecasting thermal health hazards. The combined effect of air pollution and thermal stress on morbidity was investigated by Lokys et al. (2018) along a north-south gradient in Germany, using air-pollution, meteorological and hospital admission data for obtaining correlations. Heat stress mortality was analyzed in its connection with meteorological variables expressed as UTCI units for various cities in Poland using historical meteorological series and in combination with climate change scenarios (Błażejczyk et al. 2018). Using meteorological measurements and concurrent questionnaire OTC surveys, Pantavou et al. (2016) looked at the relationship between heat-related, reported symptoms by interviewees with thermal index values throughout seasons in a Mediterranean climate (Athens, Greece). Newly, Cymes et al. (2021) correlated emergency service interventions for fainting with atmospheric data, using the UTCI as a predictor of fainting, for Poland.

VULNERABLE POPULATIONS: Morabito et al. (2014a) compared mortality data for very elderly people ($\geq$75 years) to UTCI data, establishing a causal relationship between mortality and heat stress for that population stratum. The issue of mortality in the elderly in relation to the amount of vegetation and the distance to water bodies and interactions with microclimate expressed as equivalent temperature (UTCI) was explored for Lisbon with remote sensing and GIS (Burkart et al. 2016). Decreased mortality was observed as a consequence of reduced heat in those areas, with implications to planning and climate change mitigation strategies. Larriva and Higueras (2020) evaluated health risks related to thermal and acoustic discomfort of the elderly in Madrid, Spain, using surveys and field observations. The application of the UTCI and its ability to predict thermal stress in the elderly is compared to that of other thermal comfort indices by Chindapol et al. (2017). Another vulnerable group is the homeless population. Cold-related hypothermia was correlated to weather conditions, expressed as UTCI units by Krzyżewska et al. (2017), offering useful advice for policymakers assisting the prevention of hypothermia in the most vulnerable groups of society. Romaszko et al. (2017) used UTCI predictions for assessing thermal comfort/stress conditions and mortality among the homeless in Olsztyn, Poland, a vulnerable population permanently exposed to the outdoor environment. Baruti et al. (2020) investigate the case of informal settlements in a warm-humid city in terms of OTC with SET*, UTCI, sensitizing the research community to the need for conducting more research on such settlements, with the aim to promote upgrading schemes.

MAPPING: Di Napoli et al. (2018) investigate the potential of the UTCI for predicting heat-related health risks in Europe. Using historical meteorological data, authors were able to create thermal bioclimate maps for Europe in summer thereby identifying hot spots areas with potential heat stress conditions. The predictability of the UTCI was assessed by comparing heat stress conditions verified against mortality data. Heat stress and probability of occurrence of wildfires related to meteorological conditions have been evaluated with UTCI output for mapping such hazards for decision makers contributing to the efforts to develop a European Multi Hazard-Early Warning System (MH-EWS) platform (Vitolo et al. 2019). Di Napoli et al. (2019) also developed mapping for better defining health-based, heat-stress thresholds in terms of the UTCI during heatwaves for warning systems. Authors used heat-attributable excess mortality for cities in France as a test bed against meteorological reanalysis data. By means of reanalysis, past records are fed into a numerical model in order to provide reasonable estimates of the state of the system within a given region. In this context, the ERA5-HEAT dataset is an exemplary work presented by Di Napoli et al. (2020) showing global UTCI data. Antonescu et al. (2021) used output from ERA5-HEAT reanalysis for evaluating thermal stress over Europe between 1979 and 2019 in terms of UTCI data.

## *2.4 Bioclimate*

Oxford dictionary defines bioclimate as "a climate or climatic zone considered or defined in relation to living organisms and their distribution". The main purpose of such an application of the UTCI is to assess predicted thermal conditions that might affect humans in outdoor spaces. In this case, studies might be interested in the present situation of a given location or that of a larger area, looking at the spatial variability of bioclimate and also evaluate the development of such conditions over time.

Methods here range from distributed onsite measurements, recorded data from existing meteorological stations, mobile measurements to GIS data, generally with spatial and temporal analysis, extending from very local conditions (e.g. university campuses) to entire regions, aiming at different applications, including recreational and tourist activities.

Starting with a bioclimate study of selected cities, Paris, Rome and Warsaw, which used an existing database with weather data over time (Idzikowska 2010), the majority of studies investigate bioclimate using fixed meteorological station data with estimated UTCI data: Novák (2011), Chabior (2011), Nidzgorska-Lencewicz and Mąkosza (2013), Mąkosza (2013), Dobek and Krzyżewska (2016), Błażejczyk et al. (2015), Nidzgorska-Lencewicz (2015), Araźny et al. (2016), Mąkosza and Nidzgorska-Lencewicz (2016), Zhang et al. (2018), Koźmiński and Michalska (2019), Dobek et al. (2020), Wereski et al. (2020), Błażejczyk et al. (2020a).

Jacobs et al. (2019) evaluated heat stress conditions with the UTCI in the South Asian cities Delhi, Dhaka and Faisalabad. In this study, authors analyzed the temporal evolution of heat exposure as well as intra-urban differences, using a mix of meteorological measurements from mobile and stationary devices.

A novel approach in monitoring outdoor comfort uses measured data from 'citizen scientists' in a network generated for a field study in Australia. The creation of such a network aims to circumvent difficulties arising from weather stations data, which do not consider the urban heat island effect and also from amateur networks, which may have the issue of unreliable data gathering and equipment. Rajagopalan et al. (2020) present their citizen science approach for assessing the UTCI in outdoor areas, thereby providing useful data for policy makers for urban planning. Using an existing network of citizen weather stations (CWS), Varentsov et al. (2020a) propose the use of crowdsourced meteorological observations for urban climate research and applied monitoring services in Moscow, Russia, with an application that is available at http://carto.geogr.msu.ru/mosclim/ and that will be further developed to include a real-time thermal comfort assessment based on PET and UTCI data.

Another novel approach for assessing OTC and thermal stress in urban areas is proposed by Ruedisser et al. (2020), employing 3D-data acquired from drone flights, pre-existing city models and a Monte-Carlo ray-tracing algorithm. Their method allows the calculation of the UTCI in surveyed areas.

MAPPING AND SPATIAL DISTRIBUTION OF THE UTCI: Several studies employ data from extant meteorological stations to perform a spatiotemporal analysis of greater regions. Chi et al. (2018) used 30-year historical meteorological data for analyzing trends in OTC predicted by the indices PET and UTCI for cities in mainland China thereby defining climate types. Also for China, a bioclimate mapping with spatiotemporal analysis of UTCI distribution was performed by Wu et al. (2019) using data from 591 meteorological stations comprising historical weather data. Roshan et al. (2019) utilized a statistical technique using macroscale data from meteorological stations for defining thermal comfort thresholds for four major thermal comfort indices, including the UTCI, in different climate zones in Iran. Spatiotemporal distribution of OTC levels expressed as UTCI maps were generated from historical data obtained at nine weather stations representative of the three climatic zones in Nigeria by Balogun et al. (2019). Vinogradova (2020) generated UTCI data and the spatial distribution of the UTCI for the Russian territory using observations from 500 meteorological stations in Russia.

Another approach is based on predicted and reanalysis data. The ERA5-HEAT dataset shows historical spatially gridded data for the UTCI in an hourly basis worldwide and is freely downloadable for diverse environment-health applications (Di Napoli et al. 2020). Zeng et al. (2020) presented a spatial-temporal analysis of the UTCI in the China-Pakistan economic corridor using 40-year reanalysis data from the European Centre for Medium-Range Weather Forecasts (ECMWF). Other studies adopted ERA-Interim reanalysis data for evaluating the spatial distribution of the UTCI over the Russian territory using historical data (Varentsov et al. 2020b, Konstantinov et al. 2020).

Wibowo and Salleh (2017) employed satellite imagery (Landsat land-surface temperatures 'LST') for assessing thermal conditions with UTCI distribution over the University of Indonesia campus. Wang et al. (2020) also used satellite data for running spatial analysis of UTCI distribution and mapping for over the Yangtze River Delta urban agglomeration. The relationship between LST and land cover over time was evaluated by Obiefuna et al. (2021) for the region of Lagos, Nigeria.

Leroyer et al. (2018) used the UTCI for predicting bioclimate with a Numerical Weather Prediction model in urban areas of Toronto, Canada. The index has been incorporated into the Canadian Numerical Weather Prediction (NWP) model with the aim of improving thermal conditions in urban areas.

## 2.5 Meteorological Analyses

Under this category, we include the use of the UTCI as an aid to atmospheric studies, in this case with outputs converted to predicted thermal comfort. Also included are particular aspects related to Climatology, such as the interpretation of historical data in terms of comfort parameters.

Applications of the index in Meteorology are various, ranging from understanding the relationship between large-scale atmospheric conditions on bioclimate (Więcław and Okoniewska 2015; Bartoszek et al. 2017; Kolendowicz et al. 2018; Katavoutas and Flocas 2018; Owczarek 2019; Głogowski et al. 2020b; Błażejczyk et al. 2020b; Okoniewska 2020), analysis of historical meteorological data for specific locations or regions (Bleta et al. 2014; Pecelj et al. 2017; Yan et al. 2019; Chi et al. 2018; Rozbicka and Rozbicki 2018; Katavoutas and Founda 2019) to the sensitivity of the index to particular aspects and meteorological variables such as the mean radiant temperature (Krüger et al. 2014), wind and rain patterns (Nastos et al. 2017; Hadianpour et al. 2019; Roshan and Moghbel 2020) or in combination (Fröhlich and Matzarakis 2016; Provençal et al. 2016; Du et al. 2020a).

## 2.6 Climate Change Research

Under this category, some studies looked at extreme weather conditions, interpreting those in terms of UTCI values, others utilize modeling techniques and climate change scenarios to assess biometeorological conditions in the long term.

Applications of the index with respect to impacts on human health due to extreme weather events, expressed by UTCI values or stress categories, have been explored by a few authors such as Nastos et al. (2013), Saneinejad et al. (2014a, b), Liu et al. (2015), Krzyżewska et al. (2016), Giannaros et al. (2017), Kong et al. (2017), Roshan and Nastos (2018), Mohammadi et al. (2018), Tomczyk and Owczarek (2020). Methods of assessment included meteorological records, field measurements or historical data from meteorological stations. Forecasting models have been used by some authors to obtain climate change projections, having also as relevant output UTCI data: Cheung and Hart (2014), Lokys et al. (2015), Chi et al. (2019), Brecht et al. (2020), Takane et al. (2020).

## 2.7 *Urban Climate and Planning Studies*

The most common usage of the UTCI for urban planning is by means of computer simulations, and in this case the ENVI-met package is the most frequently used software for the evaluation of rehabilitation schemes, microclimatic modifications by means of vegetation or passive techniques and so on. Some extensions, apps or plug-ins have additionally been introduced by some authors, as well as other simulation packages.

SIMULATION-BASED STUDIES: The first paper identified that employed simulation tools for evaluating urban areas with the UTCI was published by Goldberg et al. (2013), evaluating planning measures in terms of predicted OTC with simulations in ENVI-met and Rayman for Dresden, Germany. As at the time the available ENVI-met version had not yet incorporated the BioMet post-processor tool so that data from ENVI-met were post-processed in the Rayman model (Matzarakis et al. 2010), which calculates radiation-balance components as well as diverse OTC indices.

Further studies looked at specific features relevant for facilitating the assessment of the UTCI: the development of a novel method for assessing the sky-view factor, a morphological attribute used in planning that affects radiant heat exchange (Park and Tuller 2014); the use of automatic algorithms for facilitating the simulation process (Bajšanski et al. 2015; Milošević et al. 2017a); a 3-step simulation process for shading design in outdoor spaces (Mackey et al. 2015); the use of CFD for estimating OTC conditions (Manickathan et al. 2018); the use of genetic algorithms (Xu et al. 2019b, c); the development of novel micro-scale bioclimate models such as the VTUF-3D (Nice et al. 2018), the SkyHelios model, which incorporates UTCI output (Fröhlich and Matzarakis 2018; Fröhlich et al. 2019) and SOLENE-microclimat (Imbert et al. 2018); the coupling of existing models such as SOLWEIG (SOlar Long-Wave Environmental Irradiance Geometry) with energy-balance models for predicting OTC and UTCI distribution (Oswald et al. 2019); the coupling between the WRF model and an urban canopy model (Imran et al. 2018); and between ENVI-met and CFD modeling (Perini et al. 2017); development as apps such as ENVI-BUG for expanding ENVI-met input features relevant for OTC predictions (Fabbri et al. 2017); and, the utilization of a rapid fine-scale spatial model for assessing the UTCI in a 2D environment (Back et al. 2021). Recently, Fröhlich and Matzarakis (2020) compared outcomes in terms of different OTC indices including the UTCI, using different approaches relative to simulations run in a biometeorology module implemented into the Parallelized Large-Eddy Simulation Model (PALM) system. Geletič et al. (2021) used PALM-4U for investigating thermal benefits in an urban area from spatiotemporal patterns of thermal exposure.

Other studies used simulations to assess bioclimate conditions related to particular planning measures by means of human bioclimatic mapping (Park et al. 2014) and the analysis of urban planning measures and their impacts on OTC (Minella et al. 2014; Bajšanski et al. 2015; Huang et al. 2019). Evaluations based on computer simulation were carried out for achieving diverse aims: bioclimate impact from changes in urban morphology attributes (Tumini et al. 2016; Achour-Younsi and Kharrat 2016; Xu et al. 2019c, d; Hamdan and De Oliveira 2019; Aghamolaei et al. 2020; Xiong et al. 2020), effects on OTC from vegetation and urban greening schemes (Wu et al. 2016; Milošević et al. 2017b; Liang et al. 2019; Matzarakis and Fröhlich 2017; Manickathan et al. 2018; Nice et al. 2018; Shata et al. 2021; Meili et al. 2021), bioclimate effects from albedo modification of urban surfaces (Schrijvers et al. 2016; Dietrich 2018; Grifoni et al. 2017), cooling benefits from the proximity to water bodies (Rahul et al. 2020), cooling strategies and UHI mitigation measures in summer (Huang et al. 2016b; Morille and Musy 2017; Imran et al. 2018), trade-offs between solar photovoltaic (PV) roofs, vegetation and cool roofs in terms of outdoor thermal comfort and PV conversion efficiency (Berardi and Graham 2020).

Traditional urban design (Nasrollahi et al. 2017) and land-use schemes (Choe and Han 2020) or urban planning modifications over time (Barbosa et al. 2019) were also simulated for assessing bioclimate with UTCI. Such studies are in many cases supported by field measurements in order to calibrate the simulation model for numerical analyses thereby testing different scenarios in terms of urban geometry and microclimate mitigation in a given urban area (Minella et al. 2014; Battista et al. 2016; Tseliou and Tsiros 2016; Bandurski et al. 2020). Grifoni et al. (2017) evaluated microclimate and thermal comfort amelioration in terms of the UTCI with 'cool façades' by means of numerical tools and monitoring of relevant meteorological variables. Montazeri et al. (2017) evaluated numerically thermal comfort effects obtained from evaporative cooling by water spray systems during a heatwave in Rotterdam, the Netherlands. As in Grifoni et al. (2017), the study was supported by experimental data and the UTCI was used to assess the heat stress reduction due to evaporative cooling.

MEASUREMENT-SUPPORTED STUDIES: Thermal monitoring of urban areas and the post-processing of monitored data as OTC information can aid planners to know the benefits of urban redevelopment projects, to compare different urban morphologies in order to guide such processes or to test the advantages of innovative microclimate mitigation strategies.

Amongst such strategies, urban greenery has been explored in several locations as a means to improve OTC conditions. Cheung and Jim (2018a) compared the cooling effects from trees by intercepting solar radiation and evapotranspiration on human thermal comfort with the UTCI against the shading effect of an extensive concrete shelter in an urban park in Hong Kong, China. In a subsequent paper, Cheung et al. (2020) monitored seasonal and meteorological effects on the cooling magnitude of trees in the humid subtropical climate of Hong Kong with respect to air temperature and OTC output including UTCI data. Petralli et al. (2020) analyzed the thermal effects obtained from tree shading and ground material properties on OTC in terms of UTCI data using onsite meteorological measurements in an urban park in Florence, Italy. Also focusing on the usage of green spaces for urban climate mitigation, Kurban Lopez and Grasso (2017) used monitored data in diverse green areas scattered in an arid city in west-central Argentina to analyze their impact on predicted outdoor thermal comfort with the UTCI. Gómez et al. (2018) studied effects from natural elements on OTC in urban areas of Valencia, Spain. Data were recorded at different locations and converted to UTCI output for predicting the thermo-physiological state of people carrying out activities in the open space. Du et al. (2020b) studied the shading effect from vegetation and building elements on micrometeorology and OTC.

Another course of action for investigating the benefits of urban greenery as a planning strategy is based on questionnaire-surveys alongside thermal monitoring. Martini et al. (2014, 2020) carried out OTC surveys in Curitiba, Brazil, analyzing the effect of street trees on outdoor thermal sensation against UTCI data. Cheung and Jim (2018b) performed OTC surveys in urban green spaces in Hong Kong, China, and looked at possible effects from vegetation on participants' thermal perception and on attendance and usability of such spaces. A further paper on the subject evaluated the impact of Intensive Green Roofs (IGR) on UTCI conditions from microclimatic monitoring in specific sites with and without woodland IGR in Hong Kong (Lee and Jim 2019). Also using microclimatic monitoring, Petralli et al. (2020) assessed UTCI conditions in different green area settings in Florence, Italy.

Another heat stress mitigation strategy is based on the use of blue infrastructure and evaporative cooling. Hendel et al. (2016) focused on Urban Heat Island (UHI) mitigation by means of pavement watering as a feasible UHI countermeasure in Paris, France. Resulting effects were expressed as UTCI values. A subsequent paper examined pedestrian thermal comfort improvement through pavement watering in Paris, as part of mitigation measures from the City of Paris to counteract heat stress in summer (Parison et al. 2017). Parison et al. (2020) used later the database collected in 2013 for pavement watering to propose a statistical method for assessing effects of UHI countermeasures. Ulpiani et al. (2019) conducted an experimental study on an overhead water mist cooling system designed to improve OTC during summer in two Italian cities (Ancona and Rome). Measured microclimatic data were interpreted in terms of the UTCI along with surveys with local population. The same approach was employed by Oh et al. (2019) with several outdoor indices including the UTCI in a study on mist spraying to relieve thermal discomfort in hot weather involving the assessment of environmental factors, physiological responses and surveys, in Tokyo, Japan. A similar study was conducted in Antofagasta, located in the Atacama desert region of Chile, with a misting prototype and comprising subjective thermal perception and field measurements (Desert et al. 2020). Also in a hot-dry location (Tempe, Arizona), Vanos et al. (2020) evaluated the effectiveness of evaporative misters in terms of OTC (UTCI data) and business managers' motivations to use such equipment. The influence of blue infrastructure on OTC and microclimate amelioration was studied by Fung and Jim (2020) on days with different sky cover for Hong Kong, China. Onsite monitoring for sites with and without a pondside lawn were inter-compared in terms of UTCI output. Lehnert et al. (2020) investigated the combined effect of blue and green infrastructure on OTC in terms of UTCI data measured in public downtown areas.

Also with a view to planning applications, a few studies focus on particular aspects that define urban morphology such as the aspect ratio or the H/W relationship (H = average building height, W = street width), street canyon orientation and vegetation canopy. The resulting building density can bring about impacts on OTC conditions. Coutts et al. (2016) analyzed such effects in combination with urban street trees for Melbourne, Australia, using locally monitored meteorological data. Pan and Du (2017) carried out measurements in diverse urban villages in Shenzhen, China aiming to explore the correlation between urban density and OTC characteristics, examining both UHI intensity and UTCI data. Streda et al. (2014) analyzed the impact of land cover and land use on local meteorological variables and OTC in Hradec Králové, Czech Republic. Using the Local Climate Zones (LCZ) scheme (Stewart and Oke 2009a, b), Liu et al. (2018) selected locations for carrying out OTC surveys with concurrent measurements of meteorological parameters in Schenzhen, China. The LCZ scheme was also used in another study in combination with OTC data and UTCI output (Kwok et al. 2019), where authors conducted a spatiotemporal thermal climate analysis of Toulouse, France, under warm and dry summer conditions with the mesoscale atmospheric model Meso-NH. A similar temporal analysis was carried out using weather station data for two different locations, i.e. outside and inside Ankara, Turkey, for the purpose of evaluating thermophysiological stress (Nouri et al. 2021).

Finally, albedo properties have been investigated by Sen et al. (2020) with measurements of relevant OTC variables next to (pedestrian scale) reflective parking lots for microscale Urban Heat Island mitigation.

## 2.8 *Tourism*

As one of the expected areas of application of the index, the relationship between bioclimate, OTC conditions and tourism attractiveness is relevant for obvious economic reasons. Research in this area can serve as important aid to decision-making and management by the tourism industry. Spatial and temporal observations as well as the mobility of people within evaluated areas are the most common methods used for that. Climate change is also an impacting factor which has been explored.

With the aim of providing relevant information for decision makers, applications of the UTCI in international tourism, regarding the intrinsic importance of weather and climate for decision-making and their impacts on temporal and spatial patterns of tourism demand have been explored by Rutty and Scott (2014, 2015).

With the goal of supporting the tourist industry and management, some papers analyzed more locally spatial and temporal variations of the UTCI such as in

Warsaw, Poland (Lindner-Cendrowska 2013; Rozbicka and Rozbicki 2020), at Zlatibor Moutain in Serbia (Basarin et al. 2018), in particular summer destinations in China using historical weather data (Yang et al. 2018), and with GIS tools for Loznica, Serbia (Pecelj et al. 2018). In a broader spatial scale, Roshan et al. (2018) looked at implications to international tourism in Iran of historical meteorological data combined as UTCI thermal comfort data, thereby identifying biometeorological clusters which can aid tourism management in that country. A subsequent discussion under consideration of climate change influences is presented in a book chapter by Fitchett and Roshan (2020). Ge et al. (2017) investigated the spatial pattern of thermal bioclimatic conditions in China having the UTCI as relevant thermal comfort indicator for applications in tourism bioclimatology and from a tourism perspective. Hua et al. (2020) used historical data from a number of weather stations in mainland China to assess the thermal suitability of tourist attractions.

In terms of mobility, Liang and Bi (2017) correlated tourist flow, online booking and climatic information, expressed as UTCI units. Not precisely related to tourism but to inland travel, an analysis between weather variability and trip chaining and travel behavior was evaluated by Liu et al. (2016) using the UTCI as relevant indicator of meteorological conditions and explanatory variable in the models generated by the authors. Liu et al. (2020) analyzed how perception of weather conditions, expressed as UTCI values, affect decision-making for leisure activities. The method uses self-reported travel diary surveys conducted in Sweden.

Impacts from current climate and climate change projections on tourism were investigated by Miszuk et al. (2016) within the framework of the EU KLAPS project in the German-Polish border region, using in their study the UTCI as thermal bioclimate predictor. Kong et al. (2019) performed a study of climate change impacts on OTC with the UTCI as relevant indicator in diverse regions of China during summer, thereby identifying regions that are suitable for "sunbird" tourism, i.e. regions to which citizens will potentially be able to escape from heat stress in summer.

## 2.9 New Tools in OTC Research

Further applications of the index can arise from new tools developed by researchers and a few of them have been identified in the literature search. A smartphone app was developed by Heusinkveld et al. (2017) that communicates a location-specific human thermal comfort forecast expressed as UTCI units based on high-resolution numerical weather prediction (NWP) and geoinformation. A novel wind rose biometeorological data visualization tool was developed by Sadeghi et al. (2018), which takes into consideration the OTC dimension in terms of predicted UTCI data for the assessment of comfort wind resources in Sydney, Australia. The study also uses climate projections for 2030 accounting for climate change estimates. Baglia et al. (2019) propose a dynamic microclimate modeling for urban areas in China with the purpose of assessing pedestrian comfort (UTCI output), air quality and building ventilation potential.

DISCUSSION AND REVIEW PAPERS: In a review paper, Ongoma et al. (2016) describe the UTCI as a potential index to be used for thermal comfort assessment in planning with respect to growing urbanization in Nairobi City, Kenya. Hai and Feng (2018) propose an approach termed 'thermalscape' for evaluation the human-centered outdoor thermal environment performance, using the UTCI as its core indicator. Graham et al. (2020) discuss the obstacles found while implementing microclimate analysis as a design driver in an architectural office and measures that could be taken to overcome them. Basarin et al. (2020) present a systematic review of recent studies of biometeorological extremes in Europe, during extreme conditions and heatwaves, looking at influences on human health and providing an overview of heat-health warning systems (HHWS) across the European continent.

## 3 Categorization

Höppe (2002) had envisaged applications of the UTCI in weather reports and forecasts, climate change scenarios and tourism, while initial publications by the former members of the ISB expert group also listed as potential areas of application, apart from those, bioclimatological assessments, bioclimate mapping, urban planning, outdoor recreation and health, epidemiology and climate research. In the 2012 special issue of IJBM, Jendritzky et al. (2012) suggested following areas to be expected to profit from the UTCI: public weather service, public health system, precautionary planning and climate impact research in the health sector.

The literature review showed that, basically, two areas concentrate most of the applications of the index: human-centered research focused on thermal perception in outdoor spaces and studies aimed at urban planning and microclimate amelioration (Table 1), together amounting to just over 59% of all papers reviewed. It should be noted that the total amount of categorized papers has varied slightly from the total sample of publications in this review, with the exclusion of repeated items, of papers not directly related to the UTCI usage and of a few papers that were not available online. From the initial samples of 320 and 304 documents, in Scopus and WoS, respectively, the combined list of valid and non-duplicated publications amounted altogether to 317 papers.

As performance tests of the index in comparison to other existing indices can be also understood as a preliminary phase in OTC research, the bulk of research output (one third of the total) in this literature review is related to human biometeorology and its subfield human thermal comfort assessment including heat stress evaluation. Under such application, the testing of the index against field measurements and the calibration of the thermal assessment scale are necessary stages. Sensitivity tests and new methods in OTC assessment have also been explored.

A second major application is in urban climate studies and in studies related to urban planning, with about one quarter of the total. In that category, the great part of contributions is based on simulation tools, such as ENVI-met, either performing sensitivity tests with those tools or improving them by various means (coupling,

**Table 1** Publication output on the UTCI for the time frame 2000–2020, ranked by number of publications

| Topic/Application of the UTCI | Amount of peer-reviewed publications | Percentage of total (%) | References |
|---|---|---|---|
| 1. Outdoor thermal comfort (OTC) and thermal stress | 102 | 32 | Höppe (2002), Mateeva (2011), Bröde et al. (2012a, b), Krüger et al. (2012), Pantavou et al. (2013), Nidzgorska-Lencewicz and Mąkosza (2013), Bröde et al. (2013a, b), Dobek et al. (2013), Milewski (2013), Pantavou et al. (2013), Knott and Evins (2013), Langner et al. (2013), Abdel-Ghany et al. (2013), Błażejczyk et al. (2014a, b), Watanabe et al. (2014), Lai et al. (2014), Pantavou et al. (2014), Mazon (2014), Abdel-Ghany et al. (2014), Krüger et al. (2015), Maras et al. (2016), Hamanaka and Bueno-Bartholomei (2017), Vatani et al. (2015), Watanabe and Ishii (2016), Huang et al. (2016a), Hanipah et al. (2016), Nassir et al. (2017), Zamanian et al. (2017), Bjerre et al. (2017), Lam and Hang (2017), Krüger and Drach (2017), Krüger (2017), Reinhart et al. (2017), Huang et al. (2017), Hanipah et al. (2017), Krüger et al. (2017a, b), Hadianpour et al. (2018), Xu et al. (2018), Li et al. (2018), Bröde et al. (2018), Gao et al. (2018), Honjo et al. (2018), Liu and Li (2018), Ohashi et al. (2018), Lam and Lau (2018), Lam et al. (2018), Fang et al. (2018), Sharifi and Boland (2018), Pantavou et al. (2018), Adekunle (2018), Walikewitz et al. (2018), Sen and Nag (2019), Messeri et al. (2019), Galan and Guedes (2019), Szer et al. (2019), Fang et al. 2019b), Abadla et al. (2019), Haidanpour et al. (2019), Lam et al. (2019), Cheung and Jim (2019), Dhariwal et al. (2019), Jin et al. (2019), Adekunle (2019), Adekunle and Nikolopoulou (2019), Xu et al. (2019a), Zhang L. et al. (2020), Zhang Y. et al. 2020(Zhu et al. (2020), Silva and Hirashima (2020), Manavvi and Rajasekar (2020), Das and Das (2020), Chen et al. (2020), Thorsson et al. (2020), Gasparetto and Nesseler (2020), Konefał et al. (2020), Miszuk (2020), Głogowski et al. (2020a, b), Krzyżewska et al. (2020), Pecelj et al. (2020), Dong et al. (2020), Lee et al. (2020), Talhi et al. (2020), Li et al. (2020), Lam et al. (2020), Liu et al. (2020), Jin et al. (2020), Xue et al. (2020), Lam et al. (2020), He et al. (2020), Sharifi and Boland (2020), Krüger et al. (2020), Borges et al. (2020), Chen et al. (2020), Binarti et al. (2020), Adekunle (2020), Niu et al. (2020), An et al. (2021), Kuchcik et al. (2021), Lehnert et al. (2021), Gamero-Salinas et al. (2021), Wu et al. (2021), Lam et al. 2021a, b), Miszuk (2021) |

(continued)

**Table 1** (continued)

| Topic/Application of the UTCI | Amount of peer-reviewed publications | Percentage of total (%) | References |
|---|---|---|---|
| 2. Urban climate and planning studies | 85 | 27 | Matzarakis et al. (2010), Goldberg et al. (2013), Park and Tuller (2014), Park et al. (2014), Minella et al. (2014), Martini et al. (2014), Streda et al. (2014), Bajšanski et al. (2015), Mackey et al. (2015), Bajšanski et al. (2015), Tumini et al. (2016), Achour-Younsi and Kharrat (2016), Wu et al. (2016), Schrijvers et al. (2016), Huang et al. (2016b), Battista et al. (2016), Tseliou and Tsiros (2016), Hendel et al. (2016), Coutts et al. (2016), Ongoma et al. (2016), Milošević et al. (2017a, b), Perini et al. (2017), Fabbri et al. (2017), Morille and Musy (2017), Nasrollahi et al. 2017), Grifoni et al. 2017), Montazeri et al. 2017), Grifoni et al. (2017), Kurban Lopez and Grasso (2017), Parison et al. (2017), Pan and Du (2017), Manickathan et al. (2018), Nice et al. (2018), Fröhlich and Matzarakis (2018), Imbert et al. (2018), Imran et al. (2018), Matzarakis and Fröhlich (2017), Manickathan et al. (2018), Nice et al. (2018), Dietrich (2018), Imran et al. 2018), Cheung and Jim (2018a, b), Gómez et al. (2018), Liu et al. (2018), Hai and Feng (2018), Oswald et al. (2019), Hamdan and De Oliveira (2019), Liang et al. (2019), Barbosa et al. (2019), Lee and Jim (2019), Ulpiani et al. (2019), Oh et al. 2019), Kwok et al. (2019), Fröhlich and Matzarakis (2020), Xu et al. (2019b, c, d, Fröhlich et al. (2019), Huang et al. (2019), Aghamolaei et al. (2020), Rahul et al. (2020), Berardi and Graham (2020), Xiong et al. (2020), Choe and Han (2020), Bandurski et al. (2020), Cheung et al. (2020), Petralli et al. (2020), Du et al. (2020b), Martini et al. (2020), Parison et al. (2020), Desert et al. (2020), Vanos et al. (2020), Fung and Jim (2020), Lehnert et al. (2020), Nouri et al. (2021), Sen et al. (2020), Graham et al. (2020), Basarin et al. (2020), Back et al. (2021), Shata et al. (2021), Geletič et al. (2021), Meili et al. (2021) |
| 3. Climate-related impacts on human health | 29 | 9 | Idzikowska (2011), Idzikowska 2011), Hartz et al. 2013), Wieczorek et al. (2013), Urban and Kysely (2014), Morabito et al. (2014a, b), Błażejczyk et al. (2014c), Pappenberger et al. (2015), Ghalhari and Mayvaneh (2016), Pantavou et al. (2016), Burkart et al. (2016), Krzyżewska et al. (2017), Chindapol et al. (2017), Romaszko et al. (2017), Lokys et al. (2018), Ma et al. (2018b), (Błażejczyk et al. (2018), Di Napoli et al. (2018), Ma et al. (2018a), Romaszko et al. (2019), Vitolo et al. (2019), Di Napoli et al. (2019), Skutecki et al. (2019a, b), (Kuchcik (2020), Romaszko-Wojtowicz et al. (2020), Larriva and Higueras (2020), Baruti et al. (2020), Di Napoli et al. (2020), Lindner-Cendrowska and Bröde (2021), Cymes et al. (2021), Antonescu et al. (2021) |

(continued)

**Table 1** (continued)

| Topic/Application of the UTCI | Amount of peer-reviewed publications | Percentage of total (%) | References |
|---|---|---|---|
| 4. Bioclimate | 31 | 10 | Idzikowska (2010), Novák (2011), Chabior (2011), Nidzgorska-Lencewicz and Mąkosza (2013), Mąkosza (2013), Dobek and Krzyżewska (2016), Błażejczyk et al. (2015), Nidzgorska-Lencewicz (2015), Araźny et al. (2016), Mąkosza and Nidzgorska-Lencewicz (2016), Wibowo and Salleh (2017), Zhang et al. (2018), Chi et al. (2018), Leroyer et al. (2018), Koźmiński and Michalska (2019), Jacobs et al. (2019), Roshan et al. (2019), Wu et al. (2019), Balogun et al. (2019), Dobek et al. (2020), Wereski et al. (2020), Rajagopalan et al. (2020), Ruedisser et al. (2020), Vinogradova (2020), Zeng et al. (2020), Di Napoli et al. (2020), Varentsov et al. (2020a, b), Konstantinov et al. (2020), Wang et al. (2020), Błażejczyk et al. (2020a), Obiefuna et al. (2021) |
| 5.Meteorological analyses | 22 | 7 | Bleta et al. (2014), Krüger et al. (2014), Więcław and Okoniewska (2015), Liu et al. (2015), Fröhlich and Matzarakis (2016), Provençal et al. (2016), Bartoszek et al. (2017), Pecelj et al. (2017), Nastos et al. (2017), Chi et al. (2018), Katavoutas and Flocas (2018), Kolendowicz et al. (2018), Rozbicka and Rozbicki (2018), Katavoutas and Founda (2019), Hadianpour et al. (2019), Owczarek (2019), Yan et al. (2019), Błażejczyk et al. (2020b), Du et al. (2020a), Głogowski et al. (2020b), Okoniewska (2020), Roshan and Moghbel (2020) |
| 6. Comparisons with other thermal comfort indices | 21 | 7 | Nastos and Matzarakis (2012), Błażejczyk et al. (2012), Novák (2013), Matzarakis et al. (2014), Lai et al. (2014), Pantavou et al. (2014), Tumini and Perez Fargallo (2015), Provençal et al. (2016), Coccolo et al. (2016), Golasi et al. (2018), Potchter et al. (2018), Zare et al. (2018), Charalampopoulos (2019), Fang et al. (2019a), Staiger et al. (2019), Charalampopoulos and Nouri (2019), Zare et al. (2019), Katavoutas and Founda (2019), Asghari et al. (2019), Santurtún et al. (2020), Lian et al. (2020) |
| 7. Climate change research | 14 | 4 | Nastos et al. (2013), Cheung and Hart (2014), Saneinejad et al. (2014a, b), Lokys et al. (2015), Krzyżewska et al. (2016), Giannaros et al. (2017), Kong et al. (2017), Mohammadi et al. (2018), Roshan and Nastos (2018), Chi et al. (2019), Tomczyk and Owczarek (2020), Brecht et al. (2020), Takane et al. (2020) |
| 8. Tourism | 15 | 5 | Lindner-Cendrowska (2013), (Rutty and Scott (2014, 2015), Liu et al. (2016), (Miszuk et al. (2016), Ge et al. (2017), Liang and Bi (2017), Yang et al. (2018), Pecelj et al. (2018), (Roshan et al. (2018), Basarin et al. (2018), Kong et al. (2019), Fitchett and Roshan (2020), Hua et al. (2020), Rozbicka and Rozbicki (2020) |

development of apps and plugins) or by using simulations to assess bioclimate changes arising from planning measures. Field measurements were also performed, aimed at planning applications. Of particular interest are studies related to green and blue urban infrastructure.

A third worth-mentioning category is related to climate-related impacts on human health, including mortality, morbidity and diverse aspects related to human health in connection to UTCI variation. Of special interest here is the development of mapping approaches, such as the ERA5-HEAT dataset, presented by Di Napoli et al. (2020).

The application of the UTCI in studies of bioclimatic conditions in selected cities and regions has been pursued in a number of papers with temporal and spatial analyses, using various methods (weather station data, GIS data, existing models, mapping and so on).

Applications in meteorology and comparisons to other indices come next and encompass the use of the UTCI as an aid to atmospheric studies whereas comparisons to other relevant indices, particularly the PET index, have also been performed in order to evaluate the predictability of the index.

Finally, applications for tourism and climate change research could be identified, but in a relatively smaller number of publications.

## 3.1 Categorization Using VOSviewer

The VOSviewer tool version 1.6.16 (van Eck and Waltman 2013) allows users to create maps based on network data and visualize and explore these maps. It is aimed at bibliometric analyses and in the case of the two databases consulted for this literature review chapter (Scopus and WoS), it was possible to identify the main clusters of papers arising from common keywords using its functionality 'co-occurrence of author keywords'.

The procedure used was as follows: (a) saving search results as.csv and.txt files, in both Scopus and WoS samples, respectively, (b) creating maps with co-occurrences based on Author Keywords data in VOSviewer with full counting and a minimum number of 5 occurrences for a given keyword, and (c) filtering keywords that are very general or mere descriptors of the UTCI as well as names of specific locations. The maps generated could then be summarized as coincident keywords with some strength (not necessarily the highest link strength—defined as the number of publications in which terms occur together). Again coincident terms are neither just common descriptors for the UTCI nor location names, and are primarily related to areas of application of the index. Figure 2 presents the network visualizations for author keywords in both samples. Table 2 summarizes clusters found in both samples.

A similar cluster distribution was found in both search results, with some variations as the two samples are not exactly the same. Basically, clusters identified in VOSviewer bare a resemblance with UTCI applications listed in Table 1, with a distinction between human-centered studies (clusters 1 and 6, application topics 1 and 3), climate change research (cluster 2, application topic 7), microclimate studies

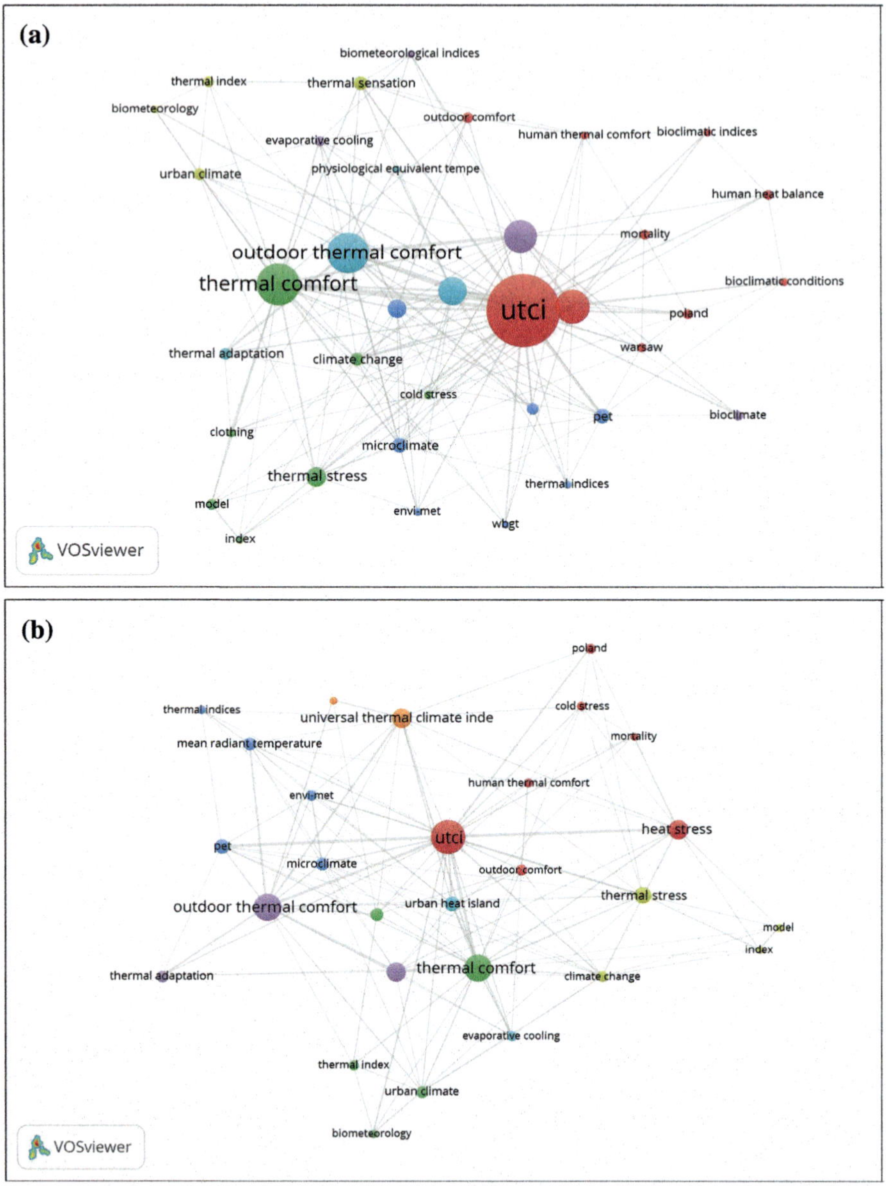


**Fig. 2** Network visualizations for author keywords: **a** Scopus search, **b** WoS search

**Table 2** Clusters defined for main author keywords in co-occurrences in Scopus and WoS, ordered by strength of link

| Scopus | WoS |
|---|---|
| Cluster 1: heat stress, outdoor comfort, mortality | Cluster 1: heat stress, outdoor comfort, cold stress, mortality |
| Cluster 2: thermal comfort, climate change, thermal stress | Cluster 2: thermal stress, climate change |
| Cluster 3: urban heat island, PET, microclimate | Cluster 3: PET, microclimate, ENVI-met |
| Cluster 4: thermal sensation, urban climate | Cluster 4: thermal comfort, thermal sensation, urban climate |
| Cluster 5: evaporative cooling, bioclimate | Cluster 5: Urban Heat Island, evaporative cooling |
| Cluster 6: outdoor thermal comfort, thermal adaptation | Cluster 6: outdoor thermal comfort, thermal adaptation |

and comparisons with other indices, predominantly with the PET index, aided or not by simulations in ENVI-met (cluster 3 and application topics 2 and 5), studies on urban climate focusing on human thermal sensation (cluster 4 and application topic 2), bioclimate and urban climate amelioration studies, particularly by means of evaporative cooling (cluster 5 and application topic 2 and 4). Application topics 6 and 8, with lower publication output and, in the case of meteorological analyses, involving a diversity of parameters of analysis, did not yield significant co-occurrences.

## 4 Conclusions

Papers on the UTCI cover different topics, but with a prevalence of OTC research and studies related to the understanding of microclimate and aimed at urban planning. Applications in meteorology, climate change and tourism are still timid, though these areas will gain in importance in the years to come. Global issues arising from climate change will tend to increase with the progress of global warming and weather extremes while worldwide tourism in the aftermath of the COVID-19 pandemic crisis will progressively rise again.

Most of the papers surveyed are still concentrated in a few countries, most evidently in Germany, Poland and China. Research output observed in Germany and Poland are in a great extent linked to studies conducted by former members of the ISB Commission responsible for the development of the UTCI. Applications in China are mounting up and currently this country is leading in outdoor thermal comfort research, as reported in a recent comprehensive review by Lai et al. (2020). With the rising, fastest growing urbanization and economic development taking place in China presently, that trend is likely to continue over the next years. Also noticed in the analysis of search results is the prominent funding of many of the studies by the National Natural Science Foundation of China, which is undoubtedly conducive to widening the application spectrum of the UTCI in that country.

The multidisciplinary character of human biometeorology when analyzing research output on UTCI applications also became evident. In the analysis of search results, the bulk of research was categorized as environmental science (28%), followed by earth and planetary sciences (25%), social sciences (16%) and by engineering (11%).

From the potential applications for the UTCI in major fields of human biometeorology, as stated by Jendritzky et al. (2012), areas such as public weather service and climate-related impacts on the health sector still need to become aware of the potentialities and practicalities of the UTCI. The next chapters will explore accuracy and regional adaptation of the UTCI to various exposure conditions and populations.

## References

Abadla Z, Schlink U, Wahab MA, Robaa SM (2019) Sea breeze climate and human thermal comfort over Gaza, Palestine. Environ Asia 12(2)

Abdel-Ghany AM, Al-Helal IM, Shady MR (2013) Human thermal comfort and heat stress in an outdoor urban arid environment: a case study. Adv Meteorol

Abdel-Ghany AM, Al-Helal IM, Shady MR (2014) Evaluation of human thermal comfort and heat stress in an outdoor urban setting in summer under arid climatic conditions. Environ Protect Eng 40(3)

Achour-Younsi S, Kharrat F (2016) Outdoor thermal comfort: impact of the geometry of an urban street canyon in a Mediterranean subtropical climate–case study Tunis, Tunisia. Proc Soc Behav Sci 216(6):689–700

Adekunle TO (2020) Occupants' comfort and stress indices in a structural timber school building in the Northeast US in different seasons. Build Res Inf 48(3):331–348

Adekunle TO, Nikolopoulou M (2019) Winter performance, occupants' comfort and cold stress in prefabricated timber buildings. Build Environ 149:220–240

Adekunle TO (2018) Summertime overheating and heat stress in multi-family Colonial Revival Style timberframe buildings. In: Proceedings of 34th PLEA international conference: PLEA 2018, pp 312–317

Adekunle TO (2019) Summer performance, comfort, and heat stress in structural timber buildings under moderate weather conditions. Smart Sustain Built Environ

Aghamolaei R, Azizi MM, Aminzadeh B, Mirzaei PA (2020) A tempo-spatial modelling framework to assess outdoor thermal comfort of complex urban neighbourhoods. Urban Clim 33:

An L, Hong B, Cui X, Geng Y, Ma X (2021) Outdoor thermal comfort during winter in China's cold regions: a comparative study. Sci Total Environ 144464

Antonescu B, Mărmureanu L, Vasilescu J, Marin C, Andrei S, Boldeanu M, Ţilea A et al (2021) A 41-years bioclimatology of thermal stress in Europe. Int J Clim (in press)

Araźny A, Uscka-Kowalkowska J, Kejna M, Przybylak R, Kunz M (2016) Diversity of biometeorological conditions in Toruń and its suburban area in 2012 | [Zróżnicowanie warunków biometeorologicznych w Toruniu i jego strefie podmiejskiej w 2012 r]. Przegląd Geograficzny 88(1):87–108

Asghari M, Teimori G, Abbasinia M, Shakeri F, Tajik R, Ghannadzadeh MJ, Ghalhari GF (2019) Thermal discomfort analysis using UTCI and MEMI (PET and PMV) in outdoor environments: case study of two climates in Iran (Arak & Bandar Abbas). Weather 74:S57–S64

Back Y, Bach PM, Jasper-Tönnies A, Rauch W, Kleidorfer M (2021) A rapid fine-scale approach to modelling urban bioclimatic conditions. Sci Total Environ 756:
Baglia V, DeKay M, Guida R, Zhu S (2019) Dynamic microclimate modelling for urban China: assessing pedestrian comfort, air quality and building ventilation potential. In: Proceedings of 34th PLEA international conference: PLEA 2018, pp 157–163
Bajšanski IV, Milošević DD, Savić SM (2015) Evaluation and improvement of outdoor thermal comfort in urban areas on extreme temperature days: applications of automatic algorithms. Build Environ 94:632–643
Balogun IA, Ntekop AA, Daramola MT (2019) Assessment of the bioclimatic conditions over some selected stations in Nigeria. SN Appl Sci 1(6):610
Bandurski K, Bandurska H, Kazimierczak-Grygiel E, Koczyk H (2020) The green structure for outdoor places in dry, hot regions and seasons—providing human thermal comfort in sustainable cities. Energies 13(11):2755
Barbosa GS, Drach PR, Corbella OD (2019) Intraurban temperature variations: urban morphologies of the densification process of Copacabana neighborhood, Brazil. Climate 7(5):65
Bartoszek K, Wereski S, Krzyżewska A, Dobek M (2017) The influence of atmospheric circulation on bioclimatic conditions in Lublin (Poland). Bull Geogr Phys Geogr Ser 12(1):41–49
Baruti MM, Johansson E, Yahia MW (2020) Urbanites' outdoor thermal comfort in the informal urban fabric of warm-humid Dar es Salaam, Tanzania. Sustain Cities Soc 62:
Basarin B, Lukić T, Matzarakis A (2020) Review of biometeorology of heatwaves and warm extremes in Europe. Atmosphere 11(12):1276
Basarin B, Lukić T, Bjelajac D, Micić T, Stojićević G, Stamenković I, Matzarakis A et al (2018) Bioclimatic and climatic tourism conditions at Zlatibor mountain (Western Serbia). ID JÁRÁS. QUARTERLY JOURNAL OF THE HUNGARIAN METEOROLOGICAL SERVICE 122(3):321–343
Battista G, Carnielo E, Vollaro RDL (2016) Thermal impact of a redeveloped area on localized urban microclimate: a case study in Rome. Energy Build 133:446–454
Berardi U, Graham J (2020) Investigation of the impacts of microclimate on PV energy efficiency and outdoor thermal comfort. Sustain Cities Soc 62:
Binarti F, Koerniawan MD, Triyadi S, Utami SS, Matzarakis A (2020) A review of outdoor thermal comfort indices and neutral ranges for hot-humid regions. Urban Clim 31:
Bjerre L, Santurette S, Jeong C-H (2017) Relationship between overall comfort and combined thermal and acoustic conditions in urban recreational spaces. In: INTER-NOISE 2017—46th international congress and exposition on noise control engineering: taming noise and moving quiet, Jan 2017
Bleta A, Nastos PT, Matzarakis A (2014) Assessment of bioclimatic conditions on Crete Island, Greece. Reg Environ Change 14(5):1967–1981
Borges VCDAL, Callejas IJA, Durante LC (2020) Thermal sensation in outdoor urban spaces: a study in a Tropical Savannah climate, Brazil. Int J Biometeorol 64(3):533–545
Brecht BM, Schädler G, Schipper JW (2020) UTCI climatology and its future change in Germany–an RCM ensemble approach. Meteorol Z 29(2):97
Bruse M (2014) Using ENVI-met Met BioMet. http://www.envi-met.info/public/files/docs/Quickstart_biomet.pdf. Accessed July 2020
Bröde P, Fiala D, Lemke B, Kjellstrom T (2018) Estimated work ability in warm outdoor environments depends on the chosen heat stress assessment metric. Int J Biometeorol 62(3):331–345
Bröde P, Krüger E, Fiala D (2013b) UTCI: validation and practical application to the assessment of urban outdoor thermal comfort. Geogr Pol 86(1):11–20
Bröde P, Krüger E, Rossi FA, Fiala D (2012a) Predicting urban outdoor thermal comfort by the Universal Thermal Climate Index UTCI—a case study in Southern Brazil. Int J Biometeorol 56(3):471–480
Bröde P, Jendritzky G, Fiala D, Havenith G (2010) The universal thermal climate index UTCI in operational use. In: Adapting to change: new thinking on comfort. Network for comfort and energy use in buildings, London, 6 pp

Bröde P, Fiala D, Błażejczyk K, Holmér I, Jendritzky G, Kampmann B, Havenith G et al (2012b) Deriving the operational procedure for the Universal Thermal Climate Index (UTCI). Int J Biometeorol 56(3):481–494
Bröde P, Błażejczyk K, Fiala D, Havenith G, Holmer I, Jendritzky G, Kampmann B et al (2013a) The Universal Thermal Climate Index UTCI compared to ergonomics standards for assessing the thermal environment. Ind Health 51(1):16–24
Bröde P, Krüger E, Fiala D (2014) Residual analysis of UTCI predictions on outdoor thermal sensation survey data. In: 8th Windsor conference: counting the cost of comfort in a changing world, 2014, Windsor. Proceedings of 8th Windsor conference. London. Network for comfort and energy use in buildings
Burkart K, Meier F, Schneider A, Breitner S, Canário P, Alcoforado MJ, Endlicher W et al (2016) Modification of heat-related mortality in an elderly urban population by vegetation (urban green) and proximity to water (urban blue): evidence from Lisbon, Portugal. Environ Health Perspect 124(7):927–934
Błażejczyk K, Baranowski J, Błażejczyk A (2014b) Heat stress and occupational health and safety–spatial and temporal differentiation. Misc Geogr 18(1):61–67
Błażejczyk K, Baranowski J, Jendritzky G, Błażejczyk A, Bröde P, Fiala D (2015) Regional features of the bioclimate of central and southern Europe against the background of the Köppen-Geiger climate classification. Geogr Pol 88(3):439–453
Błażejczyk K, Błażejczyk A, Baranowski J (2014c) Multiannual fluctuations of some climate related diseases in Poland & their possible relation to climate conditions. Prace i Studia Geograficzne 56:37–65
Błażejczyk A, Błażejczyk K, Baranowski J, Kuchcik M (2018) Heat stress mortality and desired adaptation responses of healthcare system in Poland. Int J Biometeorol 62(3):307–318
Błażejczyk K, Epstein Y, Jendritzky G, Staiger H, Tinz B (2012) Comparison of UTCI to selected thermal indices. Int J Biometeorol 56(3):515–535
Błażejczyk K, Błażejczyk M (2010) BioKlima© 2.6, Program komputerowy. http://www.igipz.pan.pl/Bioklima-zgik.html
Błażejczyk K, Bröde P, Fiala D, Havenith G, Holmér I, Jendritzky G, Kunert A et al (2010a) Principles of the new universal thermal climate index (UTCI) and its application to bioclimatic research in European scale. Misc Geogr 14(1):91–102
Błażejczyk K, Bröde P, Fiala D, Havenith G, Holmér I, Jendritzky G, Kampmann B (2010b) UTCI—new tool of bioclimatic research in different scales [UTCI - Nowe narzędzie badania warunków bioklimatycznych w różnych skalach czasowych i przestrzennych]. Przeglad Geofizyczny 55(1–2):5–19
Błażejczyk K, Kuchcik M, Błażejczyk A, Milewski P, Szmyd J (2014a) Assessment of urban thermal stress by UTCI–experimental and modelling studies: an example from Poland. DIE ERDE J Geogr Soc Berlin 145(1–2):16–33
Błażejczyk K, Nejedlik P, Skrynyk O, Halaś A, Skrynyk O, Baranowski J, Mikulova K (2020a) Thermal stress in the northern Carpathians and air circulation. Misc Geogr 1(ahead-of-print)
Błażejczyk K, Nejedlik P, Skrynyk O, Halaś A, Skrynyk O, Błażejczyk A, Mikulova K (2020b) Influence of geographical factors on thermal stress in northern Carpathians. Int J Biometeorol 1–14
Chabior M (2011) Selected aspects of the bioclimate of Szczecin. Stud Geogr 47:293–300
Charalampopoulos I (2019) A comparative sensitivity analysis of human thermal comfort indices with generalized additive models. Theoret Appl Climatol 137(1–2):1605–1622
Charalampopoulos I, Nouri AS (2019) Investigating the behaviour of human thermal indices under divergent atmospheric conditions: a sensitivity analysis approach. Atmosphere 10(10):580
Chen X, Gao L, Xue P, Du J, Liu J (2020) Investigation of outdoor thermal sensation and comfort evaluation methods in severe cold area. Sci Total Environ 749:
Cheung CSC, Hart MA (2014) Climate change and thermal comfort in Hong Kong. Int J Biometeorol 58(2):137–148

Cheung PK, Jim CY (2018a) Comparing the cooling effects of a tree and a concrete shelter using PET and UTCI. Build Environ 130:49–61
Cheung PK, Jim CY (2018b) Subjective outdoor thermal comfort and urban green space usage in humid-subtropical Hong Kong. Energy Build 173:150–162
Cheung PK, Jim CY (2019) Improved assessment of outdoor thermal comfort: 1-hour acceptable temperature range. Build Environ 151:303–317
Cheung PK, Fung CK, Jim CY (2020) Seasonal and meteorological effects on the cooling magnitude of trees in subtropical climate. Build Environ 106911
Chi X, Cubasch U, Sodoudi S (2019) Assessment of human bio-meteorological environment over the Tibetan Plateau region based on CORDEX climate model projections. Theor Appl Climatol 137(1–2):893–907
Chi X, Li R, Cubasch U, Cao W (2018) The thermal comfort and its changes in the 31 provincial capital cities of mainland China in the past 30 years. Theor Appl Climatol 132(1–2):599–619
Chindapol S, Blair J, Osmond P, Prasad D (2017) A suitable thermal stress index for the elderly in summer tropical climates. Procedia Eng 180(10):932–943
Choe SJ, Han SH (2020) Applicability of Feng Shui concepts for Korean Eum-Taek sites verified using three-dimensional digital mapping and simulations. Sustainability 12(21):8904
Coccolo S, Kämpf J, Scartezzini JL, Pearlmutter D (2016) Outdoor human comfort and thermal stress: a comprehensive review on models and standards. Urban Clim 18:33–57
Coutts AM, White EC, Tapper NJ, Beringer J, Livesley SJ (2016) Temperature and human thermal comfort effects of street trees across three contrasting street canyon environments. Theor Appl Climatol 124(1–2):55–68
Cymes I, Jalali R, Glińska-Lewczuk K, Dragańska E, Giergielewicz-Januszko, B, Romaszko J (2021) The association between the biometeorological indicators and emergency interventions due to fainting: a retrospective cohort study. Sci Total Environ 145376
Das M, Das A (2020) Exploring the pattern of outdoor thermal comfort (OTC) in a tropical planning region of eastern India during summer. Urban Clim 34:
Desert A, Naboni E, Garcia D (2020) The spatial comfort and thermal delight of outdoor misting installations in hot and humid extreme environments. Energy Build 224:
Dhariwal J, Manandhar P, Bande L, Marpu P, Armstrong P, Reinhart CF (2019) Evaluating the effectiveness of outdoor evaporative cooling in a hot, arid climate. Build Environ 150:281–288
Di Napoli C, Barnard C, Prudhomme C, Cloke HL, Pappenberger F (2020) ERA5-HEAT: a global gridded historical dataset of human thermal comfort indices from climate reanalysis. Geosci Data J
Dietrich U (2018) Urban street Canyons–impact of different materials and colours of facades and ground and different positions of persons on outdoor thermal comfort. Sustain City 95
Dobek M, Krzyżewska A (2016) Selected aspects of Lublin bioclimate I [Wybrane zagadnienia z bioklimatu Lublina]. Annales Universitatis Mariae Curie-Sklodowska, sectio B-Geographia, Geologia, Mineralogia et Petrographia 70(2):117
Dobek M, Wereski S, Krzyżewska A (2020) Bioclimatic conditions of Lublin based on the Universal Thermal Climate Index (UTCI). Misc Geogr 24(3):118–127
Dobek M, Demczuk P, Nowosad M (2013) Spatial variation of the Universal Thermal Climate Index in Lublin in specified weather scenarios/Zróznicowanie przestrzenne wskaznika UTCI w Lublinie w okreslonych scenariuszach pogodowych. In Annales Universitatis Mariae Curie-Sklodowska 68(1):21
Dong J, Peng J, He X, Corcoran J, Qiu S, Wang X (2020) Heatwave-induced human health risk assessment in megacities based on heat stress-social vulnerability-human exposure framework. Landsc Urban Plan 203:
Du J, Sun C, Xiao Q, Chen X, Liu J (2020a) Field assessment of winter outdoor 3-D radiant environment and its impact on thermal comfort in a severely cold region. Sci Total Environ 709:
Du J, Liu L, Chen X, Liu J (2020b) Field assessment of neighboring building and tree shading effects on the 3D radiant environment and human thermal comfort in summer within urban settlements in Northeast China. Adv Meteorol

Fabbri K, Di Nunzio A, Gaspari J, Antonini E, Boeri A (2017) Outdoor comfort: the ENVI-BUG tool to evaluate PMV values output comfort point by point. Energy Procedia 111:510–519

Fang Z, Lin Z, Mak CM, Niu J, Tse KT (2018) Investigation into sensitivities of factors in outdoor thermal comfort indices. Build Environ 128:129–142

Fang Z, Xu X, Zhou X, Deng S, Wu H, Liu J, Lin Z (2019b) Investigation into the thermal comfort of university students conducting outdoor training. Build Environ 149:26–38

Fang Z, Feng X, Liu J, Lin Z, Mak CM, Niu J, Xu X et al (2019a) Investigation into the differences among several outdoor thermal comfort indices against field survey in subtropics. Sustain Cities Soc 44:676–690

Fiala D, Havenith G, Bröde P, Kampmann B, Jendritzky G (2012) UTCI-Fiala multi-node model of human heat transfer and temperature regulation. Int J Biometeorol 56(3):429–441

Fitchett JM, Roshan G (2020) 14 climate change threats to cultural and heritage tourism in Iran. Cultural and heritage tourism in the Middle East and North Africa: complexities, management and practices, p 9

Fröhlich D, Gangwisch M, Matzarakis A (2019) Effect of radiation and wind on thermal comfort in urban environments-application of the RayMan and SkyHelios model. Urban Clim 27:1–7

Fröhlich D, Matzarakis A (2016) A quantitative sensitivity analysis on the behaviour of common thermal indices under hot and windy conditions in Doha, Qatar. Theor Appl Climatol 124(1–2):179–187

Fröhlich D, Matzarakis A (2018) Spatial estimation of thermal indices in urban areas—basics of the SkyHelios model. Atmosphere 9(6):209

Fröhlich D, Matzarakis A (2020) Calculating human thermal comfort and thermal stress in the PALM model system 6.0. Geosci Model Dev 13(7):3055–3065

Fung CK, Jim CY (2020) Influence of blue infrastructure on lawn thermal microclimate in a subtropical green space. Sustain Cities Soc 52:

Galan JC, Guedes JC (2019) Applicability of heat stress index in the context of military work: pilot study. Occupational and environmental safety and health. Springer, Cham, pp 313–322

Gamero-Salinas J, Kishnani N, Monge-Barrio A, López-Fidalgo J, Sánchez-Ostiz A (2021) The influence of building form variables on the environmental performance of semi-outdoor spaces. A study in mid-rise and high-rise buildings of Singapore. Energy Build 230:110544

Gao C, Kuklane K, Östergren PO, Kjellstrom T (2018) Occupational heat stress assessment and protective strategies in the context of climate change. Int J Biometeorol 62(3):359–371

Gasparetto T, Nesseler C (2020) Diverse effects of thermal conditions on performance of marathon runners. Front Psychol 11:1438

Ge Q, Kong Q, Xi J, Zheng J (2017) Application of UTCI in China from tourism perspective. Theor Appl Climatol 128(3–4):551–561

Gehl J (2011) Life between buildings: using public space. Island press

Geletič J, Lehnert M, Krč P, Resler J, Krayenhoff ES (2021) High-resolution modelling of thermal exposure during a hot spell: a case study using PALM-4U in Prague, Czech Republic. Atmosphere 12(2):175

Ghalhari GF, Mayvaneh F (2016) Effect of air temperature and universal thermal climate index on respiratory diseases mortality in Mashhad, Iran. Arch Iran Med (AIM) 19(9)

Giannaros C, Melas D, Giannaros TM (2017) Evaluation of WRF parameterization schemes during heat-wave events over the greater area of South–East Mediterranean. Perspectives on atmospheric sciences. Springer, Cham, pp 17–23

Golasi I, Salata F, de Lieto Vollaro E, Coppi M (2018) Complying with the demand of standardization in outdoor thermal comfort: A first approach to the Global Outdoor Comfort Index (GOCI). Build Environ 130:104–119

Goldberg V, Kurbjuhn C, Bernhofer C (2013) How relevant is urban planning for the thermal comfort of pedestrians? Numerical case studies in two districts of the City of Dresden (Saxony/Germany). Meteorol Z 22(6):739–751

Graham J, Berardi U, Turnbull G, McKaye R (2020) Microclimate analysis as a design driver of architecture. Climate 8(6):72

Grifoni RC, Tascini S, Cesario E, Marchesani GE (2017) Cool façade optimization: a new parametric methodology for the urban heat island phenomenon (UHI). In: 2017 IEEE international conference on environment and electrical engineering and 2017 IEEE industrial and commercial power systems Europe (EEEIC/I&CPS Europe). IEEE, pp 1–5

Gómez F, Valcuende M, Matzarakis A, Cárcel J (2018) Design of natural elements in open spaces of cities with a Mediterranean climate, conditions for comfort and urban ecology. Environ Sci Pollut Res 25(26):26643–26652

Głogowski A, Bryś K, Bryś T (2020b) Influence of NAO on forming the UTCI index in Kłodzko. Theor Appl Climatol 142(3):1555–1567

Głogowski A, Bryś K, Perona P (2020a) Bioclimatic conditions of the Lower Silesia region (South-West Poland) from 1966 to 2017. Int J Biometeorol 1–13

Hadianpour M, Mahdavinejad M, Bemanian M, Haghshenas M, Kordjamshidi M (2019) Effects of windward and leeward wind directions on outdoor thermal and wind sensation in Tehran. Build Environ 150:164–180

Hadianpour M, Mahdavinejad M, Bemanian M, Nasrollahi F (2018) Seasonal differences of subjective thermal sensation and neutral temperature in an outdoor shaded space in Tehran, Iran. Sustain Cities Soc 39:751–764

Hai Y, Feng Q (2018) Thermalscape of ecological city and its visualized evaluation. Energy Procedia 152:1139–1144

Hamanaka B, Bueno-Bartholomei CL (2017) Thermal comfort in outdoor spaces: street markets in Presidente Prudente, Brazil. In: Proceedings of 33rd PLEA international conference: design to thrive, PLEA 2017, pp 2864–2871

Hamdan DMA, De Oliveira FL (2019) The impact of urban design elements on microclimate in hot arid climatic conditions: Al Ain City, UAE. Energy Build 200:86–103

Hanipah MH, Abdullah AH, Che Sidek NA, Yunus R, Nagapan S, Yasin MN, Muhammad Yazid MNAW (2017) Wind characteristics and outdoor thermal comfort assessment in East Malaysia. Pertanika J Sci Technol 25:63–72

Hanipah MH, Abdullah AH, Sidik NAC, Yunus R, Yasin MNA, Yazid MNAWM (2016) Assessment of outdoor thermal comfort and wind characteristics at three different locations in Peninsular Malaysia. In: MATEC web of conferences, vol 47. EDP Sciences, p 04005

Hartz DA, Brazel AJ, Golden JS (2013) A comparative climate analysis of heat-related emergency 911 dispatches: Chicago, Illinois and Phoenix, Arizona USA 2003 to 2006. Int J Biometeorol 57(5):669–678

Havenith G, Fiala D, Błazejczyk K, Richards M, Bröde P, Holmér I, Jendritzky G et al (2012) The UTCI-clothing model. Int J Biometeorol 56(3):461–470

He X, An L, Hong B, Huang B, Cui X (2020) Cross-cultural differences in thermal comfort in campus open spaces: a longitudinal field survey in China's cold region. Build Environ 172:

Hendel M, Gutierrez P, Colombert M, Diab Y, Royon L (2016) Measuring the effects of urban heat island mitigation techniques in the field: application to the case of pavement-watering in Paris. Urban Clim 16:43–58

Heusinkveld BG, Sterenborg G, Steeneveld GJ, Attema JJ, Ronda RJ, Holtslag AAM (2017) Smartphone App brings human thermal comfort forecast in your hands. Bull Am Meteor Soc 98(12):2533–2541

Honjo T, Seo Y, Yamasaki Y, Tsunematsu N, Yokoyama H, Yamato H, Mikami T (2018) Thermal comfort along the marathon course of the 2020 Tokyo Olympics. Int J Biometeorol 62(8):1407–1419

Hua L, Zhang H, Liu X, Yang X, Duan H, Wu T (2020) Climate comfort evaluation of national 5A tourist attractions in the mainland of China based on universal thermal climate index. Adv Meteorol

Huang T, Li J, Xie Y, Niu J, Mak CM (2017) Simultaneous environmental parameter monitoring and human subject survey regarding outdoor thermal comfort and its modelling. Build Environ 125:502–514

Huang Q, Meng X, Yang X, Jin L, Liu X, Hu W (2016b) The ecological city: considering outdoor thermal environment. Energy Procedia 104:177–182
Huang J, Zhou C, Zhuo Y, Xu L, Jiang Y (2016a) Outdoor thermal environments and activities in open space: an experiment study in humid subtropical climates. Build Environ 103:238–249
Huang J, Hao T, Hou SS, Jones P (2019) Simulation-informed urban design: improving urban microclimate in real-world practice in a high density city. In: IOP conference series: earth and environmental science, vol 329, no 1. IOP Publishing, p 012047
Höppe P (2002) Different aspects of assessing indoor and outdoor thermal comfort. Energy Build 34(6):661–665
Idzikowska D (2010) Bioclimatic conditions of Paris, Rome and Budapest on the basis of the Universal Thermal Climate Index. Misc Geogr 14(1):103–109
Idzikowska D (2011) Relationship between mortality and UTCI in Paris, Rome, Warsaw and Budapest. Prace i Studia Geograficzne 47:311–318
Imbert C, Bhattacharjee S, Tencar J (2018) Simulation of urban microclimate with SOLENE-microclimat: an outdoor comfort case study. In: Proceedings of the symposium on simulation for architecture and urban design, pp 1–8
Imran HM, Kala J, Ng AWM, Muthukumaran S (2018) Effectiveness of green and cool roofs in mitigating urban heat island effects during a heatwave event in the city of Melbourne in southeast Australia. J Clean Prod 197:393–405
Jacobs C, Singh T, Gorti G, Iftikhar U, Saeed S, Syed A, Siderius C et al (2019) Patterns of outdoor exposure to heat in three South Asian cities. Sci Total Environ 674:264–278
Jendritzky G, de Dear R, Havenith G (2012) UTCI—why another thermal index? Int J Biometeorol 56(3):421–428
Jin H, Liu S, Kang J (2019) Thermal comfort range and influence factor of urban pedestrian streets in severe cold regions. Energy Build 198:197–206
Jin H, Liu S, Kang J (2020) Gender differences in thermal comfort on pedestrian streets in cold and transitional seasons in severe cold regions in China. Build Environ 168:
Kampmann B, Bröde P, Fiala D (2012) Physiological responses to temperature and humidity compared to the assessment by UTCI, WGBT and PHS. Int J Biometeorol 56(3):505–513
Katavoutas G, Flocas HA (2018) Universal thermal climate index (UTCI) and synoptic circulation patterns over the metropolitan city of Athens, Greece. Glob Nest J 20(3):477–487
Katavoutas G, Founda D (2019) Intensification of thermal risk in Mediterranean climates: evidence from the comparison of rational and simple indices. Int J Biometeorol 63(9):1251–1264
Knott D, Evins R (2013) Using comfort criteria and parametric analysis to drive passive building design. In: Building simulation 2013: 13th international conference of the international-building-performance-simulation-association (IBPSA), Chambery, France, 25–28 Aug 2013
Kolendowicz L, Półrolniczak M, Szyga-Pluta K, Bednorz E (2018) Human-biometeorological conditions in the southern Baltic coast based on the universal thermal climate index (UTCI). Theor Appl Climatol 134(1–2):363–379
Konefał M, Chmura P, Zacharko M, Baranowski J, Andrzejewski M, Błażejczyk K, Chmura J (2020) The influence of thermal stress on the physical and technical activities of soccer players: lessons from the 2018 FIFA World Cup in Russia. Int J Biometeorol 1–8
Kong Q, Ge Q, Xi J, Zheng J (2017) Human-biometeorological assessment of increasing summertime extreme heat events in Shanghai, China during 1973–2015. Theor Appl Climatol 130(3–4):1055–1064
Kong Q, Zheng J, Fowler HJ, Ge Q, Xi J (2019) Climate change and summer thermal comfort in China. Theor Appl Climatol 137(1–2):1077–1088
Konstantinov P, Shartova N, Varentsov M, Revich B (2020) Evaluation of outdoor thermal comfort conditions in northern Russia over 30-year period: Arkhangelsk region. Geogr Pannonica 24(4):252–260
Koźmiński C, Michalska B (2019) Assessment of bioclimatic conditions for recreation and tourism in the Polish Baltic coastal zone using the UTCI index [Ocena bioklimatycznych warunków

rekreacji i turystyki w strefie polskiego Wybrzeża Bałtyku na podstawie wskaźnika UTCI Assessment of bioclimatic conditions for recreation and tourism in the Polish Baltic coastal zone using the UTCI index]. Przegląd Geograficzny
Krzyżewska A, Dobek M, Domżał-Drzewicka R, Rząca M (2017) Emergency interventions due to weather-related hypothermia. Weather 72(1):8–12
Krzyżewska A, Bartoszek K, Wereski S (2016) The meteorological conditions during particularly severe heatwave in Lublin in August 2015 I [Warunki meteorologiczne w lublinie podczas wyjatkowo uciażliwej fali Upałów z Sierpnia 2015 r]. Przegląd Geofizyczny
Krzyżewska A, Wereski S, Dobek M (2020) Summer UTCI variability in Poland in the twenty-first century. Int J Biometeorol 1–17
Krüger E (2017) Impact of site-specific morphology on outdoor thermal perception: A case-study in a subtropical location. Urban Clim 21:123–135
Krüger E, Drach P (2017) Identifying potential effects from anthropometric variables on outdoor thermal comfort. Build Environ 117:230–237
Krüger E, Drach P, Bröde P (2015) Implications of air-conditioning use on thermal perception in open spaces: a field study in downtown Rio de Janeiro. Build Environ 94:417–425
Krüger E, Drach P, Bröde P (2017b) Outdoor comfort study in Rio de Janeiro: site-related context effects on reported thermal sensation. Int J Biometeorol 61(3):463–475
Krüger E, Minella FO, Matzarakis A (2014) Comparison of different methods of estimating the mean radiant temperature in outdoor thermal comfort studies. Int J Biometeorol 58(8):1727–1737
Krüger E, Tamura CA, Bröde P, Schweiker M, Wagner A (2017a) Short-and long-term acclimatization in outdoor spaces: exposure time, seasonal and heatwave adaptation effects. Build Environ 116:17–29
Krüger E, Bröde P, Emmanuel R, Fiala D (2012) Predicting outdoor thermal sensation from two field studies in Curitiba, Brazil and Glasgow, UK using the Universal Thermal Climate Index (UTCI). In: Proceedings of 7th Windsor conference: the changing context of comfort in an unpredictable world 2012, Windsor; United Kingdom, 12 Apr 2012 through 15 April 2012
Krüger E, Silva TJV, da Silveira Hirashima SQ, da Cunha EG, Rosa LA (2020) Calibrating UTCI'S comfort assessment scale for three Brazilian cities with different climatic conditions. Int J Biometeorol 1–10
Kuchcik M (2020) Mortality and thermal environment (UTCI) in Poland—long-term, multi-city study. Int J Biometeorol 1–13
Kuchcik M, Błażejczyk K, Halaś A (2021) Long-term changes in hazardous heat and cold stress in humans: multi-city study in Poland. Int J Biometeorol 1–12
Kurban Lopez A, Grasso C (2017) Thermal comfort in urban green spaces in arid environments. Rev Habitat Sustentable 7(1):33–43
Kwok YT, Schoetter R, Lau KKL, Hidalgo J, Ren C, Pigeon G, Masson V (2019) How well does the local climate zone scheme discern the thermal environment of Toulouse (France)? An analysis using numerical simulation data. Int J Climatol 39(14):5292–5315
Lai D, Guo D, Hou Y, Lin C, Chen Q (2014) Studies of outdoor thermal comfort in northern China. Build Environ 77:110–118
Lai D, Lian Z, Liu W, Guo C,. Liu W, Liu K, Chen Q (2020) A comprehensive review of thermal comfort studies in urban open spaces. Sci Total Environ 140092
Lam CKC, Cui S, Liu J, Kong X, Ou C, Hang J (2021b) Influence of acclimatization and short-term thermal history on outdoor thermal comfort in subtropical South China. Energy Build 231:
Lam CKC, Gallant AJ, Tapper NJ (2018) Perceptions of thermal comfort in heatwave and non-heatwave conditions in Melbourne, Australia. Urban Clim 23:204–218
Lam CKC, Gallant AJ, Tapper NJ (2019) Short-term changes in thermal perception associated with heatwave conditions in Melbourne, Australia. Theor Appl Climatol 136(1–2):651–660
Lam CKC, Gao Y, Yang H, Chen T, Zhang Y, Ou C, Hang J (2021a) Interactive effect between long-term and short-term thermal history on outdoor thermal comfort: comparison between Guangzhou, Zhuhai and Melbourne. Sci Total Environ 760:

Lam CKC, Hang J (2017) Solar radiation intensity and outdoor thermal comfort in royal botanic garden Melbourne during heatwave conditions. Procedia Eng 205:3456–3462
Lam CKC, Lau KKL (2018) Effect of long-term acclimatization on summer thermal comfort in outdoor spaces: a comparative study between Melbourne and Hong Kong. Int J Biometeorol 62(7):1311–1324
Lam CKC, Yang H, Yang X, Liu J, Ou C, Cui S, Hang J et al (2020) Cross-modal effects of thermal and visual conditions on outdoor thermal and visual comfort perception. Build Environ 186:107297
Langner M, Scherber K, Endlicher WR (2013) Indoor heat stress: an assessment of human bioclimate using the UTCI in different buildings in Berlin. DIE ERDE J Geogr Soc Berl 144(3–4):260–273
Larriva MTB, Higueras E (2020) Health risk for older adults in Madrid, by outdoor thermal and acoustic comfort. Urban Clim 34:
Lee LS, Cheung PK, Fung CK, Jim CY (2020) Improving street walkability: biometeorological assessment of artificial-partial shade structures in summer sunny conditions. Int J Biometeorol 64(4):547–560
Lee LS, Jim CY (2019) Urban woodland on intensive green roof improved outdoor thermal comfort in subtropical summer. Int J Biometeorol 63(7):895–909
Lehnert M, Brabec M, Jurek M, Tokar V, Geletič J (2021) The role of blue and green infrastructure in thermal sensation in public urban areas: a case study of summer days in four Czech cities. Sustain Cities Soc 66:
Lehnert M, Tokar V, Jurek M, Geletič J (2020) Summer thermal comfort in Czech cities: measured effects of blue and green features in city centres. Int J Biometeorol 1–13
Leroyer S, Bélair S, Spacek L, Gultepe I (2018) Modelling of radiation-based thermal stress indicators for urban numerical weather prediction. Urban Clim 25:64–81
Li J, Niu J, Mak CM, Huang T, Xie Y (2018) Assessment of outdoor thermal comfort in Hong Kong based on the individual desirability and acceptability of sun and wind conditions. Build Environ 145:50–61
Li J, Niu J, Mak CM, Huang T, Xie Y (2020) Exploration of applicability of UTCI and thermally comfortable sun and wind conditions outdoors in a subtropical city of Hong Kong. Sustain Cities Soc 52:
Lian Z, Liu B, Brown RD (2020) Exploring the suitable assessment method and best performance of human energy budget models for outdoor thermal comfort in hot and humid climate area. Sustain Cities Soc 63:
Liang C, Bi, W (2017) Seasonal variation analysis and SVR forecast of tourist flows during the year: a case study of Huangshan mountain. In: 2017 IEEE 2nd international conference on big data analysis (ICBDA). IEEE, pp 921–927
Liang X, Tian W, Li R, Niu Z, Yang X, Meng X, Yan J et al (2019) Numerical investigations on outdoor thermal comfort for built environment: case study of a Northwest campus in China. Energy Procedia 158:6557–6563
Lindner-Cendrowska K (2013) Assessment of bioclimatic conditions in cities for tourism and recreational purposes (a Warsaw case study). Geogr Pol 86(1):55–66
Lindner-Cendrowska K, Bröde P (2021) Impact of biometeorological conditions and air pollution on influenza-like illnesses incidence in Warsaw. Int J Biometeorol 1–16
Liu Y, Li S, Wang Y, Zhang T, Peng J, Li T (2015) Identification of multiple climatic extremes in metropolis: a comparison of Guangzhou and Shenzhen, China. Nat Hazards 79(2):939–953
Liu L, Lin Y, Xiao Y, Xue P, Shi L, Chen X, Liu J (2018) Quantitative effects of urban spatial characteristics on outdoor thermal comfort based on the LCZ scheme. Build Environ 143:443–460
Liu C, Susilo YO, Ahmad Termida N (2020) Weather perception and its impact on out-of-home leisure activity participation decisions. Transp B Transp Dyn 8(1):219–236
Liu C, Susilo YO, Karlström A (2016) Measuring the impacts of weather variability on home-based trip chaining behaviour: a focus on spatial heterogeneity. Transportation 43(5):843–867

Liu N, Li J (2018) A navigational approach centered on pedestrian comfort: a case study from Fuzhou University. In: 2018 1st IEEE international conference on knowledge innovation and invention (ICKII). IEEE, pp 377–380

Lokys HL, Junk J, Krein A (2018) Short-term effects of air quality and thermal stress on non-accidental morbidity—a multivariate meta-analysis comparing indices to single measures. Int J Biometeorol 62(1):17–27

Lokys HL, Junk J, Krein A (2015) Future changes in human-biometeorological index classes in three regions of Luxembourg, western-Central Europe. Adv Meteorol

Ma P, Zhou J, Wang S, Li T, Fan X, Fan J, Xie J (2018a) Differences of hemorrhagic and ischemic strokes in age spectra and responses to climatic thermal conditions. Sci Total Environ 644:1573–1579

Ma P, Wang S-G, Shang K-Z, Li T-S, Yin L (2018b) The impact of meteorological comfort conditions on respiratory disease. Zhongguo Huanjing Kexue/China Environ Sci 38(1):374–382

Mackey C, Roudsari MS, Samaras P (2015) ComfortCover: a novel method for the design of outdoor shades. In: SpringSim (SimAUD), pp 111–118

Manavvi S, Rajasekar E (2020) Semantics of outdoor thermal comfort in religious squares of composite climate: New Delhi, India. Int J Biometeorol 64(2):253–264

Manickathan L, Defraeye T, Allegrini J, Derome D, Carmeliet J (2018) Parametric study of the influence of environmental factors and tree properties on the transpirative cooling effect of trees. Agric For Meteorol 248:259–274

Maras I, Schmidt T, Paas B, Ziefle M, Schneider C (2016) The impact of human-biometeorological factors on perceived thermal comfort in urban public places

Martini A, Biondi D, Batista AC, Zamproni K, Viezzer J, Grise MM, Lima Neto EM (2014) Population's perception on thermal comfort provided by street trees of Curitiba-PR. Floresta 44(3):515–524

Martini A, Biondi D, Batista AC (2020) Thermal comfort provided by street trees in cities. Arboric J 1–12

Mateeva Z (2011) Personal factors of human thermal perception: long-lasting climate experience. Prace I Stud Geogr 47:335–344

Matzarakis A, Rutz F, Mayer H (2010) Modelling radiation fluxes in simple and complex environments: basics of the RayMan model. Int J Biometeorol 54(2):131–139

Matzarakis A, Muthers S, Rutz F (2014) Application and comparison of UTCI and PET in temperate climate conditions. Finisterra 49(98)

Matzarakis A, Fröhlich D (2017) Influence of urban green on human thermal bioclimate-application of thermal indices and micro-scale models. In: International symposium on greener cities for more efficient ecosystem services in a climate changing world, vol 1215, pp 1–10

Mazon J (2014) The influence of thermal discomfort on the attention index of teenagers: an experimental evaluation. Int J Biometeorol 58(5):717–724

McGregor GR (2012) Human biometeorology. Prog Phys Geor 36(1):93-109

Meili N, Acero JA, Peleg N, Manoli G, Burlando P, Fatichi S (2021) Vegetation cover and plant-trait effects on outdoor thermal comfort in a tropical city. Build Environ 107733

Messeri A, Morabito M, Bonafede M, Bugani M, Levi M, Baldasseroni A, Marinaccio A et al (2019) Heat stress perception among native and migrant workers in Italian industries—case studies from the construction and agricultural sectors. Int J Environ Res Public Health 16(7):1090

Milewski P (2013) Application of the UTCI to the local bioclimate of Poland's Ziemia Kłodzka region. Geogr Pol 86(1):47–54

Milošević DD, Bajšanski IV, Savić SM (2017b) Influence of changing trees locations on thermal comfort on street parking lot and footways. Urban For Urban Green 23:113–124

Milošević DD, Savić SM, Bajšanski IV (2017a) Applications of automatic algorithms for improvement of outdoor thermal comfort in cities. Procedia Eng 198:187–192

Minella FO, Krüger E, Honjo S, Goyette S, Hedjazi A (2014) Daytime microclimatic impacts of the SOVALP project in summer: a case study in Geneva, Switzerland. Simulation 90(8):857–873

Miszuk B (2021) Multi-annual changes in heat stress occurrence and its circulation conditions in the Polish-Saxon border region. Atmosphere 12(2):163
Miszuk B, Otop I, Strońska M, Schwarzak S, Surke M (2016) Tourism-climate conditions and their future development in the Polish-Saxon border area. Meteorologische Zeitschrift 421–434
Miszuk B (2020) Intensity of heat stress in 2015 and 2018 summer seasons in the region of the Lower Silesia (Poland). Misc Geogr 1(ahead-of-print)
Mohammadi B, Gholizadeh MH, Alijani B (2018) Spatial distribution of thermal stresses in Iran based on pet and UTCI indices. Appl Ecol Environ Res 16(5)
Montazeri H, Toparlar Y, Blocken B, Hensen JLM (2017) Simulating the cooling effects of water spray systems in urban landscapes: a computational fluid dynamics study in Rotterdam, The Netherlands. Landsc Urban Plan 159:85–100
Morabito M, Grifoni D, Crisci A, Fibbi L, Orlandini S, Gensini GF, Zipoli G (2014b) Might outdoor heat stress be considered a proxy for the unperceivable effect of the ultraviolet-induced risk of erythema in Florence? J Photochem Photobiol B 130:338–348
Morabito M, Crisci A, Messeri A, Capecchi V, Modesti PA, Gensini GF, Orlandini S (2014a) Environmental temperature and thermal indices: what is the most effective predictor of heat-related mortality in different geographical contexts? Sci World J
Morille B, Musy M (2017) Comparison of the impact of three climate adaptation strategies on summer thermal comfort–Cases study in Lyon, France. Procedia Environ Sci 38:619–626
Mąkosza A (2013) Bioclimatic conditions of the Lubuskie Voivodeship. Geogr Pol 86(1):37–46
Mąkosza A, Nidzgorska-Lencewicz J (2016) Contrasting bioclimatic conditions within the Polish Baltic coastal zone in the warm half-year | [Kontrastowość warunków bioklimatycznych w strefie polskiego wybrzeża Bałtyku w ciepłej połowie roku]. Przegląd Geogr 88(1):109–120
Di Napoli C, Pappenberger F, Cloke HL (2018) Assessing heat-related health risk in Europe via the Universal Thermal Climate Index (UTCI). Int J Biometeorol 62(7):1155–1165
Di Napoli C, Pappenberger F, Cloke HL (2019) Verification of heat stress thresholds for a health-based heat-wave definition. J Appl Meteorol Climatol 58(6):1177–1194
Nasrollahi N, Hatami M, Khastar SR, Taleghani M (2017) Numerical evaluation of thermal comfort in traditional courtyards to develop new microclimate design in a hot and dry climate. Sustain Cities Soc 35:449–467
Nassiri P, Monazzam MR, Golbabaei F, Dehghan SF, Rafieepour A, Mortezapour AR, Asghari M (2017) Application of Universal Thermal Climate Index (UTCI) for assessment of occupational heat stress in open-pit mines. Ind Health
Nastos PT, Bleta AG, Matsangouras IT (2017) Human thermal perception related to Föhn winds due to Saharan dust outbreaks in Crete Island, Greece. Theor Appl Climatol 128(3–4):635–647
Nastos PT, Matzarakis A (2012) The effect of air temperature and human thermal indices on mortality in Athens, Greece. Theor Appl Climatol 108(3):591–599
Nastos PT, Bleta AG, Matsangouras IT (2013) The development of Fohn winds due to Saharan Dust Outbreaks and the associated human thermal bioclimatic conditions in Crete island, Greece. In: Proceedings of the 13th international conference on environmental science and technology, Athens (Greece), vol 610
Nice KA, Coutts AM, Tapper NJ (2018) Development of the VTUF-3D v1. 0 urban micro-climate model to support assessment of urban vegetation influences on human thermal comfort. Urban Clim 24:1052–1076
Nidzgorska-Lencewicz J, Mąkosza A (2013) Assessment of bioclimatic conditions within the area of Szczecin agglomeration. Meteorol Z 22(5):615–626
Nidzgorska-Lencewicz J (2015) Variability of human-biometeorological conditions in Gdańsk. Pol J Environ Stud 24(1)
Niu J, Hong B, Geng Y, Mi J, He J (2020) Summertime physiological and thermal responses among activity levels in campus outdoor spaces in a humid subtropical city. Sci Total Environ 138757
Nouri AS, Afacan Y, Çalışkan O, Lin TP, Matzarakis A (2021) Approaching environmental human thermophysiological thresholds for the case of Ankara, Turkey. Theor Appl Climatol 143(1):533–555

Novák M (2013) Use of the UTCI in the Czech Republic. Geogr Pol 86(1):21–28
Novák M (2010) First experiences with UTCI (Universal Thermal Climate Index) in conditions of the Czech Republic. In: BIOKLIMA international conference local: Prague, Czech Republic Data, 07–09 Sept 2010
Novák M (2011) A comparison of an UTCI to some other selected indices. In: International scientific conference on bioclimate–source and limit of social development
Obiefuna JN, Okolie CJ, Nwilo PC, Daramola OE, Isiofia LC (2021) Potential influence of urban sprawl and changing land surface temperature on outdoor thermal comfort in Lagos State, Nigeria. Quaest Geogr 1(ahead-of-print)
Oh W, Ooka R, Nakano J, Kikumoto H, Ogawa O (2019) Environmental index for evaluating thermal sensations in a mist spraying environment. Build Environ 161:
Ohashi Y, Katsuta T, Tani H, Okabayashi T, Miyahara S, Miyashita R (2018) Human cold stress of strong local-wind "Hijikawa-arashi" in Japan, based on the UTCI index and thermo-physiological responses. Int J Biometeorol 62(7):1241–1250
Okoniewska M (2020) Daily and seasonal variabilities of thermal stress (based on the UTCI) in air masses typical for Central Europe: an example from Warsaw. Int J Biometeorol 1–10
Ongoma V, Muange KP, Zablon WS (2016) Potential effects of urbanization on urban thermal comfort, a case study of Nairobi city, Kenya: a review. Geogr Pannonica 20(1):19–31
Oswald SM, Revesz M, Trimmel H, Weihs P, Zamini S, Schneider A, Lindberg F et al (2019) Coupling of urban energy balance model with 3-D radiation model to derive human thermal (dis)comfort. Int J Biometeorol 63(6):711–722
Owczarek M (2019) The influence of large-scale factors on the heat load on human beings in Poland in the summer months. Theor Appl Climatol 137(1–2):855–869
Pan W, Du J (2017) Density and environmental performance: analysing diverse characteristics of outdoor thermal environment in urban villages, Shenzhen. In: Proceedings of 33rd PLEA international conference: design to thrive, PLEA 2017, pp 1829–1836
Pantavou KG, Lykoudis SP, Nikolopoulos GK (2016) Milder form of heat-related symptoms and thermal sensation: a study in a Mediterranean climate. Int J Biometeorol 60(6):917–929
Pantavou K, Lykoudis S, Nikolopoulou M, Tsiros IX (2018) Thermal sensation and climate: a comparison of UTCI and PET thresholds in different climates. Int J Biometeorol 62(9):1695–1708
Pantavou K, Santamouris M, Asimakopoulos D, Theoharatos G (2014) Empirical calibration of thermal indices in an urban outdoor Mediterranean environment. Build Environ 80:283–292
Pantavou K, Theoharatos G, Santamouris M, Asimakopoulos D (2013) Outdoor thermal sensation of pedestrians in a Mediterranean climate and a comparison with UTCI. Build Environ 66:82–95
Pappenberger F, Jendritzky G, Staiger H, Dutra E, Di Giuseppe F, Richardson DS, Cloke HL (2015) Global forecasting of thermal health hazards: the skill of probabilistic predictions of the Universal Thermal Climate Index (UTCI). Int J Biometeorol 59(3):311–323
Parison S, Hendel M, Royon L (2020) A statistical method for quantifying the field effects of urban heat island mitigation techniques. Urban Clim 33:
Parison S, Hendel M, Jurski K, Royon L (2017) The impact of different watering strategies on the cooling effects of pavement-watering during heat-waves. In: Proceedings of 33rd PLEA international conference: design to thrive, PLEA 2017, pp 120–127
Park S, Tuller SE (2014) Advanced view factor analysis method for radiation exchange. Int J Biometeorol 58(2):161–178
Park S, Tuller SE, Jo M (2014) Application of Universal Thermal Climate Index (UTCI) for microclimatic analysis in urban thermal environments. Landsc Urban Plan 125:146–155
Pecelj MM, Lukić MZ, Filipović DJ, Protić BM, Bogdanović UM (2020) Analysis of the Universal Thermal Climate Index during heatwaves in Serbia. Nat Hazards Earth Syst Sci 20(7):2021–2036
Pecelj MR, Đorđević A, Pecelj MR, Pecelj-Purković J, Filipović D, Šećerov V (2017) Biothermal conditions on Mt. Zlatibor based on thermophysiological indices. Arch Biol Sci 69(3):455–461
Pecelj M, Lukić M, Vučičević A, Uña-Álvarez D, da Silva JC, Freinkina I, Bogdanović U et al (2018) Geoecological evaluation of local surroundings for the purposes of recreational tourism. J Geogr Inst 'Jovan Cvijic' SASA 68(2)

Perini K, Chokhachian A, Dong S, Auer T (2017) Modeling and simulating urban outdoor comfort: coupling ENVI-Met and TRNSYS by grasshopper. Energy Build 152:373–384
Petralli M, Massetti L, Pearlmutter D, Brandani G, Messeri A, Orlandini S (2020) UTCI field measurements in an urban park in Florence (Italy). Misc Geogr 24(3):111–117
Potchter O, Cohen P, Lin TP, Matzarakis A (2018) Outdoor human thermal perception in various climates: a comprehensive review of approaches, methods and quantification. Sci Total Environ 631:390–406
Provençal S, Bergeron O, Leduc R, Barrette N (2016) Thermal comfort in Quebec City, Canada: sensitivity analysis of the UTCI and other popular thermal comfort indices in a mid-latitude continental city. Int J Biometeorol 60(4):591–603
Psikuta A, Fiala D, Richards M (2006) Validation of the Fiala thermo-physiological model for COST 730 using human and clothing data.Part 1. STSM Scientific Report. Karlsruhe University, Karlsruhe
Psikuta A, Fiala D, Richards M (2007a) Further validation of the Fialathermo-physiological model for COST 730 using human andclothing data. Part 2. STSM Scientific Report. Karlsruhe University, Karlsruhe
Psikuta A, Fiala D, Richards M (2007b) Further validation of theFiala thermo-physiological model for COST 730 using humanand clothing data. Part 3. STSM Scientific Report. TNO, Soesteberg
Psikuta A, Fiala D, Laschewski G, Jendritzky G, Richards M, Błażejczyk K, Havenith G (2012) Validation of the Fiala multi-node thermophysiological model for UTCI application. Int J Biometeorol 56(3):443–460
Rahul A, Mukherjee M, Sood A (2020) Impact of ganga canal on thermal comfort in the city of Roorkee, India. Int J Biometeorol 64(11):1933–1945
Rajagopalan P, Andamon MM, Paolini R (2020) Investigating thermal comfort and energy impact through microclimate monitoring-a citizen science approach. Energy Build 229:
Reinhart CF, Dhariwal J, Gero K (2017) Biometeorological indices explain outside dwelling patterns based on Wi-Fi data in support of sustainable urban planning. Build Environ 126:422–430
Romaszko-Wojtowicz A, Cymes I, Dragańska E, Doboszyńska A, Romaszko J, Glińska-Lewczuk K (2020) Relationship between biometeorological factors and the number of hospitalizations due to asthma. Sci Rep 10(1):1–9
Romaszko J, Cymes I, Dragańska E, Kuchta R, Glińska-Lewczuk K (2017) Mortality among the homeless: causes and meteorological relationships. PloS One 12(12)
Romaszko J, Skutecki R, Bocheński M, Cymes I, Dragańska E, Jastrzębski P, Glińska-Lewczuk K (2019) Applicability of the universal thermal climate index for predicting the outbreaks of respiratory tract infections: a mathematical modeling approach. Int J Biometeorol 63(9):1231–1241
Roshan G, Almomenin HS, da Silveira Hirashima SQ, Attia S (2019) Estimate of outdoor thermal comfort zones for different climatic regions of Iran. Urban Clim 27:8–23
Roshan G, Moghbel M (2020) Quantifying the cooling effect of rain events on outdoor thermal comfort in the southern coastal stations of the Caspian Sea. J Therm Biol 93:
Roshan G, Nastos PT (2018) Assessment of extreme heat stress probabilities in Iran's urban settlements, using first order Markov chain model. Sustain Cities Soc 36:302–310
Roshan G, Yousefi R, Błażejczyk K (2018) Assessment of the climatic potential for tourism in Iran through biometeorology clustering. Int J Biometeorol 62(4):525–542
Rozbicka K, Rozbicki T (2018) Variability of UTCI index in South Warsaw depending on atmospheric circulation. Theor Appl Climatol 133(1–2):511–520
Rozbicka K, Rozbicki T (2020) Long-term variability of bioclimatic conditions and tourism potential for Warsaw agglomeration (Poland). Int J Biometeorol 1–11
Ruedisser D, Weiss T, Unger L (2020) Urban thermal comfort maps based on a three-dimensional sampling algorithm and UAV-based radiometric measurements. BAUPHYSIK 42(6):335–344
Rutty M, Scott D (2014) Thermal range of coastal tourism resort microclimates. Tour Geogr 16(3):346–363

Rutty M, Scott D (2015) Bioclimatic comfort and the thermal perceptions and preferences of beach tourists. Int J Biometeorol 59(1):37–45

Sadeghi M, de Dear R, Wood G, Samali B (2018) Development of a bioclimatic wind rose tool for assessment of comfort wind resources in Sydney, Australia for 2013 and 2030. Int J Biometeorol 62(11):1963–1972

Saneinejad S, Moonen P, Carmeliet J (2014a) Comparative assessment of various heat island mitigation measures. Build Environ 73:162–170

Saneinejad S, Moonen P, Carmeliet J (2014b) Coupled CFD, radiation and porous media model for evaluating the micro-climate in an urban environment. J Wind Eng Ind Aerodyn 128:1–11

Santurtún A, Almendra R, Fdez-Arroyabe P, Sanchez-Lorenzo A, Royé D, Zarrabeitia MT, Santana P (2020) Predictive value of three thermal comfort indices in low temperatures on cardiovascular morbidity in the Iberian peninsula. Sci Total Environ 138969

Schrijvers PJC, Jonker HJJ, de Roode SR, Kenjereš S (2016) The effect of using a high-albedo material on the Universal Temperature Climate Index within a street canyon. Urban Clim 17:284–303

Sen S, Fernandèz JPRMR, Roesler J (2020) Reflective parking lots for microscale urban heat island mitigation. Transp Res Rec 2674(8):663–671

Sen J, Nag PK (2019) Human susceptibility to outdoor hot environment. Sci Total Environ 649:866–875

Sharifi E, Boland J (2018) Limits of thermal adaptation in cities: outdoor heat-activity dynamics in Sydney, Melbourne and Adelaide. Arch Sci Rev 61(4):191–201

Sharifi E, Boland J (2020) Passive activity observation (PAO) method to estimate outdoor thermal adaptation in public space: case studies in Australian cities. Int J Biometeorol 64(2):231–242

Shata RO, Mahmoud AH, Fahmy M (2021) Correlating the sky view factor with the pedestrian thermal environment in a Hot Arid University Campus Plaza. Sustainability 13(2):468

Silva TJV, Hirashima SQS (2020) Predicting urban thermal comfort from calibrated UTCI assessment scale—a case study in Belo Horizonte city, southeastern Brazil. Urban Clim 100652

Skutecki R, Cymes I, Dragańska E, Glińska-Lewczuk K, Buciński A, Drozdowski M, Romaszko J (2019b) Are the levels of lipid parameters associated with biometeorological conditions? Int J Environ Res Public Health 16(23):4636

Skutecki R, Jalali R, Dragańska E, Cymes I, Romaszko J, Glińska-Lewczuk K (2019a) UTCI as a bio-meteorological tool in the assessment of cold-induced stress as a risk factor for hypertension. Sci Total Environ 688:970–975

Staiger H, Laschewski G, Matzarakis A (2019) Selection of appropriate thermal indices for applications in human biometeorological studies. Atmosphere 10(1):18

Stewart ID, Oke TR (2009a) Newly developed "thermal climate zones" for defining and measuring urban heat island magnitude in the canopy layer. Preprints. Oke TR. Symposium eighth symposium on urban environment, 11–15 Jan, Phoenix, AZ

Stewart ID, Oke TR (2009b) Classifying urban climate field sites by "local climate zones": The case of Nagano, Japan. Preprints. In: 7th international conference on urban climate, June 29–July 3, Yokohama, Japan

Streda T, Stredova H, Sediva I (2014) Bioclimatic conditions for the recreationally important areas in the medium size city. In: Conference on public recreation and landscape protection—with man hand in hand? Krtiny, Czech Republic, 05–06 May 2014

Szer I, Szer J, Hoła B (2019) Evaluation of climatic conditions affecting workers on scaffoldings. Advances and trends in engineering sciences and technologies III—proceedings of the 3rd international conference on engineering sciences and technologies, ESaT 2018, pp 603–609

Takane Y, Ohashi Y, Grimmond CSB, Hara M, Kikegawa Y (2020) Asian megacity heat stress under future climate scenarios: impact of air-conditioning feedback. Environ Res Commun 2(1):

Talhi A, Barlet A, Bruneau D, Aichour B (2020) Towards a prediction of outdoor human thermal comfort adapted for designers of urban spaces: examining UTCI and APCI in the context of Algiers (Algeria). Int J Biometeorol 1–12

Thorsson S, Rayner D, Palm G, Lindberg F, Carlström E, Börjesson M, Holmer B (2020) Is Physiological Equivalent Temperature (PET) a superior screening tool for heat stress risk than Wet-Bulb Globe Temperature (WBGT) index? Eight years of data from the Gothenburg half marathon. Br J Sport Med

Tomczyk AM, Owczarek M (2020) Occurrence of strong and very strong heat stress in Poland and its circulation conditions. Theor Appl Climatol 139(3–4):893–905

Tseliou A, Tsiros IX (2016) Modeling urban microclimate to ameliorate thermal sensation conditions in outdoor areas in Athens (Greece). Build Simul 9(3):251–267. Tsinghua University Press

Tumini I, Higueras García E, Baereswyl Rada S (2016) Urban microclimate and thermal comfort modelling: strategies for urban renovation. Int J Sustain Build Technol Urban Dev 7(1):22–37

Tumini I, Perez Fargallo A (2015) The application of adaptive thermal comfort indices for open space assessment in Madrid. Rev Habitat Sustentable 5(2):57–67

Ulpiani G, Di Giuseppe E, Di Perna C, D'Orazio M, Zinzi M (2019) Thermal comfort improvement in urban spaces with water spray systems: field measurements and survey. Build Environ 156:46–61

Urban A, Kyselý J (2014) Comparison of UTCI with other thermal indices in the assessment of heat and cold effects on cardiovascular mortality in the Czech Republic. Int J Environ Res Public Health 11(1):952–967

Van Eck NJ, Waltman L (2013) VOSviewer manual. Leiden: Univeristeit Leiden 1(1):1–53

Vanos JK, Wright MK, Kaiser A, Middel A, Ambrose H, Hondula DM (2020) Evaporative misters for urban cooling and comfort: effectiveness and motivations for use. Int J Biometeorol 1–13

Varentsov M, Shartova N, Grischenko M, Konstantinov P (2020b) Spatial patterns of human thermal comfort conditions in russia: present climate and trends. Weather Clim Soc 12(3):629–642

Varentsov MI, Samsonov TE, Kargashin PE, Korosteleva PA, Varentsov AI, Perkhurova AA, Konstantinov PI (2020a) Citizen weather stations data for monitoring applications and urban climate research: an example of Moscow megacity. In: IOP conference series: earth and environmental science, vol 611, no 1. IOP Publishing, p 012055

Vatani J, Golbabaei F, Dehghan SF, Yousefi A (2015) Applicability of Universal Thermal Climate Index (UTCI) in occupational heat stress assessment: a case study in brick industries. Ind Health

Vinogradova V (2020) Using the Universal Thermal Climate Index (UTCI) for the assessment of bioclimatic conditions in Russia. Int J Biometeorol

Vitolo C, Di Napoli C, Di Giuseppe F, Cloke HL, Pappenberger F (2019) Mapping combined wildfire and heat stress hazards to improve evidence-based decision making. Environ Int 127:21–34

Walikewitz N, Jänicke B, Langner M, Endlicher W (2018) Assessment of indoor heat stress variability in summer and during heat warnings: a case study using the UTCI in Berlin, Germany. Int J Biometeorol 62(1):29–42

Wang C, Zhan W, Liu Z, Li J, Li L, Fu P, Jiang S (2020) Satellite-based mapping of the Universal Thermal Climate Index over the Yangtze River Delta urban agglomeration. J Clean Prod 277:123830

Watanabe S, Ishii J (2016) Effect of outdoor thermal environment on pedestrians' behavior selecting a shaded area in a humid subtropical region. Build Environ 95:32–41

Watanabe S, Nagano K, Ishii J, Horikoshi T (2014) Evaluation of outdoor thermal comfort in sunlight, building shade, and pergola shade during summer in a humid subtropical region. Build Environ 82:556–565

Weihs P, Staiger H, Tinz B, Batchvarova E, Rieder H, Vuilleumier L, Jendritzky G (2012) The uncertainty of UTCI due to uncertainties in the determination of radiation fluxes derived from measured and observed meteorological data. Int J Biometeorol 56(3):537–555

Wereski S, Krzyżewska A, Dobek M (2020) Winter UTCI variability in Poland in the 21st century. Misc Geogr 1(ahead-of-print)

Wibowo A, Salleh KO (2017) Spatial temporal analysis of urban heat hazard on education Area (University of Indonesia). Indones J Geogr 49(1):1–10

Wieczorek J, Błażejczyk K, Morita T (2013) Lighting characteristics during the polar day and their impact on changes in melatonin secretion. Geogr Pol 86(1):67–75
Więcław M, Okoniewska M (2015) Appearance of hot days in Bydgoszcz in the years 2005–2008 in different air masses and their influence on the selected biothermal indices. Sci Rev Eng Environ Sci 2015(1):67–78
Wu Z, Kong F, Wang Y, Sun R, Chen L (2016) The impact of greenspace on thermal comfort in a residential quarter of Beijing, China. Int J Environ Res Public Health 13(12):1217
Wu F, Yang X, Shen Z (2019) Regional and seasonal variations of outdoor thermal comfort in China from 1966 to 2016. Sci Total Environ 665:1003–1016
Wu P, Zhang Y, Fang Z, Gao Y (2021) Comparison of thermal comfort in different kinds of building spaces: field study in Guangzhou, China. Indoor Built Environ 1420326X20981714
Xiong K, Yang Z, Cheng C (2020) Microclimate environmental assessment and impact of mountain city pedestrian streets in summer. In: E3S web of conferences, vol 172. EDP Sciences, p 11001
Xu M, Hong B, Jiang R, An L, Zhang T (2019a) Outdoor thermal comfort of shaded spaces in an urban park in the cold region of China. Build Environ 155:408–420
Xu M, Hong B, Mi J, Yan S (2018) Outdoor thermal comfort in an urban park during winter in cold regions of China. Sustain Cities Soc 43:208–220
Xu X, Liu Y, Wang W, Xu N, Liu K, Yu G (2019b) Urban layout optimization based on genetic algorithm for microclimate performance in the cold region of China. Appl Sci 9(22):4747
Xu X, Yin C, Wang W, Xu N, Hong T, Li Q (2019c) Revealing urban morphology and outdoor comfort through genetic algorithm-driven urban block design in dry and hot regions of China. Sustainability 11(13):3683
Xu X, Wu Y, Wang W, Hong T, Xu N (2019d) Performance-driven optimization of urban open space configuration in the cold-winter and hot-summer region of China. Build Simul 12(3):411–424. Tsinghua University Press
Xue J, Hu X, Sani SN, Wu Y, Li X, Chai L, Lai D (2020) Outdoor thermal comfort at a University Campus: studies from personal and long-term thermal history perspectives. Sustainability 12(21):9284
Yan Y, Wang D, Yue S, Qu J (2019) Trends in summer air temperature and vapor pressure and their impacts on thermal comfort in China. Theor Appl Climatol 138(3–4):1445–1456
Yang J, Zhang Z, Li X, Xi J, Feng Z (2018) Spatial differentiation of China's summer tourist destinations based on climatic suitability using the Universal Thermal Climate Index. Theor Appl Climatol 134(3–4):859–874
Zamanian Z, Sedaghat Z, Hemehrezaee M, Khajehnasiri F (2017) Evaluation of environmental heat stress on physiological parameters. J Environ Health Sci Eng 15(1):24
Zare S, Hasheminejad N, Shirvan HE, Hemmatjo R, Sarebanzadeh K, Ahmadi S (2018) Comparing Universal Thermal Climate Index (UTCI) with selected thermal indices/environmental parameters during 12 months of the year. Weather Clim Extrem 19:49–57
Zare S, Shirvan HE, Hemmatjo R, Nadri F, Jahani Y, Jamshidzadeh K, Paydar P (2019) A comparison of the correlation between heat stress indices (UTCI, WBGT, WBDT, TSI) and physiological parameters of workers in Iran. Weather Clim Extrem 26:
Zeng D, Wu J, Mu Y, Deng M, Wei Y, Sun W (2020) Spatial-temporal pattern changes of UTCI in the China-Pakistan economic corridor in recent 40 years. Atmosphere 11(8):858
Zhang XY, Jiang C, Sun JX, Zhou MF (2018) Spatio-temporal variations and influencing factors of thermal comfort at different elevations. J Appl Ecol 29(9):2808–2818
Zhang Y, Liu J, Zheng Z, Fang Z, Zhang X, Gao Y, Xie Y (2020b) Analysis of thermal comfort during movement in a semi-open transition space. Energy Build 225:
Zhang L, Wei D, Hou Y, Du J, Liu ZA, Zhang G, Shi L (2020a) Outdoor thermal comfort of urban park—a case study. Sustainability 12(5):1961
Zhu Z, Liang J, Sun C, Han Y (2020) Summer outdoor thermal comfort in urban commercial pedestrian streets in severe cold regions of China. Sustainability 12(5):1876

# Chapter 4
# Sensitivity of UTCI Thermal Comfort Prediction to Personal and Situational Factors—Residual Analysis of Pedestrian Survey Data

**Peter Bröde, Claudia Di Napoli, Luísa Alcantara Rosa, Eduardo Grala da Cunha, and Eduardo L. Krüger**

**Abstract** The Universal Thermal Climate Index (UTCI) assesses the interaction of ambient temperature, wind, humidity and radiant fluxes on human physiology in outdoor environments on an equivalent temperature scale. Based upon the dynamic thermal sensation (DTS) from the UTCI-Fiala model of human thermoregulation, the UTCI allows for thermal comfort prediction. Here we compare those predictions to thermal sensation votes as recorded on the 7-unit ASHRAE scale for two Brazilian cities, Curitiba and Pelotas. Outdoor comfort surveys from 1551 respondents in Curitiba and 1148 in Pelotas, respectively, yielded negligible bias and less than one unit root-mean square error (rmse), which was similar in magnitude for both study areas. Residual analysis revealed that factors such as age, sex, body composition, site morphology (open space, street canyon), climatic state (comfort/discomfort) and clothing choice only explained a small portion of the prediction error variance, which in the total sample was dominated for over 94% by residual inter-individual variability. Adding historical weather information from the previous three days gave superior information compared to longer time lags and helped to reduce the residual variance to 88%. Those findings underpin current limitations in individual thermal comfort prediction, whereas personal and situational factors hardly affected UTCI predictive performance, which showed reasonable accuracy at the population level.

---

P. Bröde (✉)
Leibniz Research Centre for Working Environment and Human Factors at TU Dortmund (IfADo), Ardeystrasse 67, 44139 Dortmund, Germany
e-mail: broede@ifado.de

C. Di Napoli
University of Reading, Reading, UK
e-mail: c.dinapoli@reading.ac.uk

L. A. Rosa · E. G. da Cunha
Universidade Federal de Pelotas, Pelotas, Brazil
e-mail: eduardo.grala@ufpel.edu.br

E. L. Krüger
Universidade Tecnológica Federal do Paraná, Curitiba, Brazil
e-mail: ekruger@utfpr.edu.br

 
E. L. Krüger (ed.), *Applications of the Universal Thermal Climate Index UTCI in Biometeorology*, Biometeorology 4,
https://doi.org/10.1007/978-3-030-76716-7_4

**Keywords** Thermal comfort · Outdoor environment · Survey · Model · Error analysis

# 1 Introduction

The Universal Thermal Climate Index (UTCI) has become an established tool for assessing the outdoor thermal environment in the major fields of human biometeorology (Jendritzky et al. 2012; Jendritzky and Höppe 2017). The UTCI summarises the interaction of ambient temperature, wind, humidity and radiant fluxes on human physiology. The dynamic physiological responses are simulated by an advanced multi-node model of human thermoregulation (Fiala et al. 2012) coupled with a model of adaptive clothing choice in urban populations (Havenith et al. 2012). The model considers the distribution of clothing over different body parts, and the reduction of thermal and evaporative clothing resistances caused by wind and the movement of the wearer, who is assumed walking at 4 km/h (1.1 m/s) on the level. UTCI values are expressed using an equivalent temperature scale. This involved the definition of a reference environment with 50% relative humidity (but vapour pressure not exceeding 20 hPa), with still air and mean radiant temperature equalling air temperature, to which all other climatic conditions are compared.

The operational procedure (Bröde et al. 2012a) provides simplified algorithms to compute UTCI values from air temperature, wind speed, mean radiant temperature and water vapour pressure. It was supplemented by an assessment scale establishing UTCI threshold values that define different categories of thermal stress from extreme cold to extreme heat, with UTCI values from 18 to 26 °C complying with the thermal comfort zone (Bröde et al. 2012a, 2013).

Based upon the dynamic thermal sensation (DTS) from the UTCI-Fiala model of human thermoregulation (Fiala et al. 2012, 2003), the UTCI allows for thermal comfort prediction, which can be compared to observed thermal sensation votes on the ASHRAE thermal sensation scale (ASHRAE 2004). We have already presented the UTCI operational procedure and its application to outdoor thermal comfort surveys from Curitiba, Brazil (Bröde et al. 2012b, 2013). For interviews carried out in Curitiba, the observed clothing insulation was in good agreement with the UTCI-clothing model. In addition, the actual votes were well predicted by the DTS from the UTCI-Fiala model simulations carried out for UTCI reference conditions (Fiala et al. 2003, 2012). Specifically, the averaged error (bias) was found negligible and the root-mean square errors (rmse) less than one unit on the 7-unit scale (ISO 10551 1995). Detailed simulations considering the individual climatic conditions and observed clothing insulation did not further improve the predictions indicating that the assumptions underlying the UTCI model are appropriate for the surveys carried out in Curitiba (Bröde et al. 2012b).

However, we had observed larger negative bias (i.e. underestimation) and rmse for the thermal sensation votes from a survey conducted in Glasgow, UK, where the pedestrians were wearing less insulating clothing than assumed by the UTCI model (Krüger et al. 2012; Bröde et al. 2014).

Regional differences in thermal comfort, as they have been reported in field studies within Europe (Nikolopoulou and Lykoudis 2006) and in tropical regions (cf. Chap. 6: Regional adaptation of the UTCI: comparisons between different datasets in Brazil, in this book), can be partly attributed to climatic and sociocultural aspects and to some extent to personal and situational influences. These comprise psychological and physiological factors related to thermal physiology (Havenith 2001; Havenith et al. 1998; Cabanac 1971), thermal aspects of occupant behaviour (Hellwig 2015; Schweiker et al. 2013, 2016, 2020b), and non-thermal factors affecting outdoor thermal comfort (Nikolopoulou 2011; Nikolopoulou et al. 2001; Nikolopoulou and Steemers 2003; Knez et al. 2009).

It is unknown, however, whether those personal and situational factors have an impact on the accuracy of the UTCI outdoor thermal comfort predictions. To address this we have expanded our data to include a recent survey carried out in Pelotas, another Brazilian city (Krüger et al. 2020), and have:

- analysed how the residues of UTCI predictions on thermal sensation depend on personal characteristics (sex, age, body composition) and urban site morphology (open spaces vs. street canyons); as well as
- considered effects related to recent experience (Nikolopoulou et al. 2001; Nikolopoulou and Steemers 2003) by including UTCI values available from a meteorological data archive, which were calculated days, weeks and months before each survey took place.

## 2 Material and Methods

Here, we briefly review the field surveys' methodology, as detailed descriptions are available in recent publications (Bröde et al. 2012b; Krüger et al. 2020), and in the Regional Adaptation chapter of this book (Chap. 6).

### 2.1 *Outdoor Surveys*

Field measurements with concurrent administration of comfort questionnaires were carried out in Curitiba, Brazil (25°26′S, 49°16′W, 917 m amsl, subtropical climate in elevation) and in Pelotas, Brazil (31°46′18″S, 52°20′33″W, 14 m amsl, humid subtropical climate). Both field studies used similar protocols, thus ensuring compatibility of the employed procedures. In both locations, surveys were carried out in pedestrian areas during daytime (typically from 10 a.m. to 4 p.m. local time) with portable weather stations recording air temperature, relative humidity, air velocity and globe temperature, from which mean radiant temperature was calculated (ISO 7726 1998).

We applied a standard comfort questionnaire to collect personal information like age, sex, height and weight. Participants rated their thermal sensations using a symmetrical 7-unit two-pole scale ranging from $-3 =$ ‘cold’ over $0 =$ ‘neutral’ to $+3 =$ ‘hot’ (ISO 10551 1995). Intrinsic clothing thermal insulation was determined from the worn items observed on site according to standardised tables (ISO 9920 2007).

## 2.2 Data Analysis and Statistics

Only data of permanent residents (i.e. living for more than 6 months in the city) of adult age (older than 17 years) who had spent at least 15 min moving outdoors before the interview were considered eligible for the analysis. This yielded to 1148 responses from Pelotas and 1551 from Curitiba.

UTCI values were computed using the table look-up approach of the UTCI operational procedure (Bröde et al. 2012a) from measured air temperature, humidity, air velocity and mean radiant temperature. Predictions of dynamic thermal sensations (DTS) averaged over 2 h exposure time (Bröde 2019) were obtained from the output of the UTCI-Fiala model (Bröde et al. 2012a; Fiala et al. 2012) for reference clothing. In previous analyses, a more complex model incorporating actual clothing insulation did not reduce the prediction error (Bröde et al. 2012b, 2014). Therefore, we restricted our subsequent analyses on the DTS predictions for UTCI reference conditions, which are available as online dataset (Bröde 2019).

DTS prediction error was defined as the difference of DTS to the actual thermal sensation vote, with negative values indicating underestimation and positive values representing overestimation. We calculated the averaged error (bias), root-mean square error (rmse) and Pearson’s correlation coefficient (r) to assess the deviations between predicted and measured thermal sensation votes.

General additive models (Wood 2017) with locally estimated smoothing splines (LOESS) and 95%-confidence intervals (CI) were computed to describe the average course of clothing insulation, of thermal sensation and of the prediction error considering the potentially non-linear relationships with air temperature and the UTCI, respectively. For comparing models with different predictors, Akaike’s Information Criterion (AIC) was used to assess the goodness-of-fit (Zuur et al. 2009).

The influence of potential modifiers on the prediction error was assessed by computing bias, rmse and correlation coefficients for subgroups defined by city (Curitiba or Pelotas), sex and other classifying factors as described below. We calculated body mass index (BMI) from weight and height and classified the persons’ body composition as ‘underweight’, ‘normal’, ‘overweight’ or ‘obese’ according to WHO guidelines (Bröde et al. 2012b; WHO 1995), which were also applied to build age subgroups as below 25 years (young), between 25 and 64 (adult) and above 64 (elderly). Two urban site morphology groups were defined: ‘street canyons’ and ‘open spaces or crossroads’. We used the thermal state classification according to the UTCI assessment scale with the thermal comfort zone corresponding to UTCI values from 18 to 26 °C, cold discomfort below 18 °C and warm discomfort above 26 °C (Bröde et al. 2012a, 2013). The deviation of worn clothing insulation ($Icl_{obs}$) from the UTCI-clothing model ($Icl_{mod}$) was determined as percentage deviation, i.e. $(Icl_{obs} - Icl_{mod})/Icl_{mod}*100$. Percentage deviation was classified in three levels as more than 20% below (<80% $Icl_{mod}$) or above (>120% $Icl_{mod}$) or within ±20% of $Icl_{mod}$, the clothing insulation from the UTCI model.

Variance components of prediction error attributable to the factors described above were obtained separately for both cities and for the total sample, respectively, by fitting linear mixed models (Schützenmeister and Piepho 2012) considering the factors as random and using the package 'VCA' of R 4.0.2 (R Core Team 2020).

The data were supplemented by historical weather records for time periods preceding the survey campaigns, which comprised records in hourly intervals of UTCI values derived from ERA5-HEAT (Di Napoli et al. 2020). From this data, we calculated daily averages of UTCI from the recordings obtained between 10 a.m. and 4 p.m. (corresponding to the usual time frame of the surveys). For each study area, we obtained averaged historical UTCI values at 1, 3, 7, 14, 28 and 56 days prior to the actual campaign. This allows the study to cover in a logarithmic manner periods from days, weeks to almost two months. The differences between actual and historical UTCI values corresponding to the mean for the time frame of the surveys were used as predictors of the DTS prediction error by fitting cubic regression splines by generalized additive models GAM (Wood 2017). For the variance component analysis, these differences were classified in intervals ±3 °C (actually as cool/warm as in previous period), <−3 °C (actually cooler than in previous period), > +3° (actually warmer), with the thresholds corresponding to the inter-quartile range for the lagged values (Fig. 3a).

## 3 Results

### *3.1 Prediction of Clothing Insulation and Thermal Sensation*

Clothing thermal insulation showed considerable inter-individual variation, but on average dropping with increasing air temperature in both study areas (Fig. 1a). It was in good agreement with the UTCI-clothing model, especially at low temperatures in Curitiba and at higher temperatures in both cities. Although the observed insulation oscillated around the UTCI model between 12 and 25 °C, overall mean deviations (bias) from the UTCI-clothing model were negligible with a typical error (rmse) of 0.25 clo and correlation coefficients ranging between 0.7–0.8 (Table 1).

Thermal sensations also varied largely and increased with the UTCI (Fig. 1b). Generally, bias was negligible and rmse was less than one unit on the 7-unit thermal sensation scale (Table 1). These figures, including the correlation coefficients slightly above 0.6, were very similar in both study areas.

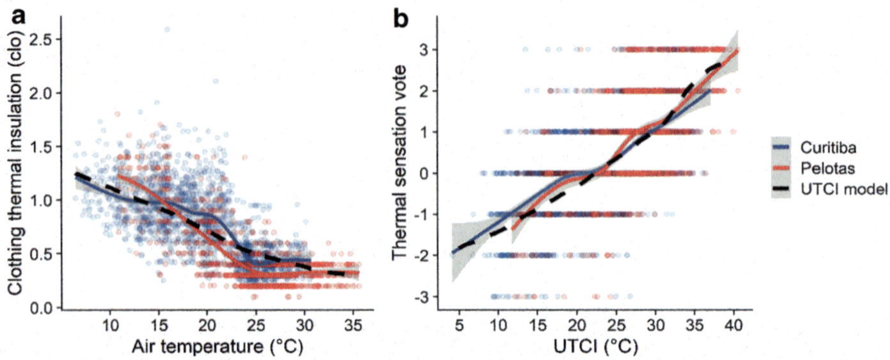


**Fig. 1** Individual recordings in Curitiba and Pelotas superimposed by smoothing splines with 95% confidence intervals for clothing thermal insulation related to air temperature (**a**) and thermal sensation votes (−3: 'cold',…,0: 'neutral',…, +3: 'hot') related to UTCI (**b**). Black dashed lines indicate predictions by the UTCI model

**Table 1** Number of respondents (n), averaged errors (bias), root-mean-square errors (rmse) and Pearson correlation coefficient (r) by study area between observed and predicted clothing insulation (clo), and for the observed thermal sensation votes (7-unit scale) compared to the dynamic thermal sensation predicted by the UTCI-Fiala model for the UTCI reference environment

| | n | Clothing insulation (clo) | | | Thermal sensation (7-unit scale) | | |
|---|---|---|---|---|---|---|---|
| | | bias | rmse | r | bias | rmse | r |
| Curitiba | 1551 | 0.02 | 0.25 | 0.69 | −0.12 | 0.95 | 0.61 |
| Pelotas | 1148 | −0.09 | 0.21 | 0.78 | −0.12 | 0.97 | 0.63 |
| Total | 2699 | −0.02 | 0.23 | 0.77 | −0.12 | 0.96 | 0.65 |

## 3.2 *Factors Influencing Thermal Sensation Prediction Error*

There were only small changes in the bias, rmse and correlation presented for the different subgroups in Table 2 compared to the overall results (Table 1). The BMI categories showed a tendency of increased underestimation error with increasing obesity. There was a small underestimation bias due to warmer sensations reported in canyons compared to open spaces (Table 2), as well as slightly increased underestimation bias with young respondents and under cold discomfort conditions more relevant in Curitiba (Fig. 2b). However, the variance component analysis (Fig. 2) revealed that all factors only accounted for a very small portion of total variance, which was dominated by residual inter-individual variability (Fig. 2a), amounting to more than 90% in relative terms (Fig. 2b).

**Table 2** Averaged thermal sensation prediction errors (bias) and root-mean-square errors (rmse) by study area comparing observed thermal sensation votes to DTS predicted by the UTCI model in relation to the modifying factors age, body composition (BMI), sex, site morphology, thermal comfort/discomfort zone according to the UTCI and deviation of worn clothing insulation from $Icl_{mod}$ of the UTCI clothing model

| Factor | Curitiba | | Pelotas | | Total | |
|---|---|---|---|---|---|---|
| | bias | rmse | bias | rmse | bias | rmse |
| Age | | | | | | |
| Young | −0.23 | 0.97 | −0.19 | 0.99 | −0.21 | 0.98 |
| Adult | −0.09 | 0.94 | −0.10 | 0.97 | −0.09 | 0.96 |
| Elderly | −0.10 | 0.96 | 0.08 | 0.84 | −0.04 | 0.92 |
| BMI category | | | | | | |
| Underweight | 0.02 | 0.95 | 0.10 | 0.96 | 0.05 | 0.95 |
| Normal | −0.09 | 0.94 | −0.10 | 1.02 | −0.09 | 0.97 |
| Overweight | −0.15 | 0.97 | −0.07 | 0.90 | −0.12 | 0.94 |
| Obese | −0.21 | 0.97 | −0.25 | 0.97 | −0.23 | 0.97 |
| Sex | | | | | | |
| Female | −0.11 | 1.01 | −0.15 | 0.99 | −0.13 | 1.00 |
| Male | −0.13 | 0.91 | −0.06 | 0.93 | −0.11 | 0.92 |
| Site morphology | | | | | | |
| Open space | 0.05 | 0.98 | −0.08 | 0.99 | −0.03 | 0.99 |
| Street canyon | −0.21 | 0.93 | −0.25 | 0.90 | −0.22 | 0.93 |
| UTCI zone | | | | | | |
| Cold discomfort | −0.30 | 0.99 | −0.24 | 0.95 | −0.29 | 0.98 |
| Comfort | −0.14 | 0.89 | 0.02 | 0.98 | −0.09 | 0.92 |
| Warm discomfort | 0.06 | 0.98 | −0.15 | 0.97 | −0.07 | 0.97 |
| Clothing insulation worn (clo) | | | | | | |
| Under-dressing (<80% $Icl_{mod}$) | −0.14 | 0.97 | −0.06 | 0.93 | −0.09 | 0.95 |
| Conformity (100 ± 20% $Icl_{mod}$) | −0.08 | 0.94 | −0.22 | 1.00 | −0.13 | 0.96 |
| Over-dressing (>120% $Icl_{mod}$) | −0.16 | 0.96 | −0.12 | 1.02 | −0.15 | 0.98 |

## 3.3 Prediction Error and Historical Weather Data

While Fig. 3a indicates only slight variation in the distribution of changes in UTCI compared to prior values with different time lags, Fig. 3b illustrates the influence of those lagged values on thermal sensation prediction error for the survey data from both study areas and the total sample, respectively, against the background of the large individual variation. The AIC values (Fig. 3c) indicate that 3 days lag information fitted better than longer lag periods, and were superior to using 1-day lag values.

The fitted spline functions in Fig. 3b for lags greater than one week exhibited monotonically decreasing prediction errors with increasing lagged values. On the

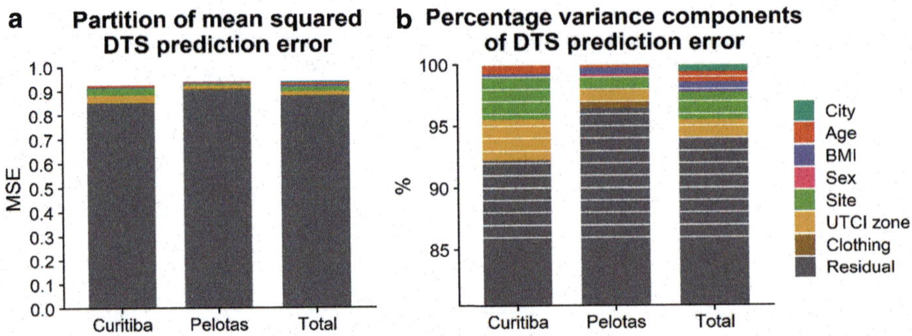


**Fig. 2** Partitioning of the variance for the DTS prediction error into factors corresponding to Table 2 shown separately for the study areas and the total sample as absolute mean square error (MSE) (**a**) and in relative terms (**b**), respectively. Note that factor "city" only applies to the total sample and that panel (**b**) does only show the range above 80%

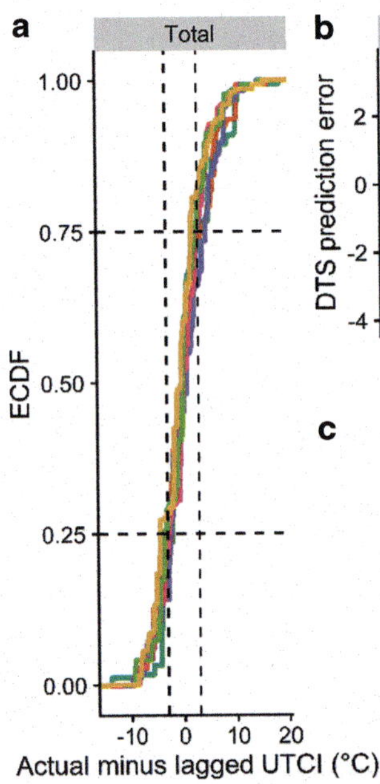


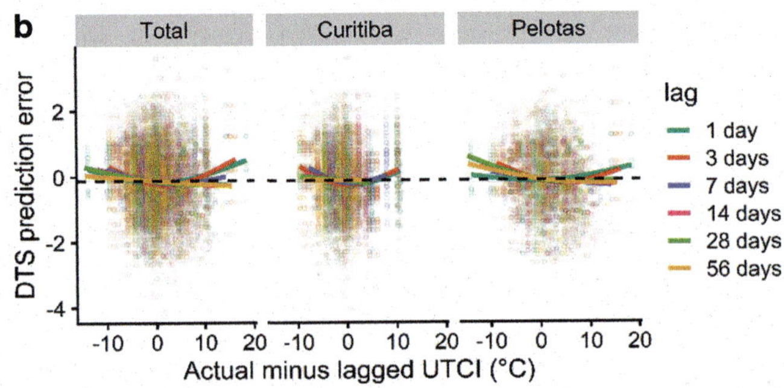


**c**

| | Total (AIC) | Curitiba (AIC) | Pelotas (AIC) |
|---|---|---|---|
| *1 day* | 7365 | 4196 | 3165 |
| *3 days* | 7332 | 4180 | 3145 |
| *7 days* | 7365 | 4201 | 3158 |
| *14 days* | 7372 | 4220 | 3158 |
| *28 days* | 7373 | 4221 | 3153 |
| *56 days* | 7386 | 4226 | 3160 |

**Fig. 3** **a** Empirical cumulative distribution function (ECDF) of the difference of actual UTCI to lagged values averaged over different lag periods from 1 day to 8 weeks for the total sample. Vertical reference lines at −3 and + 3 °C approximately intersect with the 1st and 3rd quartiles, respectively. **b** Individual DTS prediction errors in the study areas and the total sample, respectively, in relation to the difference of actual to lagged UTCI values with cubic regression spline functions fitted separately for different lag periods. Dashed horizontal reference lines indicate mean bias from Table 1. **c** Values of Akaike's Information Criterion (AIC) assessing the goodness-of-fit for the separate functions in (**b**) with lower values indicating superior fit

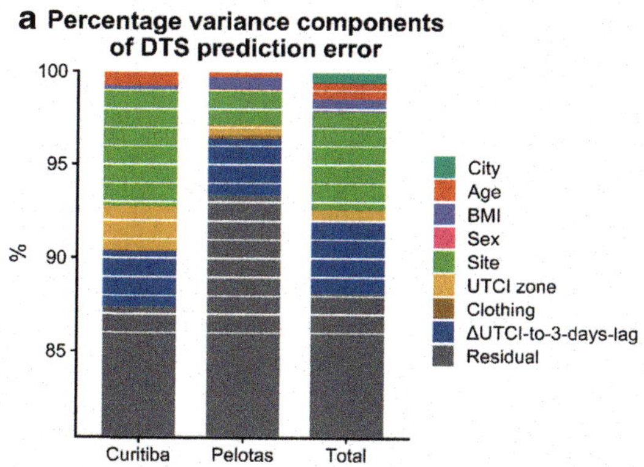


**b**

| | ΔUTCI-to-3-days-lag | bias | rmse |
|---|---|---|---|
| *Curitiba* | < -3°C | 0.02 | 0.90 |
| | ± 3°C | -0.22 | 0.96 |
| | > +3°C | -0.02 | 1.01 |
| *Pelotas* | < -3°C | 0.12 | 0.79 |
| | ± 3°C | -0.11 | 0.98 |
| | > +3°C | -0.23 | 1.02 |
| *Total* | < -3°C | 0.05 | 0.87 |
| | ± 3°C | -0.18 | 0.97 |
| | > +3°C | -0.15 | 1.02 |

**Fig. 4** **a** Relative partitioning of the variance for the error in thermal sensation votes (TSV) predicted by the UTCI model into factors considered in Fig. 2 plus the difference of actual UTCI to lagged values averaged over 3 days (ΔUTCI-to-3-days-lag) shown separately for the study areas and the total sample. Note that factor "city" only applies to the total sample and that the left panel does only show the range above 80%. **b** Bias and rmse for the categories of ΔUTCI-to-3-days-lag in the study areas and the total sample, respectively

other hand, the information from shorter previous periods (lag 1–3 days) displayed a U-shaped variation with a reduced underestimation of thermal sensation turning to slight overestimation at both negative and positive ends of the differences between actual and lagged UTCI values.

The categorized analysis for the best fitting 3 days lag period (Fig. 4b) corroborated this U-shape only in Curitiba, but not in Pelotas and the total sample, where the decreased negative bias only occurred in connection with a preceding warmer (negative difference) period. This suggests that a largely warmer period preceding the actual day resulted in diminished warm sensations of the respondents and thus a reduced negative bias. Only in Curitiba, we found this also for cooler past days (Fig. 4b).

Figure 4, which presents the variance components of DTS prediction error in relative terms, depicts that the consideration of UTCI values from the 3 previous days reduced the residual inter-individual variability by 3–4%. This was even slightly increased in combination with the other factors, e.g. for the total sample reducing residual variance from over 94% (Fig. 2) to 88% (Fig. 4).

## 4 Discussion

The dynamic thermal sensations (DTS) calculated by the UTCI-Fiala model for UTCI reference climatic conditions provided essentially unbiased predictions of actual thermal sensation votes recorded in outdoor field surveys with rmse typically less than 1 unit on the 7-unit thermal sensation scale. Given that thermal sensation votes in the range of ±1 are applied to define thermal comfort in survey studies (Rossi et al. 2012), this level of accuracy appears reasonable. Sex, age, body composition, site morphology, thermal status, and clothing choice hardly affected the prediction error with only marginal differences between Pelotas and Curitiba.

However, the large portion of 90% and more of unexplained inter-individual residual variance indicates current limitations in individual thermal comfort modelling. We had previously noted that explicitly considering individual clothing insulation in the heat exchange model did not improve the predictions (Bröde et al. 2012a, 2014). This might be explainable by inter-individual differences in human thermoregulation, probably interconnected with the clothing choice.

Recently, attempts to adapt a thermo-physiological model to Asian populations have been made by modifying the passive part of the system, i.e. anthropometry (Zhou et al. 2013). Nevertheless, given the limited influence of personal characteristics found in our study, it remains questionable, whether such alterations will sufficiently account for psychological influences (Nikolopoulou and Steemers 2003; Nasrollahi et al. 2020) or even semantic differences in perceiving thermal comfort in different cultures or regions (Tochihara et al. 2012; Schweiker et al. 2020a; Pantavou et al. 2020). Achieving a better understanding on the psychological part of thermal comfort (Lenzholzer and Nikolopoulou 2020) remains crucial for adequately considering regional (Nikolopoulou and Lykoudis 2006) and inter-cultural differences (Havenith et al. 2020).

Additional information on prior weather as a surrogate of short-term experience (Nikolopoulou and Steemers 2003) could only help reducing the large inter-individual residual variance if it refers to a short previous period, with an optimum fit for 3 days lag in our study. This concurs with earlier results for a survey carried out in Glasgow (Bröde et al. 2014), though in that study 1 day lagged values fitted best to the data. It also confirms regression analyses from earlier studies (Nikolopoulou et al. 2001) showing lower capacity of temperatures recorded at longer time lags for neutral temperature prediction. The reduced bias due to diminished warm sensations following a preceding short warmer period (Figs. 3 and 4) may be attributable to habituation or short-term acclimation (Krüger et al. 2017; Lam et al. 2021), which are considered in another chapter of this book (Chap. 5: Long and short-term acclimatization effects on outdoor thermal perception versus UTCI). Corresponding effects due to a previous cooler period only occurred in Curitiba, but had also been observed for Glasgow, UK (Bröde et al. 2014). The mechanisms underlying such regional differences still have to be elucidated, which could be facilitated by databases summarizing outdoor thermal comfort studies as described in another chapter of this book (cf. Chap. 11: Proposed framework for establishing a global database for outdoor thermal comfort research, in this book).

## 5 Conclusions

The accuracy of UTCI predictions of outdoor thermal comfort as experienced by pedestrians in two Brazilian urban areas was acceptable at the population level, thus confirming recent studies in Brazil and other regions (Xue et al. 2020; Krüger and Drach 2017). The personal and situational factors considered in this study hardly affected UTCI predictive capability.

As an outlook, recently developed personal monitoring systems have shown the potential to move forward towards an individualized assessment of thermal stress related to comfort and health in outdoor environments (Hondula et al. 2021; Sugg et al. 2020; Runkle et al. 2019; Buller et al. 2018). This may be supplemented by integrating human thermal modelling with emerging technologies (Anderson et al. 2021), with climate modelling (Brecht et al. 2020; Di Napoli et al. 2020) or weather forecasts (Petersson et al. 2019), as exemplified in the Operational Forecasting chapter of this book (Chap. 10, in this book).

**Acknowledgements** UTCI was developed within COST Action 730, the COST office is supported by the EU framework program Horizon 2020.

## References

Anderson V, Leung ACW, Mehdipoor H, Jänicke B, Milošević D, Oliveira A, Manavvi S, Kabano P, Dzyuban Y, Aguilar R, Agan PN, Kunda JJ, Garcia-Chapeton G, de França Carvalho Fonsêca V, Nascimento ST, Zurita-Milla R (2021) Technological opportunities for sensing of the health effects of weather and climate change: a state-of-the-art-review. Int J Biometeorol 65(6):779–803.https://doi.org/10.1007/s00484-020-02063-z

ASHRAE (2004) Standard 55—thermal environmental conditions for human occupancy. ASHRAE Inc., Atlanta

Brecht BM, Schädler G, Schipper JW (2020) UTCI climatology and its future change in Germany – An RCM ensemble approach. Meteorol Z 29(2):97–116. https://doi.org/10.1127/metz/2020/1010

Bröde P (2019) UTCI-to-DTS [Data set]. 1.0 edn. Zenodo. https://doi.org/10.5281/zenodo.3372231

Bröde P, Fiala D, Blazejczyk K, Holmér I, Jendritzky G, Kampmann B, Tinz B, Havenith G (2012a) Deriving the operational procedure for the universal thermal climate index (UTCI). Int J Biometeorol 56(3):481–494. https://doi.org/10.1007/s00484-011-0454-1

Bröde P, Krüger EL, Rossi FA, Fiala D (2012b) Predicting urban outdoor thermal comfort by the universal thermal climate index UTCI—a case study in Southern Brazil. Int J Biometeorol 56(3):471–480. https://doi.org/10.1007/s00484-011-0452-3

Bröde P, Krüger EL, Fiala D (2013) UTCI: validation and practical application to the assessment of urban outdoor thermal comfort. Geogr Pol 86(1):11–20. https://doi.org/10.7163/GPol.2013.2

Bröde P, Krüger EL, Fiala D (2014) Residual analysis of UTCI predictions on outdoor thermal sensation survey data. In: Cumberland Lodge, Windsor, UK, 10–13 April 2014. Network for Comfort and Energy Use in Buildings, London, pp 776–784

Buller MJ, Welles AP, Friedl KE (2018) Wearable physiological monitoring for human thermal-work strain optimization. J Appl Physiol 124(2):432–441. https://doi.org/10.1152/japplphysiol.00353.2017

Cabanac M (1971) Physiological role of pleasure. Science 173(4002):1103–1107. https://doi.org/10.1126/science.173.4002.1103
Di Napoli C, Barnard C, Prudhomme C, Cloke HL, Pappenberger F (2020) ERA5-HEAT: a global gridded historical dataset of human thermal comfort indices from climate reanalysis. Geosci Data J. https://doi.org/10.1002/gdj3.102
Fiala D, Lomas KJ, Stohrer M (2003) First principles modeling of thermal sensation responses in steady-state and transient conditions. ASHRAE Trans 109(1):179–186
Fiala D, Havenith G, Bröde P, Kampmann B, Jendritzky G (2012) UTCI-Fiala multi-node model of human heat transfer and temperature regulation. Int J Biometeorol 56(3):429–441. https://doi.org/10.1007/s00484-011-0424-7
Havenith G (2001) Individualized model of human thermoregulation for the simulation of heat stress response. J Appl Physiol 90(5):1943–1954
Havenith G, Coenen JML, Kistemaker L, Kenney WL (1998) Relevance of individual characteristics for human heat stress response is dependent on exercise intensity and climate type. Eur J Appl Physiol 77(3):231–241. https://doi.org/10.1007/s004210050327
Havenith G, Fiala D, Blazejczyk K, Richards M, Bröde P, Holmér I, Rintamaki H, Ben Shabat Y, Jendritzky G (2012) The UTCI-clothing model. Int J Biometeorol 56(3):461–470. https://doi.org/10.1007/s00484-011-0451-4
Havenith G, Griggs K, Qiu Y, Dorman L, Kulasekaran V, Hodder S (2020) Higher comfort temperature preferences for anthropometrically matched Chinese and Japanese versus white-western-middle-European individuals using a personal comfort/cooling system. Build Environ 183:107162. https://doi.org/10.1016/j.buildenv.2020.107162
Hellwig RT (2015) Perceived control in indoor environments: a conceptual approach. Build Res Inf 43(3):302–315. https://doi.org/10.1080/09613218.2015.1004150
Hondula DM, Kuras ER, Betzel S, Drake L, Eneboe J, Kaml M, Munoz M, Sevig M, Singh M, Ruddell BL, Harlan SL (2021) Novel metrics for relating personal heat exposure to social risk factors and outdoor ambient temperature. Environ Int 146:106271. https://doi.org/10.1016/j.envint.2020.106271
ISO 10551 (1995) Ergonomics of the thermal environment—assessment of the influence of the thermal environment using subjective judgement scales. International Organisation for Standardisation, Geneva
ISO 7726 (1998) Ergonomics of the thermal environment—instruments for measuring physical quantities. International Organisation for Standardisation, Geneva
ISO 9920 (2007) Ergonomics of the thermal environment—estimation of thermal insulation and water vapour resistance of a clothing ensemble. International Organisation for Standardisation, Geneva
Jendritzky G, Höppe P (2017) The UTCI and the ISB. Int J Biometeorol 61(Suppl 1):23–27. https://doi.org/10.1007/s00484-017-1390-5
Jendritzky G, de Dear R, Havenith G (2012) UTCI—why another thermal index? Int J Biometeorol 56(3):421–428. https://doi.org/10.1007/s00484-011-0513-7
Knez I, Thorsson S, Eliasson I, Lindberg F (2009) Psychological mechanisms in outdoor place and weather assessment: towards a conceptual model. Int J Biometeorol 53(1):101–111. https://doi.org/10.1007/s00484-008-0194-z
Krüger EL, Drach P (2017) Identifying potential effects from anthropometric variables on outdoor thermal comfort. Build Environ 117:230–237. https://doi.org/10.1016/j.buildenv.2017.03.020
Krüger EL, Bröde P, Emmanuel R, Fiala D (2012) Predicting outdoor thermal sensation from two field studies in Curitiba, Brazil and Glasgow, UK using the universal thermal climate index (UTCI). In: 7th Windsor conference, Cumberland lodge, Windsor, UK, 12–15 April 2012 2012. Proceedings of 7th Windsor conference. Network for Comfort and Energy Use in Buildings, London, pp 1–12
Krüger EL, Tamura CA, Bröde P, Schweiker M, Wagner A (2017) Short- and long-term acclimatization in outdoor spaces: exposure time, seasonal and heatwave adaptation effects. Build Environ 116:17–29. https://doi.org/10.1016/j.buildenv.2017.02.001

Krüger EL, Silva TJV, da Silveira Hirashima SQ, da Cunha EG, Rosa LA (2020) Calibrating UTCI'S comfort assessment scale for three Brazilian cities with different climatic conditions. Int J Biometeorol. https://doi.org/10.1007/s00484-020-01897-x

Lam CKC, Gao Y, Yang H, Chen T, Zhang Y, Ou C, Hang J (2021) Interactive effect between long-term and short-term thermal history on outdoor thermal comfort: comparison between Guangzhou, Zhuhai and Melbourne. Sci Total Environ 760:144141. https://doi.org/10.1016/j.scitotenv.2020.144141

Lenzholzer S, Nikolopoulou M (2020) Foreword to the special issue on subjective approaches to thermal perception. Int J Biometeorol 64(2):167–171. https://doi.org/10.1007/s00484-019-01857-0

Nasrollahi N, Ghosouri A, Khodakarami J, Taleghani M (2020) Heat-mitigation strategies to improve pedestrian thermal comfort in urban environments: a review. Sustainability 12(23):10000. https://doi.org/10.3390/su122310000

Nikolopoulou M (2011) Outdoor thermal comfort. Front Biosci 3:1552–1568. https://doi.org/10.2741/245

Nikolopoulou M, Lykoudis S (2006) Thermal comfort in outdoor urban spaces: analysis across different European countries. Build Environ 41(11):1455–1470. https://doi.org/10.1016/j.buildenv.2005.05.031

Nikolopoulou M, Steemers K (2003) Thermal comfort and psychological adaptation as a guide for designing urban spaces. Energy Build 35(1):95–101. https://doi.org/10.1016/S0378-7788(02)00084-1

Nikolopoulou M, Baker N, Steemers K (2001) Thermal comfort in outdoor urban spaces: understanding the human parameter. Sol Energy 70(3):227–235. https://doi.org/10.1016/S0038-092X(00)00093-1

Pantavou K, Koletsis I, Lykoudis S, Melas E, Nikolopoulou M, Tsiros IX (2020) Native influences on the construction of thermal sensation scales. Int J Biometeorol 64:1497. https://doi.org/10.1007/s00484-020-01927-8

Petersson J, Kuklane K, Gao C (2019) Is there a need to integrate human thermal models with weather forecasts to predict thermal stress? Int J Environ Res Public Health 16(22):4586. https://doi.org/10.3390/ijerph16224586

R Core Team (2020) R: a language and environment for statistical computing. R Foundation for Statistical Computing, Vienna, Austria. https://www.R-project.org/. Accessed 12 Apr 2020

Rossi FA, Krüger EL, Bröde P (2012) Definição de faixas de conforto e desconforto térmico para espaços abertos em Curitiba, PR, com o índice UTCI (Definition of thermal comfort and discomfort ranges for open spaces in Curitiba, PR, with the UCTI index). Ambiente Construído 12:41–59. https://doi.org/10.1590/S1678-86212012000100004

Runkle JD, Cui C, Fuhrmann C, Stevens S, Del Pinal J, Sugg MM (2019) Evaluation of wearable sensors for physiologic monitoring of individually experienced temperatures in outdoor workers in southeastern U.S. Environ Int 129:229–238. https://doi.org/10.1016/j.envint.2019.05.026

Schützenmeister A, Piepho H-P (2012) Residual analysis of linear mixed models using a simulation approach. Comput Stat Data Anal 56(6):1405–1416. https://doi.org/10.1016/j.csda.2011.11.006

Schweiker M, Brasche S, Bischof W, Hawighorst M, Wagner A (2013) Explaining the individual processes leading to adaptive comfort: exploring physiological, behavioural and psychological reactions to thermal stimuli. J Build Phys 36(4):438–463. https://doi.org/10.1177/1744259112473945

Schweiker M, Hawighorst M, Wagner A (2016) The influence of personality traits on occupant behavioural patterns. Energy Build 131:63–75. https://doi.org/10.1016/j.enbuild.2016.09.019

Schweiker M, Rissetto R, Wagner A (2020b) Thermal expectation: influencing factors and its effect on thermal perception. Energy Build 210:109729. https://doi.org/10.1016/j.enbuild.2019.109729

Schweiker M, André M, Al-Atrash F, Al-Khatri H, Alprianti RR, Alsaad H, Amin R, Ampatzi E, Arsano AY, Azar E, Bannazadeh B, Batagarawa A, Becker S, Buonocore C, Cao B, Choi J-H, Chun C, Daanen H, Damiati SA, Daniel L, De Vecchi R, Dhaka S, Domínguez-Amarillo S, Dudkiewicz E, Edappilly LP, Fernández-Agüera J, Folkerts M, Frijns A, Gaona G, Garg V,

Gauthier S, Jabbari SG, Harimi D, Hellwig RT, Huebner GM, Jin Q, Jowkar M, Kim J, King N, Kingma B, Koerniawan MD, Kolarik J, Kumar S, Kwok A, Lamberts R, Laska M, Lee MCJ, Lee Y, Lindermayr V, Mahaki M, Marcel-Okafor U, Marín-Restrepo L, Marquardsen A, Martellotta F, Mathur J, Mino-Rodriguez I, Montazami A, Mou D, Moujalled B, Nakajima M, Ng E, Okafor M, Olweny M, Ouyang W, Papst de Abreu AL, Pérez-Fargallo A, Rajapaksha I, Ramos G, Rashid S, Reinhart CF, Rivera MI, Salmanzadeh M, Schakib-Ekbatan K, Schiavon S, Shooshtarian S, Shukuya M, Soebarto V, Suhendri S, Tahsildoost M, Tartarini F, Teli D, Tewari P, Thapa S, Trebilcock M, Trojan J, Tukur RB, Voelker C, Yam Y, Yang L, Zapata-Lancaster G, Zhai Y, Zhu Y, Zomorodian Z (2020a) Evaluating assumptions of scales for subjective assessment of thermal environments—do laypersons perceive them the way, we researchers believe? Energy Build 211:109761.https://doi.org/10.1016/j.enbuild.2020.109761

Sugg MM, Fuhrmann CM, Runkle JD (2020) Perceptions and experiences of outdoor occupational workers using digital devices for geospatial biometeorological monitoring. Int J Biometeorol 64(3):471–483. https://doi.org/10.1007/s00484-019-01833-8

Tochihara Y, Lee JY, Wakabayashi H, Wijayanto T, Bakri I, Parsons K (2012) The use of language to express thermal sensation suggests heat acclimatization by Indonesian people. Int J Biometeorol 56(6):1055–1064. https://doi.org/10.1007/s00484-011-0519-1

WHO (1995) Physical status: the use and interpretation of anthropometry. Report of a WHO Expert Committee. WHO Technical Report Series. Geneva. http://whqlibdoc.who.int/trs/WHO_TRS_854.pdf. Accessed 20 Jul 2010

Wood SN (2017) Generalized additive models: an introduction with R, 2nd edn. Chapman & Hall/CRC, Boca Raton, Florida

Xue J, Hu X, Sani SN, Wu Y, Li X, Chai L, Lai D (2020) Outdoor thermal comfort at a university campus: studies from personal and long-term thermal history perspectives. Sustainability 12(21):9284. https://doi.org/10.3390/su12219284

Zhou X, Lian Z, Lan L (2013) An individualized human thermoregulation model for Chinese adults. Build Environ 70:257–265. https://doi.org/10.1016/j.buildenv.2013.08.031

Zuur AF, Ieno EN, Walker NJ, Saveliev AA, Smith GM (2009) Things are not always linear: additive modelling. In: Zuur AF, Ieno EN, Walker NJ, Saveliev AA, Smith GM (eds) Mixed effects models and extensions in ecology with R. Statistics for Biology and Health. Springer, New York, pp 35–69. https://doi.org/10.1007/978-0-387-87458-6_3

# Chapter 5
# Long and Short-Term Acclimatization Effects on Outdoor Thermal Perception Versus UTCI

**Cho Kwong Charlie Lam, Eduardo L. Krüger, Ivan Julio Apolonio Callejas, and Andreas Wagner**

**Abstract** The chapter looks at human acclimatization in the short and in the long term with respect to intercations between reported thermal perception, monitored meteorological variables and calculated UTCI data. Two different perspectives are employed to investigate both issues: responses from 16 participants in a longitudinal study using a controlled thermal environment (climate chamber) and responses from a cross-sectional study with local population. In both cases, outdoor data were monitored and compared to participants' thermal responses. In the longitudinal study, a 5-h exposure in a controlled thermal environment preceded a 30-min period outdoors, during which standard standard thermal comfort questionnaires were administered at three time stamps along with the monitoring of meteorological data for assessing potential short-term acclimatization effects. In the cross-sectional study, reported thermal history was used as an indicator of short-term acclimatization. For the evaluation of long-term acclimatization, two approaches have been used: in the longitudinal study, changes in thermal votes were compared over three seasons of the year; in the cross-sectional study, we compared thermal responses from two populations living under different climatic conditions but subject to summer conditions (Guangzhou/Zhuhai and Melbourne). The UTCI was used as a relevant thermal index for assessing acclimatization effects. The method of analysis also differed between

C. K. C. Lam (✉)
School of Atmospheric Sciences, Sun Yat-sen University, Zhuhai, China
e-mail: linzug@mail.sysu.edu.cn

E. L. Krüger
Department of Civil Construction, Federal University of Technology of Parana, Curitiba, Brazil
e-mail: ekruger@utfpr.edu.br

I. J. A. Callejas
Department of Architecture and Urban Planning, Universidade Federal de Mato Grosso (UFMT), Cuiabá, Brazil
e-mail: ivancallejas@ufmt.br

A. Wagner
Institute for Building Design and Technology, Karlsruhe Institute of Technology (KIT), Karlsruhe, Germany
e-mail: andreas.wagner2@kit.edu

 
E. L. Krüger (ed.), *Applications of the Universal Thermal Climate Index UTCI in Biometeorology*, Biometeorology 4,
https://doi.org/10.1007/978-3-030-76716-7_5

studies: in the longitudinal study, a dynamic approach was used with trend analysis over time; in the cross-sectional study, comparisons were made using right-here-right-now responses performing cluster analysis, followed by t-test and chi-square test of independence.

**Keywords** Acclimatization · Dynamic thermal sensation · Multisensory interactions · Thermal history · UTCI

## 1 Introduction

The way outdoor thermal comfort (OTC) surveys are configured usually involves the administration of right-here-right-now comfort assessment questionnaires to a given population at a particular point in time. At that particular moment, microclimate will have been affected dynamically in the course of the preceding hours in terms of radiant and convective heat exchanges to and from surrounding surfaces (Pearlmutter et al. 1999, 2007). Local microclimate will exhibit a thermal environment strongly related to the present weather conditions in conjunction with local urban morphology (Lin et al. 2017). A person's thermal assessment, by its turn, will evaluate thermal conditions as he found them at that specific moment. However, the extent of time the subject has been exposed to the outdoor environment prior to that given moment will play a role in his thermal assessment (Nikolopoulou and Steemers 2003). Analogously, the previous thermal environment the subject came from, whether it was warmer or cooler than the present environment he encounters, will also affect his judgment. In a similar fashion, the long-term memories of previous thermal environments and past climate types experienced by the subject will additionally affect the way he perceives the present conditions (Lenzholzer 2010).

In terms of thermophysiological parameters, Höppe (2002) shows that the extent of time spent in outdoor environments can be within the range of minutes, yet variations in skin and core temperature, modeled with the Munich Energy-balance Model for Individuals (MEMI) Model, are expected to last up to three hours and 43 h, respectively, until the human body reaches thermal equilibrium with the thermal environment on a cold winter day. Under hot conditions, thermal adaptation happens in a faster pace, with a bit less than half an hour for the skin surface temperature and just lower than one hour for the core temperature. In this respect, Höppe (2002) advocated for a dynamic modeling of outdoor thermal comfort as opposed to steady-state models, thereby introducing the UTCI, which was at the time under development.

In physiology research, short-term acclimatization is an adaptation process through which the human body adjusts to new thermal conditions involving physiological strain (de Freitas and Grigorieva 2014). Such adaptation takes place firstly in the respiratory tract through heat exchanges elicited by respiration, and independently from clothing habits and behavioural adjustments. The process involves the reduction of physiological strain under repeated stress exposure. Heat exchanges involved in the respiratory and thermoregulatory systems and the accompanying thermal strain tend to drop with acclimatization. Such a process can last up to several weeks (de Freitas and Grigorieva 2014; Hanna and Tait 2015). In physiological terms, acclimatization is accounted for by appropriate indices that assess thermal strain, such as the Acclimatization Thermal Strain Index (ATSI) proposed by de Freitas and Grigorieva (2009).

Likewise, coping mechanisms such as behavioural and technological adaptations can greatly diminish thermal stress imposed, for instance, by progressing global warming (Hanna and Tait 2015). In the context of climate change, and in purely thermophysiological terms, three types of acclimatization can occur, roughly: low acclimatization, which will essentially affect those who remain unchallenged by weather extremes by a permanent exposure to air-conditioned spaces; partial acclimatization, which is the case of the population majority, with interchanging exposure to heat (i.e. frequently moving from indoors to outdoors); and high acclimatization, which applies to outdoor workers and those engaged in outdoor activities.

In terms of thermal perception, the cognitive interpretation of the aforementioned physiological reactions will greatly influence a person's adaptive behaviour in seeking immediate relief from thermal stress conditions. Yet, despite the strong relationship between measurable physiological changes and interpreted thermal sensation (or thermal perception), changes in relevant physiological markers such as skin temperature alone are not a requirement for the initiation of thermoregulatory behaviour in humans. Perceptual mechanisms can be in many cases more reliable behavioural controllers (Schlader et al. 2011). In that sense, the concept of alliesthesia (Cabanac 1971), i.e. seeking pleasantness from acting in favour of or against a given stimulus, applies, thereby mediating behavioural responses (de Dear 2011). Alliesthesia can be applied in a short time scale such as in transient thermal environments (Parkinson et al. 2012; Parkinson and de Dear 2015; Shooshtarian 2019; Lam et al. 2021a), and in a longer time scale such as seasonal alliesthesia (Spagnolo and de Dear 2003; Schweiker et al. 2020c).

Thermal discomfort might greatly affect behavioural strategies to reduce heat and cold stress (Shooshtarian et al. 2018), dictating to which extent one undergoes acclimatization in outdoor spaces (Taylor 2014). In such context, the way a person perceives the thermal environment is influenced by seasons, expectations towards that particular outdoor location and their immediate thermal history (Nikolopoulou and Steemers 2003). Moreover, thermal sensitivity is also affected by the 'range effect', as discussed by Humphreys et al. (2007, p. 72), which refers to a psychological condition where 'people accustomed to a wide range of temperature might be less sensitive to change than those accustomed to smaller variations'. The range effect has been noticed in outdoor comfort studies (Heng and Chow 2019; Lam et al. 2021a) and is observed in comparisons shown for Brazilian cities in Chap. 5 of this book (Ref. Chap. 5: Regional adaptation of the UTCI: comparisons between different datasets in Brazil).

Seasonal influence on subjective thermal preferences is well demonstrated by the work done by de Dear and Brager (2001), which culminated with the ANSI/ASHRAE Standard 55: Thermal Environmental Conditions for Human Occupancy (ASHRAE 2017). According to the adaptive comfort approach, indoor design temperatures or acceptable temperature ranges in naturally ventilated buildings should be dictated by outdoor meteorological parameters. In this case, prevailing mean outdoor temperatures define the range of indoor operative temperature within given thermal acceptability limits. The adaptive comfort approach thus accounts for seasonal effects on preferred thermal conditions, based on the adaptation potential of humans throughout different seasons of the year as observed from a global database of 21,000 measurements conducted primarily in office buildings (ASHRAE 2017). The approach follows a trend in thermal comfort research that aims at non-steady-state thermal conditions, where occupants can attain comfort through their physiological and behavioural temperature regulatory mechanisms instead of relying on technological adaptation (de Dear 2011). Similar findings of seasonal effects on outdoor thermal comfort have also been reported in terms of thermal adaptation (Lin et al. 2011; Kántor et al. 2016; Zhou et al. 2020).

However, perceptual mechanisms expressed by subjective thermal responses additionally entail psychological aspects such as thermal expectation (Becker et al. 2003; Nikolopoulou and Steemers 2003; Schweiker et al. 2020b). The role of expectation is demonstrated by Nasrollahi et al. (2017) in a study on thermal comfort conditions experienced by tourists in different sites in Iran. Under hot conditions, most of the tourists felt warm at the sites despite being satisfied with the thermal conditions, which can be attributed to their expectation towards the visited sites. More strikingly, Rutty and Scott (2015) present a field study where expectation with the thermal environment in Caribbean beaches became evident when respondents cast their preference for no change or even warmer conditions at 39 °C in the UTCI scale. This temperature corresponds to 'very strong heat stress' according to the index's thermal comfort/stress assessment scale. Expectations towards a given outdoor location can also be verified by attitudes of the subjects with regard to that environment and its overall appearance, particularly with respect to urban morphology (Knez et al. 2009; Lenzholzer and Koh 2010; Klemm et al. 2017; Krüger and Costa 2019).

Then, there is the issue of thermal history (Nikolopoulou and Steemers 2003; Lam et al. 2021b), which can affect how a person perceives a recently entered thermal environment. Lenzholzer and de Vries (2020) propose that studies dealing with human thermal perception (the interpretation of sensory signals) should focus on two time scales: the short-term and the long-term scales. Both scales are in this case related to weather parameters: momentary and accumulated weather experience. Thus, in OTC surveys based on right-here-right-now questionnaires, the short-term acclimatization can roughly lie within minutes to a couple of hours before the interview takes place (Ji et al. 2019). The long-term acclimatization takes place over seasons (Zhang et al. 2016; Krüger et al. 2017). As an example of acclimatization in the short term, de Dear et al. (1993) investigated a step-change exposure from a thermally controlled environment. They showed that reported thermal sensation throughout the 90-min exposure in the new environment varied over time, with a marked cold overshoot in thermal sensation for down-steps, lasting for a few minutes, in short duration.

Using the UTCI as a relevant thermal index, as described in Chap. 2 of this book (Ref. Chap. 2: Literature review on UTCI applications), a number of studies have been conducted on thermal history and air-conditioning usage (Cândido et al. 2010, Krüger et al. 2015), as well as on short- and long-term acclimatization effects (Krüger et al. 2017; Lam and Lau 2018; Lam et al. 2021a). Moreover, past studies have examined heat-wave episodes and their implications on perceived thermal comfort (Lam et al. 2018, 2019) and exposure time length on OTC (Cheung and Jim 2019).

This chapter will look at human acclimatization in the short term (activities performed and thermal environments occupied by respondents prior to their subjective thermal assessment evaluation) and in the long term (seasonal effects, or original climatic region of the respondent) with respect to outdoor thermal perception. Specifically, this chapter addresses two research questions:

(1) To what extent do short-term thermal history and length of outdoor exposure time affect subjective thermal perception?
(2) To what extent does long-term acclimatization (seasonal effects and climatic region of subjects) affect subjective thermal perception?

## 2 Materials and Methods

Two different perspectives are employed here to investigate acclimatization influences: (a) responses from 16 participants in a longitudinal study using a controlled thermal environment; and, (b) responses from a cross-sectional study with local population. Each study had a different experimental setup and followed an appropriate research protocol for data collection and analysis. In the longitudinal study, a 5-h exposure in a controlled thermal environment preceded a 30-min period outdoors, during which standard OTC questionnaires were administered at three time stamps alongside the monitoring of meteorological data. In the cross-sectional study, informed thermal history was used as an indicator of acclimatization effects.

For the evaluation of long-term acclimatization, two approaches have been used: in the longitudinal study, changes in thermal votes were compared over three seasons of the year; in the cross-sectional study, we compared thermal responses from populations living under different climatic conditions, both in China and in Australia. The following sections describe the local climate, experimental setup and research protocols of both studies.

### *2.1 Longitudinal Study Carried Out in Karlsruhe, Germany*

In the longitudinal study, short and long-term acclimatization effects are investigated from subjective responses to outdoor conditions of participants who took part in a controlled field experiment over three seasons in a temperate climate in Karlsruhe, Baden-Württemberg, Germany (49°0′24.81″N, 8°24′13.16″E).

Local climate can be defined as warm and temperate, 'Cfb' type according to the Köppen-Geiger climate classification (Kottek et al. 2006). The average annual temperature is 10.5 °C, with a daily average in the coldest month (January) of

1.1 °C and 19.7 °C in the hottest month (July) (climate-data.org). Karlsruhe has been impacted by recurring heat wave episodes such as many other cities in Western-Central Europe over the last 20 years (Basarin et al. 2020), one of which was noticed during our measurements in summer 2015 (Krüger et al. 2017) and also documented by Muthers et al. (2017) while focusing on South-West Germany (where the state of Baden-Württemberg is located).

The field study was developed in the outside premises of a climate chamber located at the Karlsruhe Institute of Technology, termed 'Laboratory for Occupant Behaviour, Satisfaction, Thermal Comfort and Environmental Research' (LOBSTER). The facility comprises two adjoining 24 $m^2$ offices, designed as semi-controllable thermal environments with operable windows and shading devices. In each office, two workstations with internet access allow participants to engage in common office activities while participating in several indoor comfort surveys. Thermal control is achieved with thermally activated surfaces equipped with a capillary tube system while the ventilation system is based on floor convectors inlets for cooled and heated air and exhaust fan-driven outlets (Schweiker et al. 2014). The one façade with glazing (visible light transmittance: VLT 70) of each test environment has a window-to-wall ratio of 73%. Thanks to a rotating mechanism at the bottom of the facility, the glazed façade can assume diverse solar orientations. A thorough description of the facility is presented by Wagner et al. (2018).

The sample consists of thermal votes obtained from 16 German males with an average height of 1.80 m (SD 0.06 m), weight of 80 kg (SD 8.9 kg), and 24.9 years old (SD 3.6). These participants occupied both offices in two groups of two participants per session, for a consecutive period of 5 h under controlled thermal conditions, after which they were led to the outdoor environment for further 30 min. Clothing was standardized as much as possible: indoors, participants wore sneakers, t-shirt and jeans; outdoors, depending on season and weather conditions on a particular day, an additional fleece pullover and/or a jacket were worn, according to a group decision. Clothing insulation thus ranged between 0.11 $m^2$ K/W (indoors) and, outdoors, between 0.11 and 0.16 $m^2$ K/W in summer, 0.20 $m^2$ K/W in spring and 0.26 $m^2$ K/W in winter (ISO 9920 2007). Assumed metabolic rate for indoors corresponded to a seated position, reading and doing light work (1.2 Met); outdoors, a short walking circuit was completed a couple of times around the facility (Fig. 1) at a regular pace close to 4 km/h (2.8 Met) (ISO 7730 2005), so as to match assumptions made for the UTCI's state-of-the-art adaptive clothing model (Richards and Havenith 2007). Food intake and beverages were standardized for all participants: only still water and neutral, sugarless biscuits and fruits were provided by the research team during sessions.

Conditions monitored indoors and outdoors comprise the relevant thermal comfort variables air temperature ($T_a$), relative humidity (RH) and speed ($v_a$) and the mean radiant temperature ($T_{mrt}$), the latter calculated according to ISO 7726 (2001) from globe temperature readings ($T_g$). For that, two Ahlborn comfortmeters ALMEMO 2690 were used, one at each office. They were continuously monitored by the researchers, who would promote slight changes in the air-conditioning system to ensure quasi-steady-state thermal conditions within the lower and upper limits of the

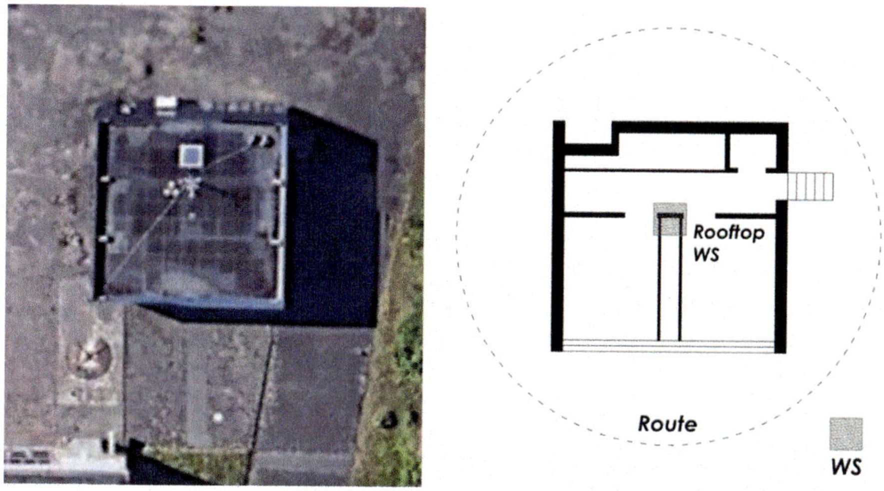


**Fig. 1** Walking route outside the climate chamber, Google image and schematic view

thermal comfort zone defined for the thermal index 'PMV' for 'Class B' thermal environments (ISO 7730 2005). During the 5-h period in the chamber, other aspects regarding indoor comfort were evaluated in a parallel study focusing on non-visual aspects of daylight and well-being (Tamura et al. 2017). For that parallel study, window façade orientation was tested as well as the absence of natural light. The two daylight situations were the solar orientation of the glazed façade to south or southwest ('equatorial orientation' or 'EQ') and, alternatively, to north-northwest ('non-equatorial orientation' or 'NEQ'). For test days with only artificial lighting 'ART', the light source was from four Osram L36W/840 fluorescent lamps, with luminaires facing the ceiling, resulting in diffuse light at desk level.

After the 5-h acclimatization period indoors, subjects were led to the walking circuit outside the chamber, stopping at times next to a hand-made weather station consisting of two HOBO U12-011 dataloggers, one of which was placed within a plastic 15-mm globe painted grey for recording globe temperature; and the other logger hung on a string inside a 50-cm-long PVC tube acting as a naturally ventilated solar radiation shield. Spot measurements of wind speed were additionally taken using a hand-held anemometer (Testo 416 Mini-Vane anemometer), which was attached to the tripod. Back-up data from the roof-top Thies weather station, located at 6 m from ground level on the rooftop of the climate chamber were also used. Measurements at pedestrian level comprised of air temperature and humidity at 1.30 m, globe temperature at 1.2 m and wind speed at approximately 1.6 m.

A standard comfort questionnaire was administered to the subjects at three time stamps: immediately after leaving the facility (or at time stamp 1); after 15 min of light walk around it (time stamp 2); and after a further 15 min (time stamp 3). Subjective thermal perception was assessed according to the German version of the 7-point perceptual judgment scale with a neutral point (ISO 10551 2019) for

**Table 1** Breakdown of experimental sessions

| | | | | | | | | | | | | |
|---|---|---|---|---|---|---|---|---|---|---|---|---|
| *Winter (January 12 through February 6)* | | | | | | | | | | | | |
| Test day | 1 | 2 | 3 | 4 | 5 | 6 | 7 | 8 | 9 | 10 | 11 | 12 |
| Light source | EQ | ART | NEQ | ART | NEQ | EQ | EQ | NEQ | ART | NEQ | EQ | ART |
| Sky conditions | C | N/A | OC | N/A | OC | OC | C | OC | N/A | C | OC | N/A |
| *Spring (April 13 through May 8)* | | | | | | | | | | | | |
| Test day | 13 | 14 | 15 | 16 | 17 | 18 | 19 | 20 | 21 | 22 | 23 | 24 |
| Light source | EQ | ART | NEQ | NEQ | EQ | ART | EQ | NEQ | ART | ART | NEQ | EQ |
| Sky conditions | PC | N/A | C | C | C | N/A | C | C | N/A | N/A | PC | PC |
| *Summer (June 22 through July 17)* | | | | | | | | | | | | |
| Test day | 25 | 26 | 27 | 28 | 29 | 30 | 31 | 32 | 33 | 34 | 35 | 36 |
| Light source | ART | EQ | NEQ | NEQ | ART | EQ | EQ | ART | NEQ | NEQ | EQ | ART |
| Sky conditions | N/A | PC | C | C | N/A | C | C | N/A | OC | OC | C | N/A |

Exposure: *EQ* equatorial; *NEQ* non-equatorial; *ART* electric light source
Sky conditions: *N/A* not applicable (external blinds drawn); *C* clear sky; *PC* partly overcast/cloudy; OC = overcast

thermal sensation and preference. Groups of four were tested each day and for three consecutive days per season, so that each individual would follow three times the same experimental procedure, hence a total of 9 test days per subject.

Outdoor thermal conditions were post-processed with the UTCI using Bioklima 2.6 (Błażejczyk and Błażejczyk 2010). Air temperature and humidity, as well as globe temperature readings at pedestrian level, and the wind speed measured on the rooftop, rescaled for the height of 10 m according to a logarithmic formula (Bröde et al. 2012) were used for calculations of the UTCI.

In total, 36 sessions took place, including winter, spring and summer periods in 2015, and accounting for the three different light exposures as previously mentioned (Table 1).

### 2.1.1 Analysis Protocol

The analysis focussed on reported thermal sensation and preference votes cast over the various exposure conditions. We employed descriptive statistics, significance tests, ANOVA and multiple regression analysis for evaluating individual responses from a same subject over three time stamps and under different conditions (previous exposure as regards window configuration and seasonal changes), encompassing 414 thermal votes (missing data) and at grouped responses for each of the 108 sessions. As a relevant index for analysis, UTCI's Dynamic Thermal Sensation (DTS), more specifically its prediction bias (DTS minus the actual thermal sensation vote), was used as a variable of interest. Improved models for DTS were tested by adding stepwise exposure time, previous exposure conditions and season.

## 2.2 Cross-Sectional Study

### 2.2.1 Regional Climate of Study Areas

For the cross-sectional study, our study areas are Guangzhou and Zhuhai in Southeast China and Melbourne in Australia (Fig. 2). In these cities, surveys were conducted under warm and hot conditions, so the long-term acclimatization is not related to the season but associated with long-term thermal history. Table 2 shows the 1981–2010 average air temperature ($T_a$), relative humidity (RH) and wind speed ($v_a$) in September in Guangzhou and Zhuhai, as well as in January and February in Melbourne. These months are the same months during which

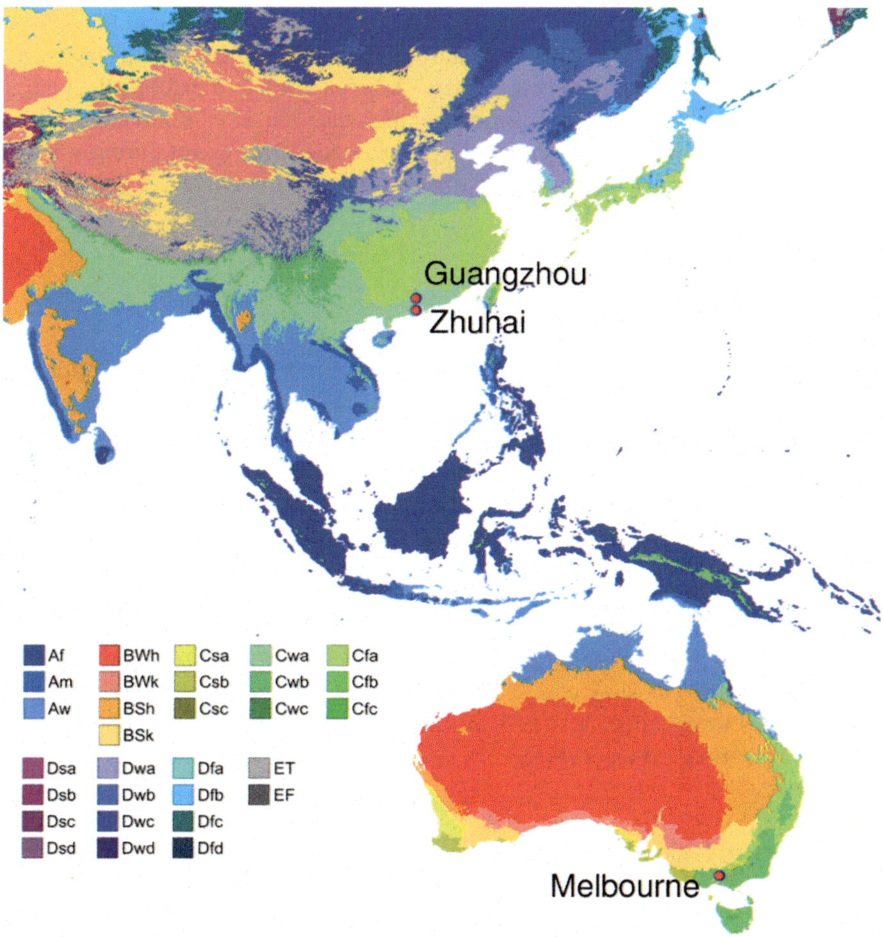


**Fig. 2** Survey study sites according to the Köppen climate classification. The colour scheme was adapted from Peel et al. (2007)

**Table 2** Mean air temperature, relative humidity and wind speed in Guangzhou, Zhuhai and Melbourne for the period 1981–2010. Guangzhou and Zhuhai data are derived from China Meteorological Administration (2012), whereas Melbourne data are derived from the Bureau of Meteorology (2020). We chose the long-term monthly average corresponding to the same months where we conducted our surveys

| City | $T_a$ (°C) | RH (%) | $v_a$ (m/s) |
|---|---|---|---|
| Guangzhou[a] | 32.1 | 72 | 1.5 |
| Zhuhai[a] | 30.6 | 80 | 3.1 |
| Melbourne[b] | 21.1 | 65 (9 a.m.) and 47 (3 p.m.) | 2.4 (9 a.m.) and 3.5 (3 p.m.) |

[a]September average, [b]January and February average

we conducted the surveys in the three cities. Both Guangzhou (Köppen Cfa) and Zhuhai (Köppen Cwa) have a humid subtropical climate. From May to September, summer is characterized by high temperature and humidity affected by the monsoon. Melbourne has a temperate oceanic climate (Köppen Cfb), and exhibits fluctuating weather conditions with a large diurnal temperature range in summer (Sturman and Tapper 2006). Melbourne occasionally experiences heat waves during summer (December–February), which is related to persistent incursions of hot, dry continental air.

### 2.2.2 Site Description

In Melbourne, we chose the Melbourne Gardens (37°50′S, 144°58′E; Fig. 3a) and Cranbourne Gardens (38°7′S 145°16′E; Fig. 3b) as our study sites. Moreover, outdoor thermal comfort surveys and concurrent meteorological measurements were conducted in three Chinese university campuses in Guangzhou (23°05′N 113°17′E; Fig. 3c), Panyu (a peri-urban district in Guangzhou, 23°03′N 113°23′E; Fig. 3d), and Zhuhai (22°21′N 113°34′E; Fig. 3e), which is near the South-eastern coast of China. These sites represent open spaces and urban parks. Overall, we have selected different sites in three cities to cover a range of sky view factors (SVF) in green and open sites (from shaded, partially shaded to fully exposed). SVF refers to 'the amount of sky that can be seen from a given point', ranging from 0 (fully obstructed) to 1 (fully visible)' (Gosling et al. 2014). Moreover, selecting similar landscapes in different cities would allow us to evaluate thermal history's effect on outdoor thermal comfort in a similar urban context.

### 2.2.3 Meteorological Measurements

In Melbourne, we used Campbell Scientific CR211X loggers and Kestrel 4400 heat stress trackers to measure $T_a$, RH, $v_a$, and globe temperature ($T_g$). The $T_g$ was measured by black globe thermometers (150-mm for Campbell Scientific station and 25-mm for Kestrel 4400 heat stress trackers). Campbell Scientific CR211X loggers

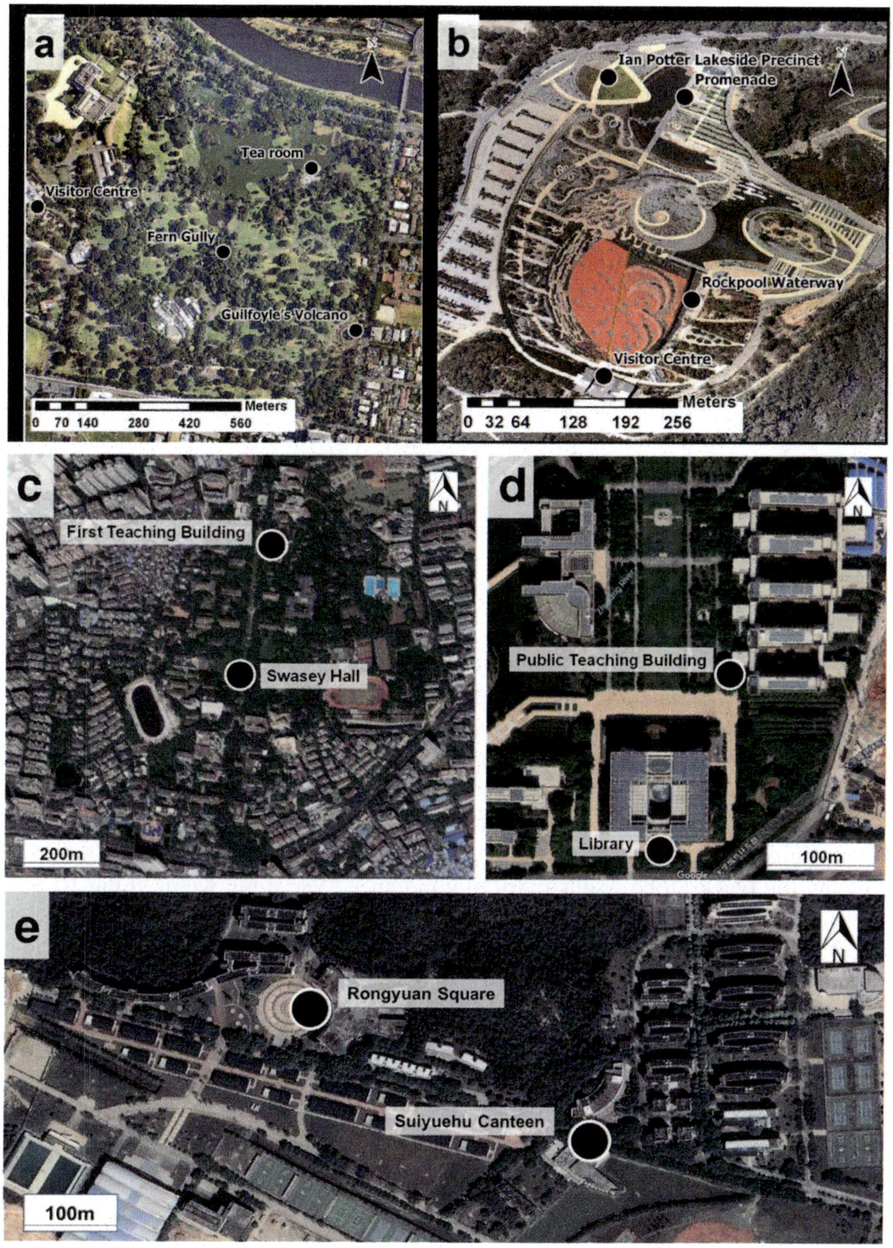


**Fig. 3** Selected locations of weather stations and survey sites in Melbourne, showing: **a** Melbourne Gardens and **b** Cranbourne Gardens (Nearmap aerial imagery: January 2014 and February 2014), **c** Guangzhou campus, **d** Panyu campus, **e** Zhuhai campus (Google Maps: September 2019)

(with 150-mm globe thermometers) were employed in the Visitor Centre and the Tea room in the Melbourne Gardens (Fig. 3a), as well as the Visitor Centre and Rockpool Waterway in the Cranbourne Gardens (Fig. 3b). We used Kestrel 4400 heat stress trackers (with 25-mm globe thermometers) at Fern Gully and Guilfoyle's Volcano in the Melbourne Gardens (Fig. 3a) and Promenade and Ian Potter Lakeside Precinct in the Cranbourne Gardens (Fig. 3b). The same meteorological instrument was used at each location during the entire survey period in Melbourne. The meteorological data were based on 1-min measurements and later averaged into a 10-min interval.

We used the Rainwise Portlog in the First teaching building and Swasey Hall in Guangzhou (Fig. 3c) and Rongyuan Square and Suiyuehu Canteen in Zhuhai (Fig. 3e). Furthermore, we used the Davis Vantage Pro 2 $v_a$ in the public teaching building and library in Panyu (Fig. 3d). All weather stations were set up at 1.5 m above ground. Similar to Melbourne, our weather stations measured $T_a$, RH and $v_a$. We used a 150 mm black globe thermometer (TJHY HQZY-1) to measure $T_g$. The weather data were averaged into a 1-min interval for both Rainwise and Davis weather stations.

### 2.2.4 UTCI Calculation

Using meteorological measurements collected and physiological observations made during the survey campaigns in the three cities, we calculated UTCI. To calculate the UTCI, we input $T_a$, $RH$, $v_a$ at 10 m ($V_{10m}$), as well as $T_{mrt}$, which was derived from the globe temperature. We used the logarithmic formula (Bröde et al. 2012; Fiala et al. 2012) to convert our 1.5 m wind speed to $V_{10m}$.

In Guangzhou, Zhuhai and Melbourne, we used $T_a$, $RH$, $v_a$ to calculate the $T_{mrt}$ from the globe thermometers based on the equation defined in ISO 7726 (2001). For reducing the black globe's sensitivity to wind speed variation, we used the 10-min average $T_a$, $v_a$, and $T_g$ to calculate $T_{mrt}$ in both Guangzhou, Zhuhai, and Melbourne.

In Melbourne, the $T_{mrt}$ of a 150-mm black globe ($T_{mrt150mm}$) was calibrated against integral radiation measurements by Kipp and Zonen CNR1 (Coutts et al. 2016). We also calibrated the $T_{mrt}$ of Kestrel weather station's 25-mm black globe against the 150-mm black globe for Melbourne (Lam and Lau 2018). In Guangzhou and Zhuhai, the $T_{mrt150mm}$ was calibrated for size and precision against the readings of three Kipp & Zonen CNR4 net radiometers, which measured the radiation fluxes from six perpendicular directions concurrently (Lam et al. 2021b).

### 2.2.5 Thermal Comfort Survey

We conducted thermal comfort surveys in two botanic gardens in Melbourne ($n$ = 3293) from 8th to 19th January 2014 and from 5th to 16th February 2014 (within 24 days). Surveys were also conducted on university campuses during summer (3rd–15th September 2018, i.e. within 13 days) in Guangzhou and Panyu ($n$ = 2882), as well as Zhuhai ($n$ = 1422). The survey questions were posed by the interviewers and had two parts. The first part included respondents' demographic background, origin, clothing insulation, as well as their activity and exposure 10–15 min before the survey. Based on the tick box of clothing in the survey, the respondents' clothing insulation values were estimated from the tables provided in ISO 9920 (2007).

In the second part, survey questions were designed according to the guidelines of ISO 10551 (2019). Both Guangzhou and Melbourne surveys included questions regarding thermal sensation vote (TSV: the ASHRAE 7-point scale) and thermal preference vote (TPV). The Melbourne survey used the 3-point McIntyre scale (McIntyre 1980), whereas the 7-point preferred TSV scale was used to indicate thermal preference. Details on the harmonisation method used for TSV and McIntyre scales are given by Lam et al. (2021b). Further details of the survey questions are shown in Lam et al. (2021b).

We derived the Köppen climate zone classifications of survey respondents in Guangzhou, Zhuhai and Melbourne based on their reported origin. Our survey in Guangzhou and Zhuhai recorded the city of origin, whereas our survey in Melbourne recorded the country of origin and postcode for Australian respondents. For the Melbourne survey, certain samples only reported country of origin and we could not define their exact climate zone. These samples (404 out of 3293 votes) were discarded for the associated analysis.

### 2.2.6 Analysis Protocol

We first conducted a cluster analysis to form different groups from the survey with a similar TSV and TPV characteristic. After that, we conducted t-tests and chi-square tests of independence to compare the long-term and short-term acclimatization parameters between these clusters.

For the cross-sectional study, the raw survey data were analysed and grouped through a clustering procedure to classify different clusters (i.e. relatively homogenous groups). We compared these different groups of survey participants in terms of their proportion of long-term acclimatization (e.g. reported climate zones of origin) and short-term thermal history (prior exposure environments and activities, as reported during surveys). We adopted two steps to the cluster analysis. First, hierarchical cluster analysis (Bridges 1966) was used to determine the number of clusters required based on the dendrogram and observing the changes in the distance coefficients. Second, we used k-means cluster analysis (Hartigan and Wong 1979) to form the clusters according to a priori fixed k number of clusters, which has been used in previous outdoor thermal comfort studies (Pigliautile and Pisello 2020; Acero et al. 2020). In this study, the number of clusters was determined by the hierarchical clustering algorithm. Hierarchical clustering is an algorithm that groups similar objects into clusters. The endpoint is a set of clusters, where each cluster is distinct from each other, and the objects within each cluster are broadly similar to each other.

For the k-means method, we calculate the squared Euclidean distance (SED) between two data objects, which are comprised of different variables. The iteration process begins with selecting the cluster centroids (CC). Then, the k-means method calculates the SED for each grouping of CC and data objects. The cluster is formed by assigning all data objects to their nearest CC. After that, a new CC is provided by averaging the data objects of each cluster. When no data require reassignment to another CC, the process of reassigning CC and calculating SED stops. Hence, the k-means method groups data objects according to their similarity. The final CC represents unique values of the selected variables that characterize the clusters' data objects. There are two reasons for using the k-means clustering. First, this approach can classify all data without removing data that could have been recognized as outliers by other algorithms. Second, k-means clustering is suitable for large sample sizes and good at segmenting large data sets.

The clustering input parameters are (i) thermal sensation vote (TSV) and (ii) thermal preference vote (TPV) in Guangzhou, Zhuhai and Melbourne. These two parameters are suitable to summarize outdoor thermal comfort in these cities. After forming the new clusters, scatterplots were used to visualize the cluster distribution (e.g. distribution of thermal sensation vote and thermal preference vote). Then, the clusters were evaluated by short-term thermal history variables (reported activity and exposure environment before the survey), as well as long-term acclimatization indicators (e.g. city or country of origin reported by the subject).

Following the clustering analysis, we conducted an independent samples t-test to compare the mean thermal sensation vote (MTSV) between the two clusters in each city across different UTCI thermal stress categories (e.g. moderate to extreme heat stress). For the t-test, we used eta squared ($\eta^2$) as our effect size. Moreover, we conducted a chi-square test of independence to investigate whether there is an association between long- and short-term acclimatization and outdoor thermal perception in Guangzhou/Zhuhai and Melbourne. The long-term acclimatization indicator is the respondents' climate zone, whereas short-term acclimatization indicators are activity and exposure environment before the survey. The effect size is shown by Cramer's V ($\varphi_c$). After that, we used a Bonferroni-corrected z test to examine whether the column proportion between different clusters is significantly different (Lomax and Hahs-Vaughn 2020; Sharpe 2015).

# 3 Results

## *3.1 Longitudinal Study Carried Out in Karlsruhe, Germany*

According to the physiological concept of Alliesthesia (Cabanac 1971) "a given stimulus will arouse either pleasure or displeasure according to the internal state of the stimulated subject". Therefore, stepping from thermal homogeneity to transient outdoor conditions should create immediate responses that would then diminish with the time of exposure. When a subject experiences a thermally static environment for a long time, with no opportunity for the body to interpret the 'usefulness' of a stimulus for thermoregulation, there is a greater chance that he will more effectively experience thermal pleasure under sudden transient conditions. In the climate chamber study, such long-term exposure in a thermally homogeneous environment was granted before each subject was exposed to the outdoors.

Outdoor conditions measured over the three seasons are varied, including mild days in winter, unexpected warm days in spring and heat wave episodes in summer. Nevertheless, post-processed climatic conditions represented as UTCI units (Fig. 4) differed significantly between seasons ($p < 0.0001$).

A side-by-side comparison of the same individual's thermal response (sensation and preference) over three time stamps (0 min, 15 min, and 30 min) after leaving the chamber against UTCI changes along the course of those sessions is shown in Table 3. Accounting for missing data, altogether 414 votes have been evaluated. No statistical significance has been verified, neither in terms of exposure conditions nor in obtained thermal responses throughout the three time stamps. Yet, although UTCI remained quite unchanged, mean values for thermal sensation and preference varied slightly over the course of outdoor exposure (Table 3). Locally estimated smoothing splines (LOESS) with 95%-confidence intervals (CI) were used considering the potentially

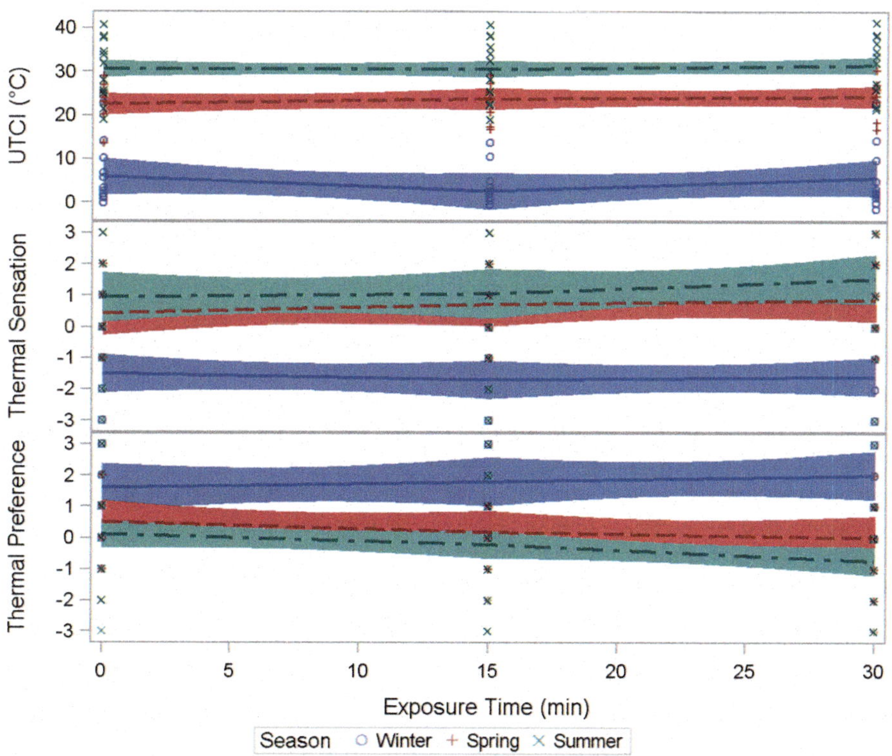


**Fig. 4** UTCI, thermal sensation and preference related to season and exposure time. Individual data and smoothing function (LOESS) with 95% CI for time effect

**Table 3** One-way ANOVA with repeated measures results with F- and p-values for UTCI, thermal sensation (TS) and thermal preference (TP) accounting for exposure time—all seasons

| | UTCI | | | TS | | | TP | | |
|---|---|---|---|---|---|---|---|---|---|
| | F | p | F stat | F | p | F stat | F | p | F stat |
| Stats | 0.040 | 0.9607 | 3.018 | 0.689 | 0.5025 | 3.018 | 1.710 | 0.1821 | 3.018 |
| Exposure time (min) | 0 | 15 | 30 | 0 | 15 | 30 | 0 | 15 | 30 |
| Mean | 19.7 | 19.6 | 20.0 | −0.09 | 0.00 | 0.15 | 0.72 | 0.57 | 0.41 |
| Standard deviation | 11.9 | 12.2 | 12.7 | 1.64 | 1.75 | 1.88 | 1.31 | 1.44 | 1.54 |
| Confidence level (95%) | 2.01 | 2.06 | 2.13 | 0.28 | 0.29 | 0.32 | 0.22 | 0.24 | 0.26 |

**Table 4** Mean prediction error (bias) and root-mean-squared error (rmse) of DTS (Dynamic Thermal Sensation) related to season and exposure time

| Season | bias | | | rmse | | |
|---|---|---|---|---|---|---|
| | 0 min | 15 min | 30 min | 0 min | 15 min | 30 min |
| Winter | −0.08 | −0.02 | −0.01 | 0.84 | 0.49 | 0.41 |
| Spring | −0.23 | −0.43 | −0.53 | 0.74 | 0.63 | 0.75 |
| Summer | 0.34 | 0.25 | 0.04 | 0.59 | 0.56 | 0.67 |
| *Total* | *0.01* | *−0.07* | *−0.17* | *0.73* | *0.56* | *0.63* |

non-linear patterns of the UTCI and thermal perception votes over the course of the three time stamps.

Table 4 illustrates the time course of UTCI's Dynamic Thermal Sensation (DTS) bias, which represents the difference between the estimator and the actual thermal vote. Moreover, root-mean-square error (rmse) of prediction for the different seasons showed that bias and rmse were reduced after 15 min in winter and after 30 min in summer, respectively. For spring, there was not much change in rmse and even a slightly increased negative bias, meaning underestimation of thermal sensation. In order to harmonize data, as there were on occasion less than four participants per session (missing data, comprising 6% of the total of thermal perception votes), grouped thermal responses for each session were computed. An overshoot of cool responses in summer lead to increased bias (i.e. to an overestimation of DTS), which was attenuated with temperature and time of exposure. Such effect has been reported in past research by de Dear et al. (1993) while conducting temperature step changes in a thermally controlled environment. In that study, however, overshoot in cold responses was noticed for down-steps, lasting for a few minutes; in our case, the cold overshoot was noticed more evidently in summer and extended to 15 min.

In Table 5, the effect of previous window exposure is represented. The possibility of having a view to the outdoor environment before leaving the climate chamber or to have a more or less accentuated access to daylight could be analysed, again, by looking at aggregate responses over the 108 sessions. Overall, bias is low and changes in responses can be noticed most pronouncedly for a previous non-equatorial view

**Table 5** Mean prediction error (bias) and root-mean-square error (rmse) of DTS (Dynamic Thermal Sensation) related to previous window exposure and exposure time

| Exposure | bias | | | rmse | | |
|---|---|---|---|---|---|---|
| | 0 min | 15 min | 30 min | 0 min | 15 min | 30 min |
| ART | 0.03 | 0.05 | 0.01 | 0.79 | 0.72 | 0.73 |
| EQ | 0.06 | 0.04 | −0.15 | 0.83 | 0.38 | 0.53 |
| NEQ | −0.05 | −0.29 | −0.36 | 0.54 | 0.54 | 0.60 |
| *Total* | *0.01* | *−0.07* | *−0.17* | *0.73* | *0.56* | *0.63* |

Exposure: *EQ* equatorial; *NEQ* non-equatorial; *ART* electric light source

orientation (NEQ), however with rising underestimation of thermal sensation by DTS over the three time stamps.

Three-factor ANOVAs performed for the three factors (exposure time, season and previous window exposure) indicated that differences are only found for DTS bias with respect to season, but no statistically significant interactions were found between previous exposure conditions, season and exposure time.

Nevertheless, Fig. 5a shows that, in spring, a high overestimation by DTS occurs for all previous exposure conditions. In winter, previous exposure to electric lighting (ART) yielded an underestimation of thermal sensation outdoors, whereas previous exposures to natural light (EQ, NEQ) led to an overestimation of it, yet with a reduced bias for an equatorial window orientation, which might suggest that visual cues of outdoor conditions in that season are important for a more accurate thermal perception before leaving to outdoors. In summer, all previous exposures led to an overestimation of thermal sensation by DTS with a more accentuated bias for 'ART' and then for 'EQ'. Patterns found for spring and summer might indicate a relationship between illuminance level subjects were exposed to prior to their outdoor residency and outdoor thermal sensation. Measured illuminance at desk level performed during indoor sessions with a JETI Specbos 1201 spectroradiometer showed that overall, for the 'EQ' Mean = 1165 lx (SD = 448 lx), for 'NEQ' Mean = 1100 lx (SD = 170 lx) and for 'ART' a constant value of 1506 lx (SD = 9 lx).

With respect to exposure time (Fig. 5b), for those who had been previously devoid of the outside view, bias remained low and unchanged over the three time stamps. There was a change in reported thermal sensation over time for previous exposures to natural light. Again, this seems to indicate a lasting relationship between previous light exposure and outdoor thermal perception.

The interaction between season and exposure time (Fig. 5c) confirms the findings of Table 4, with subtle differences in votes in winter and more accentuated changes in reported thermal sensation votes over time in the other two seasons, in spring with a rising underestimation of thermal sensation and in summer with an initial overestimation by DTS for thermal sensation, with a clear overshooting response at the first time stamp. Such patterns denote a seasonal acclimatization effect, with a more precise thermal response in the coldest period of the year, an uncertain perception of the weather in spring and diminishing bias in summer over time, perhaps due to

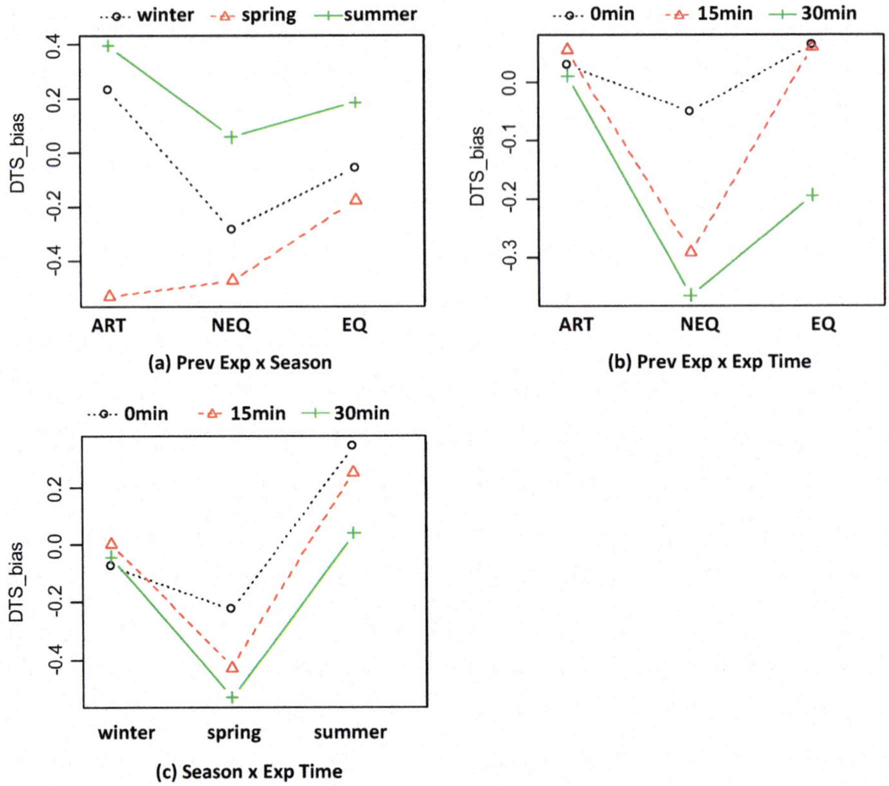


**Fig. 5** Interactions between variables previous exposure ('Prev Exp'), season ('Season') and exposure time ('Exp Time') derived from three-factor ANOVAs

short-term acclimatization. It should be stressed that heat wave episodes took place during the summer sessions, significantly affecting the way subjects perceived the outdoor thermal environment over the three time stamps (Krüger et al. 2017).

Since seasonal effects reached statistical significance, one-way ANOVAs have been carried out separately for winter, spring and summer, in order to evaluate whether prior exposure in a given season is relevant for defining thermal sensation and preference. Statistically significant differences only occurred under winter conditions, particularly between prior exposure to electric lighting (ART) and natural light exposure (EQ and NEQ). For sensation, 'ART' showed increased cold sensation (mean ± SE: $-2.2 \pm 0.12$) whereas 'NEQ' and 'EQ' both had diminished cold sensation, $-1.3 \pm 0.13$ and $-1.5 \pm 0.16$, respectively. For thermal preference, a reverse relationship is observed: 'ART' had $2.2 \pm 0.12$ in terms of thermal preference vote, meaning that participants voted for warmer conditions, whereas for 'EQ' and 'NEQ', the intensity of preference for warmer conditions was reduced: $1.6 \pm 0.13$ and $1.6 \pm 0.14$, respectively.

### 3.1.1 Effect Size and Model Building

The several variables tested, relative to their contribution to diminishing prediction bias and to promoting a better fit between DTS and thermal sensation were integrated stepwise in multiple regression models. Starting with DTS and adding one by one the variables exposure time ('Exp Time'), season ('Season') and previous exposure from the window during the 5-h stay in the indoor environment prior to outdoors ('Prev Exp'), the effect size (Pearson's r-value), mean bias of prediction, Willmott's index of agreement (d) (Willmott 1981), root-mean-square error (rmse) and features of the regression model were determined (Table 6) for the whole ensemble and in Table 7 for the grouped data per questionnaire round. For variations in 'Exp Time' according to the three increasing time stamps, at 0 min, 15 min and 30 min, the values 1, 2, 3 have been used. Changes in 'Season' have been assigned as follows: winter = 1, spring = 2, summer = 3, so as to follow the rhythm of rising temperatures over the course of seasonal changes. For 'Prev Exp', assigned values were, for ART = 1, for NEQ = 2, for EQ = 3, thus in a rising rank order in terms of previous exposure to natural light.

Increases in the predictive power of UTCI's Dynamic Thermal Sensation (DTS) for reported thermal sensation are minimal when new variables are included. The complete model (model 6) showed a somewhat better fit to reported data but with only two variables having any statistical significance. Predicted thermal sensation according to that model rises with exposure time (positive slope coefficient) and drops (negative slope coefficient) with the course of seasonal changes, from winter to summer and with previous exposure to natural light, which could suggest the influence of acclimatization in the short and in the long term and of visual cues on

**Table 6** Improved predictive models for DTS, from individually reported thermal votes (N = 414)

| | DTS | Model 1 | Model 2 | Model 3 | Model 4 | Model 5 | Model 6 |
|---|---|---|---|---|---|---|---|
| N | 414 | 414 | 414 | 414 | 414 | 414 | 414 |
| R | 0.840 | 0.841 | 0.844 | 0.840 | 0.845 | 0.844 | 0.845 |
| R-squared | 0.706 | 0.708 | 0.712 | 0.706 | 0.713 | 0.712 | 0.713 |
| Mean bias | −0.122 | −0.027 | −0.024 | −0.027 | −0.024 | −0.024 | −0.024 |
| RMSE | 0.957 | 0.949 | 0.943 | 0.950 | 0.940 | 0.943 | 0.940 |
| Wilmott | 0.908 | 0.908 | 0.910 | 0.908 | 0.911 | 0.910 | 0.911 |
| *Equation parameters* | | | | | | | |
| Intercept | – | −0.096 | 0.711* | 0.062 | 0.521* | 0.730* | 0.539 |
| Slope-DTS | – | 0.983** | 1.123** | 0.982** | 1.120** | 1.124** | 1.122** |
| Slope-Exp Time | – | 0.095 | – | – | 0.091 | – | 0.091 |
| Slope-Season | – | – | −0.306* | – | −0.301* | −0.308* | −0.303* |
| Slope-Prev Exp | – | | – | 0.016 | – | −0.007 | −0.007 |

*Significant at $p < 0.05$ level
**Significant at $p < 0.001$ level

**Table 7** Improved predictive models for DTS, from grouped thermal votes (N = 108)

| | DTS | Model 1 | Model 2 | Model 3 | Model 4 | Model 5 | Model 6 |
|---|---|---|---|---|---|---|---|
| N | 108 | 108 | 108 | 108 | 108 | 108 | 108 |
| R | 0.917 | 0.917 | 0.918 | 0.921 | 0.922 | 0.921 | 0.922 |
| R-squared | 0.840 | 0.841 | 0.843 | 0.848 | 0.850 | 0.848 | 0.850 |
| Mean bias | −0.07 | 0.00 | 0.00 | 0.00 | 0.00 | 0.00 | 0.00 |
| RMSE | 0.645 | 0.6368 | 0.6359 | 0.6240 | 0.6199 | 0.6248 | 0.6206 |
| Wilmott | 0.955 | 0.956 | 0.956 | 0.958 | 0.958 | 0.958 | 0.958 |
| *Equation parameters* | | | | | | | |
| Intercept | – | 0.015 | −0.111 | 0.712* | 0.529 | 0.728* | 0.544 |
| Slope-DTS | – | 0.981** | 0.982** | 1.126** | 1.123** | 1.127** | 1.124** |
| Slope-Exp Time | – | | 0.092 | | 0.088 | | 0.088 |
| Slope-Season | – | | | −0.314* | −0.311* | −0.317* | −0.313* |
| Slope-Prev Exp | – | 0.028 | | | | −0.006 | −0.005 |

*Significant at $p < 0.05$ level
**Significant at $p < 0.001$ level

predicted thermal sensation. However, with respect to acclimatization impacts on reported thermal sensation, only the long-term effect is noticeable and consistent for ‘Season’ in models 2, 4–6, with a drop in reported thermal sensation throughout seasons having a slight contribution to the predictive power of DTS. In this case, subjects become adapted to warmer conditions with the approach of summer and will likely report diminished warm sensation.

For the aggregated data (Table 7) with the mean thermal responses of each group of respondents, per time stamp, session and season, correlations rise as dispersion drops, i.e. data spread is reduced by the averaging procedure. Mean bias for all models, including one or more of the added variables of interest, drops virtually to zero. Again, improvements from introducing new variables are negligible. In terms of significance, apart from DTS, only ‘Season’ plays a role in the models, with an inverse relationship to reported thermal sensation.

The longitudinal study examined the effect of acclimatization and seasonal effect on outdoor thermal comfort and DTS using a panel study design. The following study investigates long-term acclimatization and short-term thermal history on outdoor thermal comfort in a cross-sectional manner.

**Table 8** Meteorological variables and UTCI ranges (showing maximum and minimum values) for Guangzhou, Zhuhai and Melbourne during the survey period. The meteorological variables are air temperature ($T_a$), relative humidity (RH), wind speed ($v_a$) and mean radiant temperature ($T_{mrt}$)

| Location | $T_a$ (°C) | RH (%) | $v_a$ (m/s) | $T_{mrt}$ (°C) | UTCI (°C) |
|---|---|---|---|---|---|
| Guangzhou | 26.2–34.6 | 44–93 | 0–2.5 | 26.3–66.8 | 28.0–43.0 |
| Zhuhai | 25.8–36.1 | 52–95 | 0–3.7 | 26.8–73.7 | 28.7–47.8 |
| Melbourne | 15.8–41.3 | 14.6–99.9 | 0–3.7 | 18.5–71.9 | 16.0–48.4 |

## 3.2 *Cross-Sectional Study (Guangzhou, Zhuhai and Melbourne)*

### 3.2.1 Meteorological Conditions

Table 8 summarizes the meteorological conditions in Guangzhou, Zhuhai and Melbourne during the survey period. In Guangzhou and Zhuhai, the daily maximum $T_a$ reached over 34 °C, and maximum UTCI could reach beyond 46 °C in Zhuhai (extreme thermal stress). During the survey period, the maximum $T_a$ in Guangzhou and Zhuhai was 2.5 °C and 5.5 °C higher than the 1981–2010 average maximum $T_a$ (32.1 °C and 30.6 °C, respectively). Sunny days were selected for our survey in Guangzhou and Zhuhai.

In Melbourne, the range of $T_a$ was greater than Guangzhou and Zhuhai during the survey period. There were heat waves in Melbourne from 14th to 17th January 2014 and from 7th to 9th February 2014. The maximum $T_a$ reached 41.3 °C and maximum UTCI was 48.4 °C during the Melbourne heat wave. Melbourne was experiencing an atypical year in 2014, with maximum $T_a$ (14.7 °C) above the maximum average $T_a$ (26.6 °C) for the summer period (1981–2010). In 2014, Melbourne was experiencing below-average rainfall, with most areas in Melbourne recording about 75% of their long-term average rainfall (Bureau of Meteorology 2015). In Melbourne, rainfall was absent during the survey period apart from 17th January 2014 (1.2 mm) and 16th February 2014 (7.2 mm).

### 3.2.2 Respondents' Characteristics (Guangzhou, Zhuhai and Melbourne)

Table 9 presents the respondents' characteristics in Guangzhou, Zhuhai and Melbourne. In Guangzhou, the proportions of male and female respondents were approximately 50%, whereas there were 15.3% and 15.5% more female respondents than male respondents in Zhuhai and Melbourne, respectively. In Melbourne, 74% of respondents were walking 15 min before the survey, whereas walking and sitting were the most common activities before the survey in Guangzhou (83.2%) and Zhuhai (84.2%). In Guangzhou, 33.8% of respondents were in an air-conditioned

**Table 9** Respondents characteristics from Guangzhou, Zhuhai and Melbourne

| | | Guangzhou | | Zhuhai | | Melbourne | |
|---|---|---|---|---|---|---|---|
| | | *n* | % | *n* | % | *n* | % |
| Gender | Female | 1425 | 49.4 | 816 | 57.4 | 1872 | 56.8 |
| | Male | 1420 | 49.3 | 598 | 42.1 | 1359 | 41.3 |
| | Missing | 37 | 1.3 | 8 | 0.6 | 62 | 1.9 |
| Age | 17–24 | 2389 | 82.9 | 1306 | 91.8 | 315 | 9.6 |
| | 25–44 | 395 | 13.7 | 78 | 5.5 | 1260 | 38.3 |
| | 45–64 | 46 | 1.6 | 18 | 1.3 | 1078 | 32.7 |
| | 65+ | 2 | 0.1 | 2 | 0.1 | 619 | 18.8 |
| | Missing | 50 | 1.7 | 18 | 1.3 | 21 | 0.6 |
| Activity 10–15 min before the survey | Lying down | 33 | 1.1 | 34 | 2.4 | 18 | 0.5 |
| | Sitting | 842 | 29.2 | 751 | 52.8 | 596 | 18.1 |
| | Standing | 229 | 7.9 | 94 | 6.6 | 198 | 6.0 |
| | Walking | 1555 | 54.0 | 447 | 31.4 | 2435 | 74.0 |
| | Running | 132 | 4.6 | 9 | 0.6 | 24 | 0.7 |
| | Other | 91 | 3.2 | 87 | 6.1 | 18 | 0.5 |
| Exposure 10–15 min before the survey | Outdoor (exposed) | 962 | 33.4 | 235 | 16.5 | 1928 | 58.6 |
| | Outdoor (shaded) | 746 | 25.9 | 246 | 17.3 | 1108 | 33.7 |
| | Indoor (no AC) | 198 | 6.9 | 218 | 15.3 | 112 | 3.4 |
| | Indoor (AC) | 975 | 33.8 | 723 | 50.8 | 143 | 4.3 |

(AC) environment 15 min before the survey, compared with 50.8% of Zhuhai respondents. In contrast, only 4.3% of Melbourne respondents were in an AC environment 15 min before the survey, because most Melbourne respondents were in an outdoor environment in the botanic gardens.

### 3.2.3 Cluster Analysis Results (Guangzhou, Zhuhai and Melbourne)

Based on the hierarchical clustering, it is possible to form either two or three clusters in our case. When three clusters were formed in the k-means cluster analysis, the sample size ratio between the largest and smallest cluster was large (10.69 for Guangzhou and Zhuhai and 7.85 for Melbourne). Therefore, we chose to form two clusters for both Guangzhou/Zhuhai and Melbourne, which had a sample size ratio below 3.26 between the larger and smaller cluster. These two clusters are the warm and neutral groups. For Guangzhou and Zhuhai, the warm group ($n = 3170$) had a mean thermal sensation vote (MTSV) of 2 (warm) and mean thermal preference vote (MTPV) of −1 (prefer cooler), whereas the neutral group ($n = 1134$) had a

MTSV of 0 and MTPV of 0 (no change) (Fig. 6a). When the UTCI thermal stress level was moderate to very strong, our t-test results showed that the warm group had a significantly higher MTSV than the neutral group ($p < 0.001$) (Table 10). As the UTCI increased from 27 to 48 °C, the warm group's MTSV increased from +2 to + 3, but the MTSV of the neutral group remained at 0 (Fig. 6c). Under the same UTCI range, the MTPV of the warm group was −1, while the MTPV of the neutral group

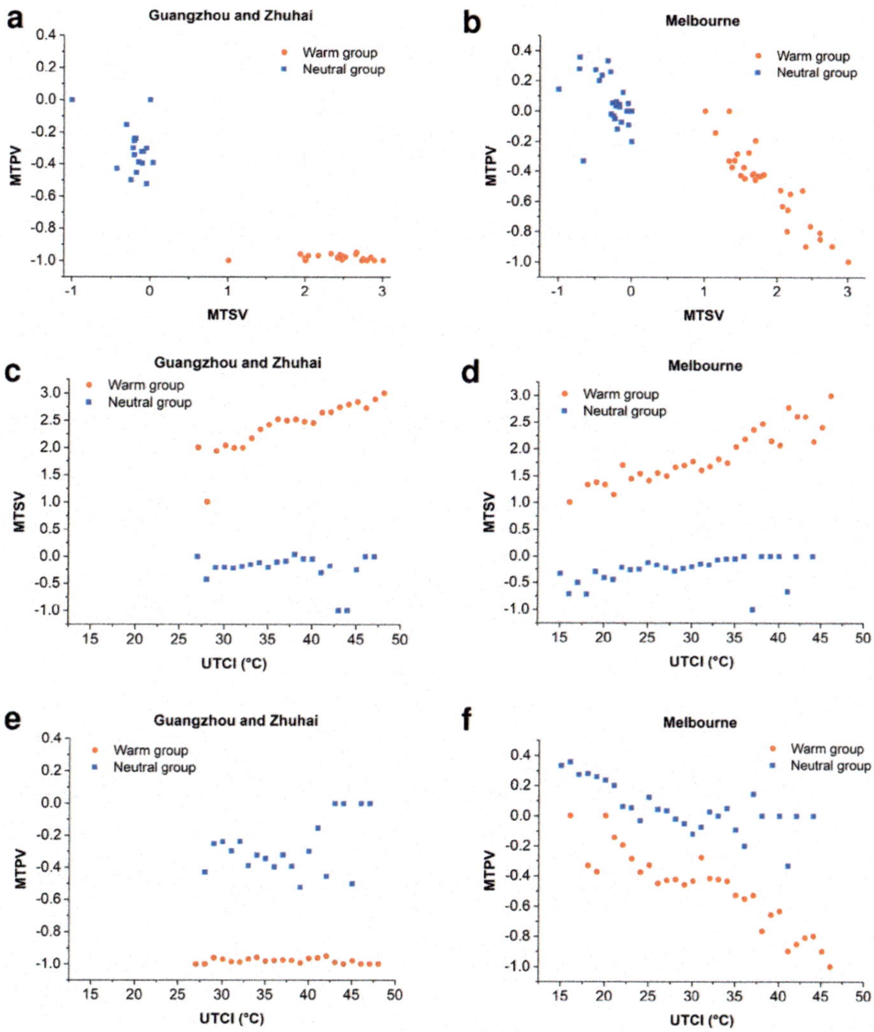


**Fig. 6** Cluster analysis results, showing mean thermal sensation vote (MTSV) versus mean thermal preference vote (MTPV) in **a** Guangzhou and Zhuhai and **b** Melbourne; UTCI versus MTSV in **c** Guangzhou and Zhuhai and **d** Melbourne; UTCI versus MTPV in **e** Guangzhou and Zhuhai and **f** Melbourne

**Table 10** Independent samples t-test for thermal sensation vote (TSV) for different cluster groups in Guangzhou, Zhuhai and Melbourne. The Table shows the mean (M) and standard deviation (SD) of TSV, the t-statistic (t), p-value (p) and eta squared ($\eta^2$). The extreme thermal stress results for Guangzhou and Zhuhai are not shown as n < 10 for group 2. Similarly, very strong thermal stress results for Melbourne are not shown for the same reason

| UTCI thermal stress level | Warm group | | Neutral group | | *df* | *t* | *p* | $\eta^2$ |
|---|---|---|---|---|---|---|---|---|
| | *M* | *SD* | *M* | *SD* | | | | |
| *Guangzhou and Zhuhai* | | | | | | | | |
| Moderate | 1.98 | 0.89 | −0.20 | 0.56 | 1231.9 | 53.73 | <0.001** | 0.690 |
| Strong | 2.39 | 0.82 | −0.16 | 0.52 | 1169.7 | 80.80 | <0.001** | 0.757 |
| Very strong | 2.61 | 0.70 | −0.14 | 0.73 | 109.1 | 34.11 | <0.001** | 0.576 |
| *Melbourne* | | | | | | | | |
| Moderate | 1.64 | 0.69 | −0.20 | 0.60 | 1007.7 | 52.10 | <0.001** | 0.653 |
| Strong | 1.93 | 0.73 | −0.11 | 0.45 | 303.4 | 41.83 | <0.001** | 0.698 |

*Note* **p < 0.001
Warm group—Guangzhou and Zhuhai: TSV = 2, prefer cooler; Melbourne: TSV = 2, no change
Neutral group—Guangzhou and Zhuhai: TSV = 0, no change; Melbourne: TSV = 0, no change

was between 0 and −0.4 (Fig. 6e). The warm group of Guangzhou/Zhuhai preferred to feel cooler regardless of the UTCI thermal stress levels.

In Melbourne, the warm group ($n = 2266$) had a MTSV of 2 and MTPV of 0, whereas the neutral group ($n = 1011$) had a MTSV of 0 and MTPV of 0 (Fig. 6d). Moreover, the MTPV of the warm group decreased from 0 (no change) to −1 (prefer cooler) when MTSV increased from +1 to +3 (Fig. 6b). At moderate to very strong UTCI thermal stress levels, the warm group reported a significantly higher MTSV than the neutral group ($p < 0.001$) (Table 10). For UTCI between 15 and 46 °C, the MTSV of the warm group increased from +1 to +3. In contrast, the MTSV of the neutral group was between −0.5 and 0 (Fig. 6d). Under the same UTCI range, the MPTV of the warm group decreased from 0 to −1. In contrast, the MTPV of the neutral group was between 0.4 and −0.2 (Fig. 6f). Overall, the neutral group of Guangzhou/Zhuhai and Melbourne exhibited a similar pattern of MTSV and MTPV. However, the MTPV trend of the warm group differed between Guangzhou/Zhuhai and Melbourne.

### 3.2.4 Long-Term Acclimatization and Thermal Perception

A chi-square test of independence was conducted to compare the proportion of respondents' climatic origins between the two cluster groups, which represented different thermal perceptions (warm and neutral). Further examination of column properties was done by the Bonferroni-corrected z test. In Guangzhou and Zhuhai, the relation between respondents' Köppen climate zones and thermal perception was significant, $\chi^2$ (3, $n = 4252$) = 3.6, $p = 0.003$, $\varphi_c = 0.06$. Based on the Bonferroni-corrected z test, the warm and neutral groups had significantly different proportions of respondents from Köppen climate zone C and D ($p < 0.05$) (Fig. 7a). Within Köppen climate zone C, there was no significant association between climatic origins and thermal perception, $\chi^2$ (4, $n = 3616$) = 8.0, $p = 0.09$, $\varphi_c = 0.05$. Moreover, the proportion of respondents from Köppen Cfa, Cfb, Cwa, Cwb, and Cwc was similar between the warm and neutral groups in Guangzhou and Zhuhai (Fig. 7b).

In Melbourne, the relation between Köppen climate zones and the two cluster groups was insignificant, $\chi^2$ (4, $n = 2873$) = 4.6, $p = 0.33$, $\varphi_c = 0.04$ (Fig. 7c). Within Köppen climate zone C, there was also no significant association between climatic origins and thermal perception, $\chi^2$ (4, $n = 2686$) = 6.2, $p = 0.19$, $\varphi_c = 0.05$. The proportion of respondents from Köppen Cfa, Cfb, Csa, Csb, and Cwa was similar between the warm and neutral groups (Fig. 7d). Most Melbourne survey respondents

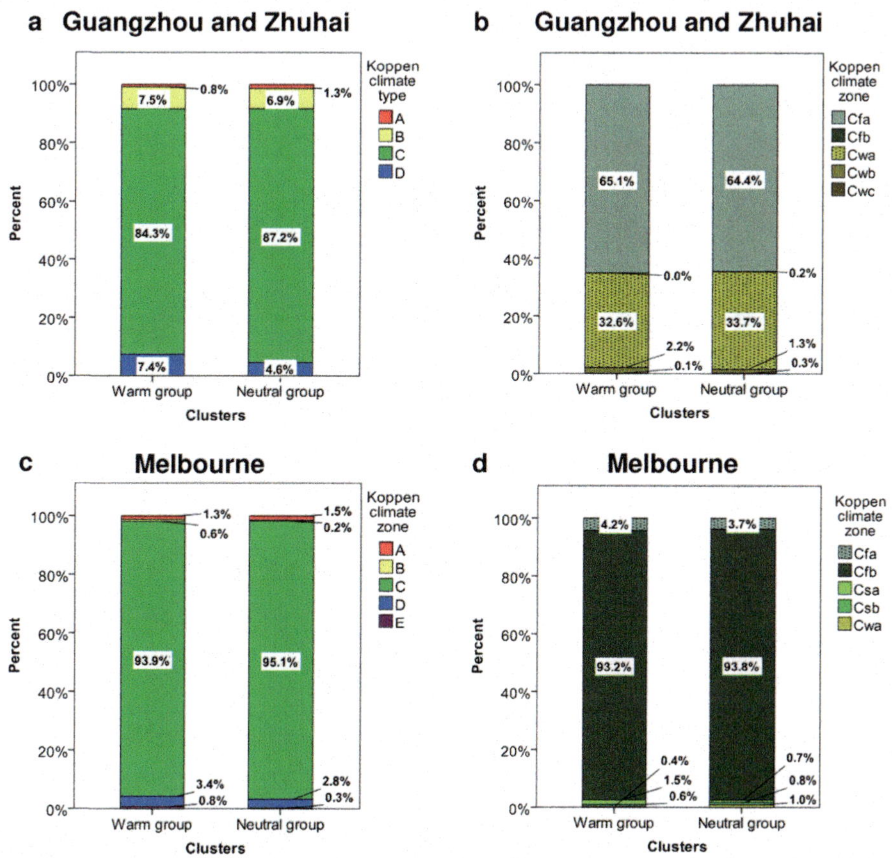


**Fig. 7** The percentage of **a** major Köppen climate zones, **b** Köppen climate zone C subgroups. The percentage of **c** major Köppen climate zones and **d** Köppen climate zone C subgroups for the warm and neutral groups in Melbourne

were from Cfb ($n = 2508$). Therefore, the differences in thermal perception between the warm and neutral groups would likely be due to factors other than climatic origins, such as short-term thermal history.

### 3.2.5 Short-Term Thermal Exposures, Activity Levels and Thermal Perception

We performed a chi-square test of independence to compare the proportion of exposure environments and activities before taking the survey between the two cluster groups (warm and neutral). There was a significant association between short-term thermal exposures and thermal perception in Guangzhou and Zhuhai, $\chi^2$ (3, $n$ = 4303) = 89.1, $p < 0.001$, $\varphi_c = 0.14$. The warm and neutral groups had significantly different proportions of people who had been outdoor (exposed) and indoor (AC and non-AC) ($p < 0.05$) (Fig. 8a). Furthermore, the relation between activity levels

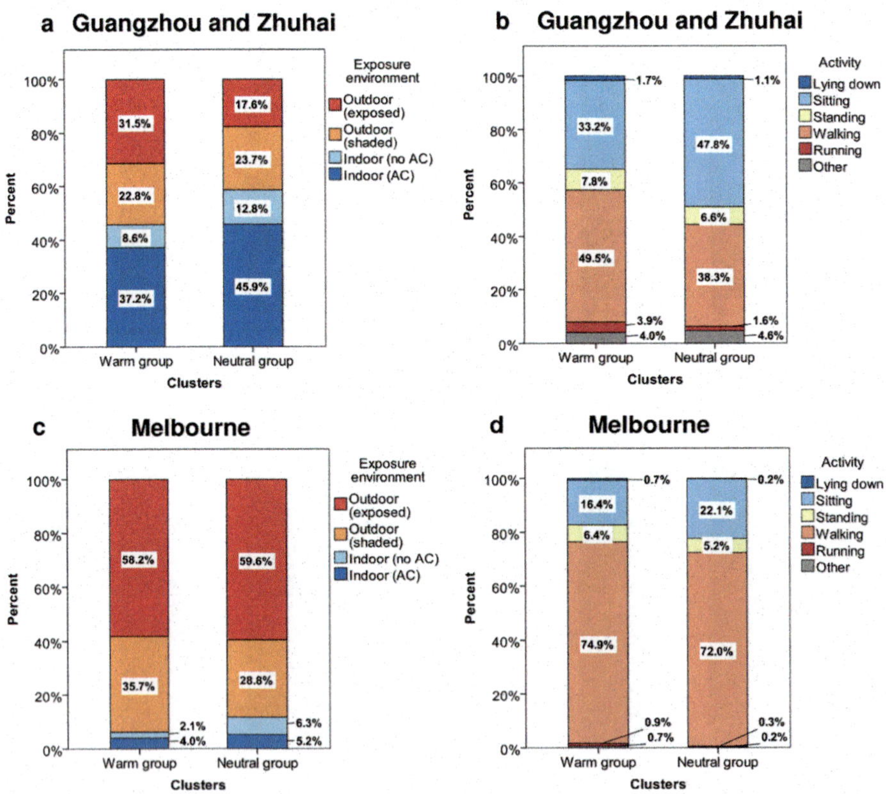


**Fig. 8** The percentage of exposure environments before taking the survey for the warm and neutral groups in **a** Guangzhou and Zhuhai, and **c** Melbourne. The percentage of activity levels prior to the survey for the warm and neutral groups in **b** Guangzhou and Zhuhai, and **d** Melbourne

and thermal perception was significant, $\chi^2$ (5, $n = 4304$) = 88.3, $p < 0.001$, $\varphi_c = 0.14$. Compared with the warm group, the neutral group had significantly different proportions of who had been sitting, walking, or running before taking the survey ($p < 0.05$) (Fig. 8b). The warm and neutral groups in Guangzhou and Zhuhai differed in both their prior exposure environments and types of activities.

In Melbourne, our analysis indicated a significant relation between short-term exposure environments and thermal perception, $\chi^2$ (3, $n = 3275$) = 49.0, $p < 0.001$, $\varphi_c = 0.12$. Compared with the warm group, the neutral group had significantly different proportions of people who had been outdoors (shaded) and indoors with no AC ($p < 0.05$) (Fig. 8c). Furthermore, the relation between activity levels and thermal perception was significant, $\chi^2$ (5, $n = 3274$) = 24.7, $p < 0.001$, $\varphi_c = 0.09$. The warm and neutral groups had significantly different proportions of people who had been sitting before taking the survey ($p < 0.05$) (Fig. 8d). Similar to Guangzhou and Zhuhai, the warm and neutral groups in Melbourne also had different percentages of thermal exposures and activity levels. These results highlight the influence of short-term thermal history on people's outdoor thermal perception.

## 4 Discussion and Conclusions

This chapter demonstrates the use of longitudinal and cross-sectional study design to investigate the impact of long- and short-term acclimatization on thermal comfort perception. The longitudinal study showed that even though changes in thermal responses could be observed over progressing exposure times, those differences were not statistically significant. Similar non-significant effects on reported thermal sensation, analysed against predicted thermal sensation with UTCI's derived 'Dynamic Thermal Sensation' DTS, also became evident. However, relationships between precedent visual comfort and reported thermal sensation were observed, which reinforce the need for further investigation on such interactions as recently suggested by a few authors (Chinazzo et al. 2019; Ko et al. 2020; Schweiker et al. 2020a). From results obtained, long-term acclimatization or the seasonal influence on subjective thermal assessment was found to be a relevant and statistically significant variable capable of improving the goodness of fit of DTS predictions for thermal sensation.

For the cross-sectional study, most Melbourne survey respondents are from Cfb (Fig. 7), representing Melbourne residents who are acclimatized to the local climate. Therefore, it is unsurprising that there was no significant relation between climatic origins and cluster groups in Melbourne. The differences in the thermal perception between Melbourne's cluster groups are more likely to be due to differences in exposure environment and activity levels before taking the survey (Fig. 8). However, we could not determine the Köppen climate zones of certain overseas visitors in Melbourne as we only knew their country of origin. In the Guangzhou and Zhuhain samples, long-term acclimatization and short-term thermal history (thermal exposures and activities) influenced subjective thermal perception, which supports the findings of previous studies (Lai et al. 2020; Xue et al. 2020; Ji et al. 2019).

The cross-sectional study demonstrates that clustering analysis can be used as an unsupervised technique to reveal various thermal comfort patterns. Each 'thermal comfort personality' is characterized by different thermal sensation and preference. Our chi-square test results of different clusters indicate that these differences in thermal sensation and preference are affected by climatic background, short-term activity, and exposure environment on outdoor thermal comfort. These findings agree with our past studies that used pre-defined groups in data analysis (Lam et al. 2020, 2021a, b). Clustering analysis has been previously employed in thermal comfort studies in both indoor (Gauthier et al. 2020; Bennetts et al. 2020; Sood et al. 2020) and outdoor environments (Jayathissa et al. 2019; Pigliautile and Pisello 2020), resulting in various thermal comfort patterns. Instead of a one-size-fits-all approach, deriving thermal comfort patterns presents a human-centric view on outdoor thermal comfort (Nazarian and Lee 2020), which has implications for designing urban space and mitigation strategies tailored to different population needs.

Future studies could combine physiological measurements by wearable sensors and surveys (Nazarian et al. 2021) to better understand the influence of physiological and psychological adaptation on thermal comfort in real-time. These approaches could provide further insight into how acclimatization affects thermal comfort and the associated comfort patterns.

**Acknowledgements** We acknowledge the Brazilian research funding agency CAPES, the European Union 7th Framework Programme (FP7/2007–2013) grant agreement no. PIRG08-GA-2010-277061 and the Fachgebiet Bauphysik & Technischer Ausbau/Karlsruher Institut für Technologie (fbta/KIT), in special Marcel Schweiker and Andreas Wagner. For the cross-sectional study in Guangzhou and Zhuhai, we acknowledge the support by the National Natural Science Foundation of China (No. 41905005, No. 41875015, No. 51811530017), and the Natural Science Foundation of Guangdong Province, China (No. 2018A030310307, 2018B030311068). For the Melbourne survey, we acknowledge the CRC for Water Sensitive Cities and the Royal Botanic Garden Victoria (RBGV) for their support. We also thank Cintia Tamura for the data collection in Karlsruhe and Yanping Gao for her assistance in analysing the survey data in Guangzhou, Zhuhai and Melbourne.

## References

Acero JA, Koh EJ, Pignatta G, Norford LK (2020) Clustering weather types for urban outdoor thermal comfort evaluation in a tropical area. Theoret Appl Climatol 139(1):659–675

ASHRAE (2017) ANSI/ASHRAE Standard 55—thermal environmental conditions for human occupancy. ASHRAE, Atlanta, GA

Basarin B, Lukić T, Matzarakis A (2020) Review of biometeorology of heatwaves and warm extremes in Europe. Atmosphere 11(12):1276

Becker S, Potchter O, Yaakov Y (2003) Calculated and observed human thermal sensation in an extremely hot and dry climate. Energy Build 35(8):747–756

Bennetts H, Arakawa Martins L, van Hoof J, Soebarto V (2020) Thermal personalities of older people in South Australia: a personas-based approach to develop thermal comfort guidelines. Int J Environ Res Public Health 17(22):8402

Błażejczyk K, Błażejczyk M (2010) BioKlima© 2.6, Program komputerowy. http://www.igipz.pan.pl/Bioklima-zgik.html

Bridges CC Jr (1966) Hierarchical cluster analysis. Psychol Rep 18(3):851–854
Bröde P, Fiala D, Błażejczyk K, Holmér I, Jendritzky G, Kampmann B, ... Havenith G (2012) Deriving the operational procedure for the Universal Thermal Climate Index (UTCI). Int J Biometeorol 56(3):481–494
Bureau of Meteorology (2015) Melbourne in 2014: another warm year with below average rainfall. Bureau of Meteorology. http://www.bom.gov.au/climate/current/annual/vic/archive/2014.melbourne.shtml. Accessed 12 Oct 2020
Bureau of Meteorology (2020) Climate statistics for Australian locations: summary statistics Melbourne regional office. Bureau of Meteorology. http://www.bom.gov.au/climate/averages/tables/cw_086071.shtml. Accessed 8 July 2020
Cabanac M (1971) Physiological role of pleasure. Science 173(4002):1103–1107
Cândido C, De Dear R, Lamberts R, Bittencourt L (2010) Cooling exposure in hot humid climates: are occupants 'addicted'? Architect Sci Rev 53(1):59–64
Cheung PK, Jim CY (2019) Improved assessment of outdoor thermal comfort: 1-hour acceptable temperature range. Build Environ 151:303–317
China Meteorological Administration (2012) Chinese surface weather station monthly climate dataset (1981–2010). National Meteorological Information Center. http://data.cma.cn/data/cdc detail/dataCode/A.0029.0004.html. Accessed 9 Oct 2020
Chinazzo G, Wienold J, Andersen M (2019) Daylight affects human thermal perception. Sci Rep 9(1):1–15
Coutts AM, White EC, Tapper NJ, Beringer J, Livesley SJ (2016) Temperature and human thermal comfort effects of street trees across three contrasting street canyon environments. Theoret Appl Climatol 124(1–2):55–68
De Dear R (2011) Revisiting an old hypothesis of human thermal perception: alliesthesia. Build Res Inf 39(2):108–117
De Dear R, Brager GS (2001) The adaptive model of thermal comfort and energy conservation in the built environment. Int J Biometeorol 45(2):100–108
De Dear RJ, Ring JW, Fanger PO (1993) Thermal sensations resulting from sudden ambient temperature changes. Indoor Air 3(3):181–192
De Freitas CR, Grigorieva EA (2009) The Acclimatization Thermal Strain Index (ATSI): a preliminary study of the methodology applied to climatic conditions of the Russian Far East. Int J Biometeorol 53(4):307–315
De Freitas CR, Grigorieva EA (2014) The impact of acclimatization on thermophysiological strain for contrasting regional climates. Int J Biometeorol 58(10):2129–2137
Fiala D, Havenith G, Bröde P, Kampmann B, Jendritzky G (2012) UTCI-Fiala multi-node model of human heat transfer and temperature regulation. Int J Biometeorol 56(3):429–441
Gauthier S, Bourikas L, Al-Atrash F, Bae C, Chun C, de Dear R, ... Wagner A (2020) The colours of comfort: from thermal sensation to person-centric thermal zones for adaptive building strategies. Energy Build 216:109936
Gosling SN, Bryce EK, Dixon PG, Gabriel KM, Gosling EY, Hanes JM, ... Wanka ER (2014) A glossary for biometeorology. Int J Biometeorol 58(2):277–308
Hanna EG, Tait PW (2015) Limitations to thermoregulation and acclimatization challenge human adaptation to global warming. Int J Environ Res Public Health 12(7):8034–8074
Hartigan JA, Wong MA (1979) AK-means clustering algorithm. J Roy Stat Soc Ser C (Appl Stat) 28(1):100–108
Heng SL, Chow WT (2019) How 'hot' is too hot? Evaluating acceptable outdoor thermal comfort ranges in an equatorial urban park. Int J Biometeorol 63(6):801–816
Höppe P (2002) Different aspects of assessing indoor and outdoor thermal comfort. Energy Build 34(6):661–665
Humphreys MA, Nicol JF, Raja IA (2007) Field studies of indoor thermal comfort and the progress of the adaptive approach. Adv Build Energy Res 1(1):55–88
ISO 7726 (2001) Ergonomics of the thermal environment. Instruments for measuring physical quantities. International Organization for Standardization (ISO), Geneva

ISO 7730 (2005) Moderate thermal environment—determination of the PMV and PPD indices and specification of the conditions for thermal comfort. International Organization for Standardization, Geneva

ISO 9920 (2007) Ergonomics of the thermal environment—estimation of thermal insulation and water vapour resistance of a clothing ensemble. International Organization for Standardization, Geneva

ISO 10551 (2019) Ergonomics of the thermal environment—assessment of the influence of the thermal environment using subjective judgement scales. International Organization for Standardization, Geneva

Jayathissa P, Quintana M, Sood T, Nazarian N, Miller C (2019) Is your clock-face cozie? A smartwatch methodology for the in-situ collection of occupant comfort data, in Journal of Physics: Conference Series, vol. 1343, no. 1. IOP Publishing, p. 012145

Ji W, Cao B, Geng Y, Zhu Y, Lin B (2019) A study on the influences of immediate thermal history on current thermal sensation. Energy Build 198:364–376

Kántor N, Kovács A, Takács Á (2016) Seasonal differences in the subjective assessment of outdoor thermal conditions and the impact of analysis techniques on the obtained results. Int J Biometeorol 60(11):1615–1635

Klemm W, van Hove B, Lenzholzer S, Kramer H (2017) Towards guidelines for designing parks of the future. Urban For Urban Green 21:134–145

Knez I, Thorsson S, Eliasson I, Lindberg F (2009) Psychological mechanisms in outdoor place and weather assessment: towards a conceptual model. Int J Biometeorol 53(1):101–111

Ko WH, Schiavon S, Zhang H, Graham LT, Brager G, Mauss I, Lin YW (2020) The impact of a view from a window on thermal comfort, emotion, and cognitive performance. Build Environ 175:106779

Kottek M, Grieser J, Beck C, Rudolf B, Rubel F (2006) World map of the Köppen-Geiger climate classification updated

Krüger E, Drach P, Bröde P (2015) Implications of air-conditioning use on thermal perception in open spaces: a field study in downtown Rio de Janeiro. Build Environ 94:417–425

Krüger EL, Tamura CA, Bröde P, Schweiker M, Wagner A (2017) Short-and long-term acclimatization in outdoor spaces: exposure time, seasonal and heatwave adaptation effects. Build Environ 116:17–29

Krüger EL, Costa T (2019) Interferences of urban form on human thermal perception. Sci Total Environ 653:1067–1076

Lai D, Lian Z, Liu W, Guo C, Liu W, Liu K, Chen Q (2020) A comprehensive review of thermal comfort studies in urban open spaces. Sci Total Environ 140092

Lam CKC, Lau KKL (2018) Effect of long-term acclimatization on summer thermal comfort in outdoor spaces: a comparative study between Melbourne and Hong Kong. Int J Biometeorol 62(7):1311–1324

Lam CKC, Gallant AJ, Tapper NJ (2018) Perceptions of thermal comfort in heatwave and non-heatwave conditions in Melbourne, Australia. Urban Climate 23:204–218

Lam CKC, Gallant AJ, Tapper NJ (2019) Short-term changes in thermal perception associated with heatwave conditions in Melbourne, Australia. Theo Appl Climatol 136(1):651–660

Lam CKC, Yang H, Yang X, Liu J, Ou C, Cui S, … Hang J (2020) Cross-modal effects of thermal and visual conditions on outdoor thermal and visual comfort perception. Build Environ 186:107297

Lam CKC, Cui S, Liu J, Kong X, Ou C, Hang J (2021a) Influence of acclimatization and short-term thermal history on outdoor thermal comfort in subtropical South China. Energy Build 231:110541

Lam CKC, Gao Y, Yang H, Chen T, Zhang Y, Ou C, Hang J (2021b) Interactive effect between long-term and short-term thermal history on outdoor thermal comfort: comparison between Guangzhou, Zhuhai and Melbourne. Sci Total Environ 760:144141

Lenzholzer S (2010) Engrained experience—a comparison of microclimate perception schemata and microclimate measurements in Dutch urban squares. Int J Biometeorol 54(2):141–150

Lenzholzer S, Koh J (2010) Immersed in microclimatic space: microclimate experience and perception of spatial configurations in Dutch squares. Landscape Urban Plan 95(1–2):1–15

Lenzholzer S, de Vries S (2020) Exploring outdoor thermal perception—a revised model. Int J Biometeorol 64(2):293–300
Lin P, Gou Z, Lau SSY, Qin H (2017) The impact of urban design descriptors on outdoor thermal environment: a literature review. Energies 10(12):2151
Lin TP, De Dear R, Hwang RL (2011) Effect of thermal adaptation on seasonal outdoor thermal comfort. Int J Climatol 31(2):302–312
Lomax RG, Hahs-Vaughn DL (2020) Inferences about proportions. In: An introduction to statistical concepts, 4th edn. Routledge, New York, pp 291–338
McIntyre DA (1980) Indoor climate. Applied Science Publishers, London
Muthers S, Laschewski G, Matzarakis A (2017) The summers 2003 and 2015 in south-west Germany: heat waves and heat-related mortality in the context of climate change. Atmosphere 8(11):224
Nasrollahi N, Hatami Z, Taleghani M (2017) Development of outdoor thermal comfort model for tourists in urban historical areas; a case study in Isfahan. Build Environ 125:356–372
Nazarian N, Lee JKW (2020) Personal assessment of urban heat exposure: a systematic review. Environ Res Lett
Nazarian N, Liu S, Kohler M, Lee JK, Miller C, Chow WT, … Norford LK (2021) Project Coolbit: can your watch predict heat stress and thermal comfort sensation? Environ Res Lett 16(3):034031
Nikolopoulou M, Steemers K (2003) Thermal comfort and psychological adaptation as a guide for designing urban spaces. Energy Build 35(1):95–101
Parkinson T, De Dear R (2015) Thermal pleasure in built environments: physiology of alliesthesia. Build Res Inf 43(3):288–301
Parkinson T, De Dear R, Candido C (2012) Perception of transient thermal environments: pleasure and alliesthesia. In: Proceedings of 7th Windsor Conference, Windsor, UK, April 2012
Pearlmutter D, Bitan A, Berliner P (1999) Microclimatic analysis of "compact" urban canyons in an arid zone. Atmos Environ 33(24–25):4143–4150
Pearlmutter D, Berliner P, Shaviv E (2007) Integrated modeling of pedestrian energy exchange and thermal comfort in urban street canyons. Build Environ 42(6):2396–2409
Peel MC, Finlayson BL, McMahon TA (2007) Updated world map of the Köppen-Geiger climate classification. Hydrol Earth Syst Sci 11(5):1633–1644
Pigliautile I, Pisello AL (2020) Environmental data clustering analysis through wearable sensing techniques: new bottom-up process aimed to identify intra-urban granular morphologies from pedestrian transects. Build Environ 171:106641
Richards M, Havenith G (2007) Progress towards the final UTCI model. In: Mekjavic I, Kounalakis S, Taylor N (eds) Environmental Ergonomics XII, Biomed, Ljubljana, Piran Slovenia, 19–24 August 2007, pp 521–524
Rutty M, Scott D (2015) Bioclimatic comfort and the thermal perceptions and preferences of beach tourists. Int J Biometeorol 59(1):37–45
Schlader ZJ, Simmons SE, Stannard SR, Mündel T (2011) The independent roles of temperature and thermal perception in the control of human thermoregulatory behavior. Physiol Behav 103(2):217–224
Schweiker M, Brasche S, Hawighorst M, Bischof W, Wagner A (2014) Presenting LOBSTER, an innovative climate chamber, and the analysis of the effect of a ceiling fan on the thermal sensation and performance under summer conditions in an office-like setting. In: Windsor conference: counting the cost of comfort in a changing World, Windsor, UK, April 2014
Schweiker M, Ampatzi E, Andargie MS, Andersen RK, Azar E, Barthelmes VM, … Zhang S (2020a) Review of multi-domain approaches to indoor environmental perception and behaviour. Build Environ 176:106804
Schweiker M, Rissetto R, Wagner A (2020b) Thermal expectation: influencing factors and its effect on thermal perception. Energy Build 210:109729
Schweiker M, Schakib-Ekbatan K, Fuchs X, Becker S (2020c) A seasonal approach to alliesthesia. Is there a conflict with thermal adaptation? Energy Build 212:109745

Sharpe D (2015) Chi-square test is statistically significant: now what? Pract Assess Res Evaluat 20(1):8

Shooshtarian S (2019) Theoretical dimension of outdoor thermal comfort research. Sustain Cities Soc 47:101495

Shooshtarian S, Rajagopalan P, Sagoo A (2018) A comprehensive review of thermal adaptive strategies in outdoor spaces. Sustain Cities Soc 41:647–665

Sood T, Janssen P, Miller C (2020) Spacematch: using environmental preferences to match occupants to suitable activity-based workspaces. Front Built Environ 6(113)

Spagnolo J, De Dear R (2003) A field study of thermal comfort in outdoor and semi-outdoor environments in subtropical Sydney Australia. Build Environ 38(5):721–738

Sturman AP, Tapper NJ (2006) The weather and climate of Australia and New Zealand, 2nd edn. Oxford University Press, Melbourne, Australia

Tamura C, Krüger E, Schweiker M, Moosman C, Wagner A (2017) Visual perception and daylight variables: experiments in a climate chamber. In: 33rd PLEA international conference-design to Thrive, Edinburgh, July 2017, pp 3396–3403

Taylor NAS (2014) Human heat adaptation. Comprehen Physiol 4:325–365

Wagner A, O'Brien W, Dong B (2018) Exploring occupant behavior in buildings—methods and challenges. Springer International Publishing, Cham, Switzerland

Willmott CJ (1981) On the validation of models. Phys Geogr 2(2):184–194

Xue J, Hu X, Sani SN, Wu Y, Li X, Chai L, Lai D (2020) Outdoor thermal comfort at a University Campus: studies from personal and long-term thermal history perspectives. Sustainability 12(21):9284

Zhang Y, Chen H, Wang J, Meng Q (2016) Thermal comfort of people in the hot and humid area of China—impacts of season, climate, and thermal history. Indoor Air 26(5):820–830

Zhou Z, Deng Q, Yang W, Zhou J (2020) Effect of seasonal adaptation on outdoor thermal comfort in a hot-summer and cold-winter city. Adv Build Energy Res 14(2):202–217

# Chapter 6
# Regional Adaptation of the UTCI: Comparisons Between Different Datasets in Brazil

**Eduardo L. Krüger, Ivan Julio Apolonio Callejas, Luísa Alcantara Rosa, Eduardo Grala da Cunha, Linccon Carvalho, Solange Leder, Thiago Vieira, Simone Hirashima, and Patricia Drach**

**Abstract** Bands of UTCI values have been translated into comfort and stress categories according to a thermal stress assessment scale proposed for the index. Those categories were originally defined based on strain reactions observed in lab tests, thus having a pronounced human-physiological component. Apart from the 10 proposed comfort/stress categories for the UTCI, the sub-interval between 18 and 26 °C UTCI

E. L. Krüger (✉)
Programa de Pós-Graduação em Engenharia Civil, Departamento de Construção Civil, Universidade Tecnológica Federal do Paraná - UTFPR, Campus Curitiba - Sede Ecoville, Rua Deputado Heitor Alencar Furtado, 4900, Curitiba 81280-340, Brazil
e-mail: ekruger@utfpr.edu.br

I. J. A. Callejas
Universidade Federal de Mato Grosso, Cuiabá, Brazil
e-mail: ivancallejas@ufmt.br

L. A. Rosa · E. G. da Cunha
Universidade Federal de Pelotas, Pelotas, Brazil
e-mail: luisa.alcantara@ufpel.edu.br

E. G. da Cunha
e-mail: eduardo.grala@ufpel.edu.br

L. Carvalho · S. Leder
Universidade Federal da Paraíba, João Pessoa, Brazil
e-mail: linccon.carvalho@academico.ufpb.br

S. Leder
e-mail: solange.leder@academico.ufpb.br

T. Vieira · S. Hirashima
Centro Federal de Educação Tecnológica de Minas Gerais, Belo Horizonte, Brazil
e-mail: vieira.thiago@gmail.com

S. Hirashima
e-mail: simonehirashima@cefetmg.br

P. Drach
Universidade Estadual do Rio de Janeiro, Petrópolis, Brazil
e-mail: patricia.drach@uerj.br



E. L. Krüger (ed.), *Applications of the Universal Thermal Climate Index UTCI in Biometeorology*, Biometeorology 4,
https://doi.org/10.1007/978-3-030-76716-7_6

was assumed to comply with the thermal comfort zone definition of the Commission for Thermal Physiology of the International Union of Physiological Sciences. Yet, several studies have suggested that such thermal comfort zone might fluctuate due to adaptation to regional climatic conditions by local populations. This occurs primarily as a result of long-term thermal adaptation and also due to other influencing factors, such as demographics and economic and sociocultural aspects. There is thus a need for adjusting the proposed thermal comfort zone for the UTCI when using it in very diverse climatic contexts. Such adjustments can be best gauged by reported thermal perception of local populations when employing questionnaire-based surveys alongside on-site monitoring of relevant meteorological variables. The aim of this chapter is to investigate the need for regional adaptation of the UTCI's thermal comfort zone to different populations within a large range of latitudes and climatic conditions in Brazil. Results suggest that subjective responses follow the patterns of exposure conditions during surveys, which directly affect obtained neutral temperatures and thermal sensitivity. Weather patterns and seasons might, however, not be representative of local climate. We thus advocate for standardized protocols for conducting outdoor thermal comfort surveys, wherein samples should reflect as much as possible local demographics and exposure conditions should be fine-tuned with the range and extent of thermal comfort/stress conditions in the surveyed location.

**Keywords** Urban climate · Outdoor thermal comfort · Questionnaire-based surveys · Regional adaptation

## 1 Introduction

As pointed out in the literature review chapter of this book (Chap. 3), the very first publication dealing with calibration procedures of the thermal comfort assessment scale of the UTCI was a study carried out by Mateeva (2011) for three different locations in Bulgaria. The aim was to draw a relationship between human thermal perception of weather and index values calculated for those weather conditions, with a long-term goal of establishing an "exact quantitative expression of the corresponding differences between both parameters as a function of the continuous adaptation of an organism to the local climatic conditions" (Mateeva 2011, p. 336). In that study, the issue of regional adaptation was explored by Mateeva in terms of the thermal sensitivity of people living in three different locations in Bulgaria with relatively small deviations in latitude yet with remarkable differences in climate. Surveys were carried out with local population concurrently with onsite measurements. The aim of the study as well as its method has been pursued in other investigations since then.

The thermal stress assessment scale originally proposed for the UTCI allows the interpretation of index results in terms of 10 different thermal stress categories. The assessment scale was introduced so as to allow a correspondence between UTCI values and their categorization in terms of thermal stress for given applications of the index (Bröde et al. 2012a). Those categories were defined based on strain reactions observed in lab tests, thus having a pronounced human-physiological component. The sub-interval between 18 and 26 °C of the 'no thermal stress' category, which ranges from 9 to 26 °C (UTCI scale), was assumed to comply with the thermal comfort zone definition of the Commission for Thermal Physiology of the International Union of Physiological Sciences (2003). The Commission states: "The range of ambient temperatures, associated with specified mean radiant temperature, humidity, and air movement, within which a human in specified clothing expresses indifference to the thermal environment for an indefinite period".

Nevertheless, when we account for human thermal sensitivity and regional adaptation to local climatic conditions, the thermal comfort zone might shift downward or upward, broadening or narrowing depending on local climate exposure. As in Pantavou et al. (2018), those regional variations in thermal sensitivity suggest the influence of local climate as a factor of long-term thermal experience which strongly defines the thermal comfort range.

Comparisons have been drawn between the original UTCI thermal stress assessment scale and empirical data obtained from surveys with local population, allowing for a direct calibration of the original 'no thermal stress' and 'thermal comfort' zones (e.g. Bröde et al. 2012b; Hadianpour et al. 2019; Jin et al. 2019; Krüger et al. 2020; Borges et al. 2020). Other researchers compare field data to thresholds obtained from previous studies (Kántor et al. 2012; Golasi et al. 2018; Potchter et al. 2018). Comparisons to previously published material have however the drawback that the methods used during surveys and for the collection of local microclimatic data might not follow exact same protocols. Kántor et al. (2012) provide evidence of intrinsic differences in the instruments and methods of data collection, scales and semantics used to assess thermal perception which might or not be linked to the research objectives behind each study. In their study, those differences were in great part responsible for the difficulties encountered by the authors to harmonize third-party data to their own field data. Bringing outcomes from different studies that used different protocols to an equal base for comparisons is a relevant issue in this case.

A thorough calibration of the UTCI using data collected and presented in a standardized way was performed by Pantavou et al. (2018). In their work, authors used the extensive Outdoor Thermal Comfort (OTC) database from the RUROS research initiative (**R**ediscovering the **U**rban **R**ealm and **O**pen **S**paces) encompassing seven cities in three European countries (Nikolopoulou and Lykoudis 2006). The need for standardization of protocols in OTC research has been stressed by Johansson et al. (2014) and such effort more recently became a research initiative towards the development of a data repository for OTC research (Lau and Krüger 2020a, b).

Despite intrinsic limitations arising from comparisons of subjective responses to climatic data gathered by different researchers following not the exact same protocols, the present chapter looks at the effects of regional adaptation of local population to very diverse climatic conditions in Brazil. As in Kántor et al. (2012), such effects will play a role in the results obtained due to both physiological acclimatization and psychological adaptation. In the case of this study, the uniformity in semantics, language and overall features of the samples, all obtained within Brazil, should contribute to minimize sociocultural differences in subjective thermal assessments.

Deviations found in the thermal comfort zones obtained for each location are likely to be more strongly correlated to prevailing climatic features. A broader aim of this research is to investigate the sensitivity of various populations to the thermal environment, with the understanding that regional adaptation to local climate should be the focus of climate-responsive urban planning.

The method of comparison is based on regression analysis and uses an averaging procedure for variations in UTCI units for subjective votes and concurrent objective parameters. This procedure not only reduces the scattering of data but also allows us to assess sensitivity of different populations (the slope of the fitted regression function), neutral UTCI conditions and the derived thermal comfort zone for each location.

## 2 Field Data

The field studies were carried out in six different Brazilian cities, spanning a latitude range between 7° S and 30° S (Fig. 1). Climatological Normals[1] for 1981–2010 are summarized in Table 1. The Table presents annual temperature ranges along with the total number of days with ambient temperatures below 10 °C (daily minima) and above 25 °C (daily maxima) using data from the network of official meteorological stations that belongs to the Brazilian National Meteorological Service (INMET). The Köppen-Geiger's classification for each location is also informed in the Table.

In each location, surveys were conducted at selected spots alongside meteorological measurements with portable equipment. As the locations were surveyed by different groups of researchers, the equipment used as well as the format of the comfort questionnaire may diverge somewhat. Monitoring points used for the surveys were also varied, with distinct types of vegetation cover, paving materials and building morphological compositions. In João Pessoa, the interviews were conducted in a particular spot in a public park, which was shaded by native vegetation. Those exposure conditions might have affected both thermal perception and thermal sensitivity of the participants, as described later. General characteristics of the surveys are presented in Table 2.

For all datasets, the calculation of the mean radiant temperature ($T_{mrt}$) for the different data sets was performed using the equation for forced convection according to ISO 7726 (1998). The calculation of the UTCI was done in BioKlima 2.6 (Błażejczyk and Błażejczyk 2010), using $T_a$, RH, $v_a$ and the post-processed $T_{mrt}$ data. The wind speed measured on site was converted to that of the required height of 10 m above ground for UTCI calculations, using a logarithmic formula (Bröde et al. 2012a).

[1]Climatological Normals are defined by the World Meteorological Organization (WMO) as averages of climatological data computed for consecutive periods of 30 years, according to WMO's Technical Regulations.

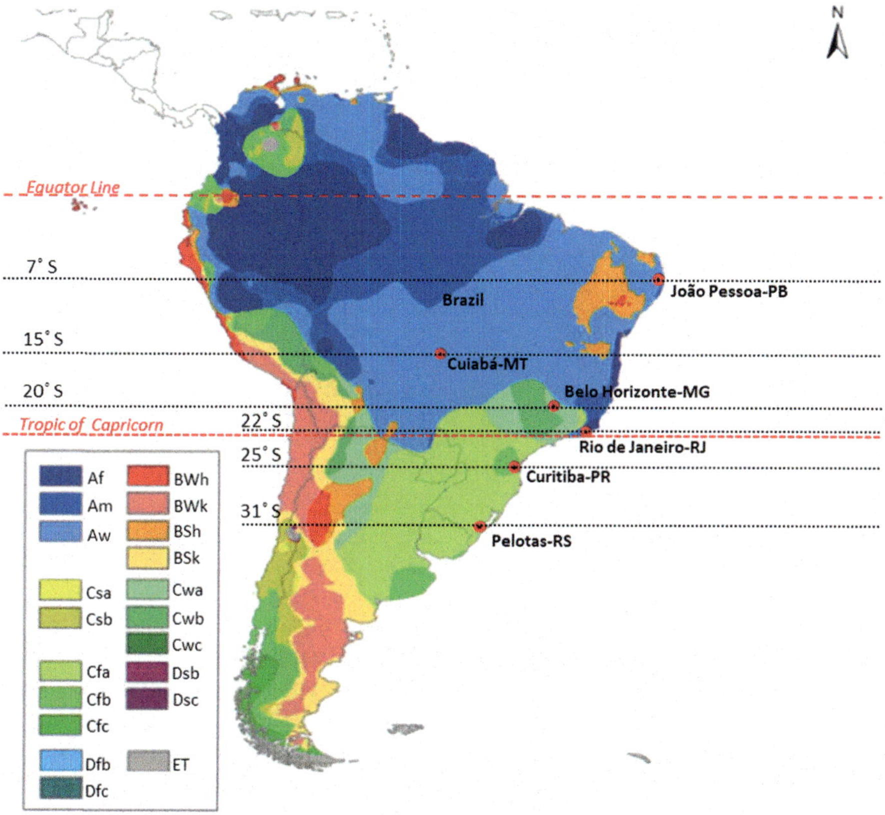


**Fig. 1** Location of the field studies (adapted from Peel et al. 2007)

Table 3 presents overall characteristics of the samples as well as the range of values, the mean and the dominant thermal comfort/stress assessment class of the UTCI obtained in the field datasets. Age groups have been defined according to the World Health Organisation (WHO) as follows: below 25 years ('young'), between 25 and 64 ('adult') and above 64 ('elderly'). The Body Mass Index (BMI) was calculated as a function of body mass (weight) and height of an individual, defined as the body mass divided by the square of the body height, given in kg/m$^2$ units. Four different classes were then assigned to the calculated BMI values according to WHO thresholds. Obesity classes I–III have been grouped as one. In the case of very young persons, up to 19 years of age, the thresholds recommended by WHO for this age group, broken down to boys and girls, were applied (De Onis and Lobstein 2010).

In João Pessoa and Belo Horizonte, biometric data have not been adequately assessed so that a clear distribution between age groups and BMI classes was not feasible. In João Pessoa, such information was not part of the survey questionnaire whereas in Belo Horizonte, too coarse age and height/weight groups were used in the interviews (e.g. 20–29 years of age, 180–189 cm tall, 90–99 kg).

**Table 1** General climatic and geographical features of the locations chosen for analysis (http://www.inmet.gov.br, https://dateandtime.info/citycoordinates.php, https://en.climate-data.org)

| Location | Latitude | Longitude | Elevation (m a.s.l.) | Annual temp range (monthly means) | Number of days with Tmin ≤ 10 °C | Number of days with Tmax ≥ 25 °C | Köppen-Geiger's climate type |
|---|---|---|---|---|---|---|---|
| João Pessoa | 7° 06′ 54″ S | 34° 51′ 47″ W | 45 | 25.1–27.9 °C | 0 | 365 | Am |
| Cuiabá | 15° 35′ 45″ S | 56° 05′ 48″ W | 193 | 23.4–27.9 °C | 2 | 352 | Aw |
| Belo Horizonte | 19° 55′ 14″ S | 43° 56′ 16″ W | 872 | 19.1–23.8 °C | 1 | 279 | Cwa |
| Rio de Janeiro | 22° 54′ 10″ S | 43° 12′ 27″ W | 5 | 22.3–28.1 °C | 0 | 299 | Am |
| Curitiba | 25° 25′ 40″ S | 49° 16′ 23″ W | 924 | 13.3–21.0 °C | 74 | 156 | Cfb |
| Pelotas | 31° 46′ 18″ S | 52° 20′ 33″ W | 14 | 12.2–23.5 °C | 84 | 149 | Cfa |

There are noticeable differences in demographic data among the datasets, partly explained by the locations where survey took place in each city. Discrepancies can be more evidently noticed between Rio and Pelotas, with an inverse proportion between male and female interviewees and a much less percentage of young interviewees in Rio, possibly due to the predominantly CBD (**C**entral **B**usiness **D**istrict) character of downtown Rio. In general, half of the samples consist of individuals with normal weight, followed by overweight and obese subjects.

Wider variations in UTCI exposure are observed in the two southern cities of Curitiba and Pelotas, and also in Cuiabá, characterized by a Tropical Savannah climate type with marked wet and dry seasons, followed by Belo Horizonte. Dominant UTCI classes for the coldest locations (Curitiba and Pelotas) differ by one UTCI thermal stress category, with Pelotas having the majority of measured conditions in 'moderate heat stress'.

## 3 Methods

The method of comparison comprised of analyses of raw and grouped data for reported thermal sensation, adopting 1 °C bins in the UTCI scale for averaging subjective thermal responses. As grouping criterion, binned data should contain at least three thermal sensation votes (TSV). For that reason, there was a slight reduction of the original samples, when discarding UTCI bins with less than three votes. Methods of analysis used in the comparisons encompassed:

**Table 2** General characteristics of the surveys

| Location | Equipment | Amount of points covered | Seasons covered | Variables measured | Questionnaires items |
|---|---|---|---|---|---|
| João Pessoa | Davis Vantage Pro2 and TGD-300 m | 1 location | Dry/humid | $T_a$, RH, $v_a$, $T_G$, $I_g$ | TS, TP, TC, thermal history, wind perception, clothing insulation |
| Cuiabá | HOBO U30-NRC micro weather station | 3 locations | All seasons | $T_a$, RH, $v_a$, $T_G$, $I_g$ | TS, TP, TC, thermal history*, time spent outdoors, clothing insulation, biometrics |
| Belo Horizonte | HOBO U12 loggers: thermohygrometer and globe thermometer; and ALNOR anemometer | 8 locations | All seasons | $T_a$, RH, $v_a$, $T_G$ | TS, TP, TC, thermal history*, time spent outdoors*, wind perception*, clothing insulation, biometrics |
| Rio de Janeiro | Davis Vantage Pro2 weather station | 7 locations | Summer/spring | $T_a$, RH, $v_a$, $T_G$, $I_g$ | TS, TP, TC, thermal history, time spent outdoors, clothing insulation, biometrics |
| Curitiba | HOBO weather station (H21-001) | 15 locations | Summer/fall/winter | $T_a$, RH, $v_a$, $T_G$, $I_g$ | TS, TP, TC, time spent outdoors, clothing insulation, biometrics |

(continued)

**Table 2** (continued)

| Location | Equipment | Amount of points covered | Seasons covered | Variables measured | Questionnaires items |
|---|---|---|---|---|---|
| Pelotas | TGD-400 m | 5 locations | Winter/spring/summer | $T_a$, RH, $v_a$, $T_G$ | TS, TP, TC, thermal history, time spent outdoors, wind perception, clothing insulation, biometrics |

*Notes* * only part of the sample, variables: $T_a$ (air temperature), RH (relative humidity), $v_a$ (wind speed),$T_G$ (globe temperature), $I_g$ (global solar radiation); questions: Thermal Sensation (TS), Thermal Preference (TP), Thermal Comfort assessment (TC)

**Table 3** General characteristics of the samples

| Location | Sample size | Sex | Age | BMI | UTCI range (°C) | UTCI $_{mean}$ (°C) | UTCI class |
|---|---|---|---|---|---|---|---|
| João Pessoa | 900 | 46% male<br>54% female | – | – | 26.6–33.9 | 30.4 | Moderate heat stress |
| Cuiabá | 685 | 45% male<br>55% female | 32% young<br>61% adult<br>7% elderly | 3% underweight<br>48% normal<br>33% overweight<br>16% obese | 14.0–45.4 | 30.4 | Strong heat stress |
| Belo Horizonte | 3630 | 55% male<br>45% female | – | – | 12.2–39.7 | 26.6 | Moderate heat stress |
| Rio de Janeiro | 1328 | 59% male<br>41% female | 16% young<br>75% adult<br>9% elderly | 2% underweight<br>42% normal<br>38% overweight<br>18% obese | 18.5–44.6 | 31.1 | Moderate heat stress |
| Curitiba | 1685 | 57% male<br>43% female | 30% young<br>61% adult<br>9% elderly | 3% underweight<br>52% normal<br>34% overweight<br>12% obese | 4.7–37.6 | 22.4 | No thermal stress |
| Pelotas | 1974 | 38% male<br>62% female | 34% young<br>61% adult<br>5% elderly | 1% underweight<br>45% normal<br>36% overweight<br>18% obese | 12.6–40.4 | 26.6 | Moderate heat stress |

(1) Generation of trend lines for binned TSV data over variations of the UTCI by means of linear regression;
(2) Representation of the percentages of TSV for different thermal comfort/stress categories of the UTCI; and,
(3) Consideration of the Dynamic Thermal Sensation (DTS) for UTCI data and generation of biases between predicted and reported thermal sensation.

## 4 Results

Overall conditions of the entire sample are presented in Fig. 2. A wide spectrum of climatic conditions and thermal perception votes were assessed with a greater concentration of situations in heat stress as defined for the UTCI, ranging from moderate to very strong heat stress, though the bulk of conditions evaluated lies within the 'moderate heat stress' category. As expected, thermal sensation votes for cool and comfort decrease for higher UTCI heat stress categories while there is an increase in warm and hot sensations. For thermal preference, an inverse relationship is found. Thermal comfort votes are highest for the 'no thermal stress' category (it should be stressed that the 'slightly cool' class has only 9 thermal votes, assessed only in Curitiba). A more refined breakdown of the 'no thermal stress' category into the thermal comfort range suggested by Bröde et al. (2012a), with 18 °C ≤ UTCI ≤

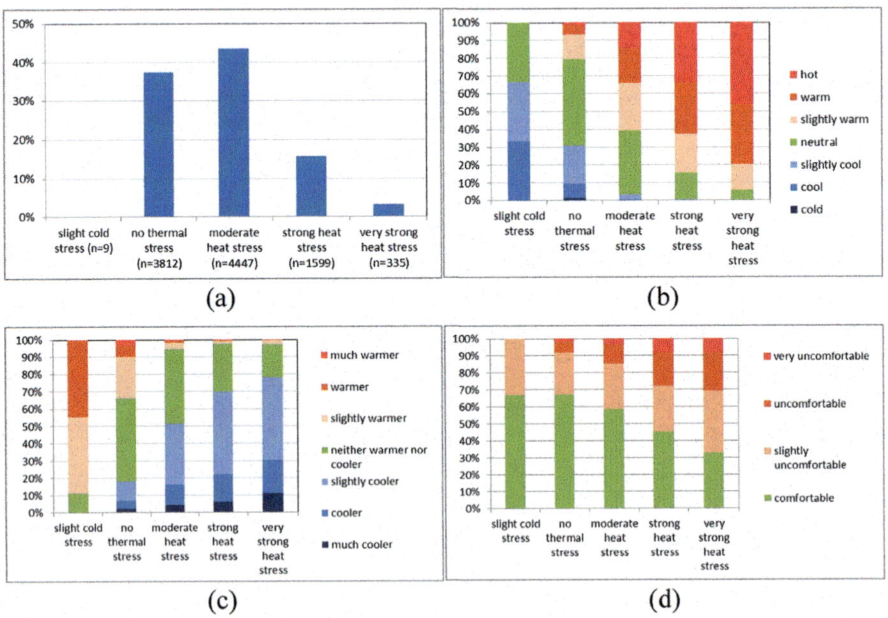


**Fig. 2** Overall exposure conditions (**a**) and thermal perception votes for the complete dataset (n = 10,202): thermal sensation (**b**), thermal preference (**c**), thermal comfort assessment (**d**)

26 °C, yielded a slight improvement in the percentage of comfort votes: 68%, versus 67% for the wider spectrum of the 'no thermal stress' range.

In general, survey participants' responses appear consistent with expected thermal sensation and thermal preference, with a higher concentration of comfort votes around the proposed thermal comfort range for the UTCI.

A first comparison between locations can be visualized in Fig. 3, which represents the binned TSV data for the different ranges of the UTCI at each location. Figure 4 zooms in on the common UTCI range observed between studies, roughly 27–34 °C UTCI, constrained by the reduced fluctuation in the UTCI as measured in João Pessoa.

In the overall comparison, the slope of the curves suggest a similar thermal sensitivity of local population to changes in UTCI units in most of the cities, except for João Pessoa, which stands out more evidently in the graph that depicts a common base for thermal exposure (Fig. 4). Surveys in João Pessoa were conducted in a shaded spot, which restricted thermal exposure variations and thus affected thermal responses. Table 4 presents properties of the trend line equations generated for each city, along with neutral UTCI temperatures, calculated for MTSV = 0 in the corresponding regression equation.

In general, MTSV were well correlated with the UTCI with r-squared from 0.87. Slope coefficients ranged between 0.10 and 0.15, except for João Pessoa, which had a much higher slope of 0.21. Neutral temperatures obtained demonstrate some consistency with local climatic features, with the high-latitude cities and the high-elevation Belo Horizonte having the lowest values and the warmer locations showing

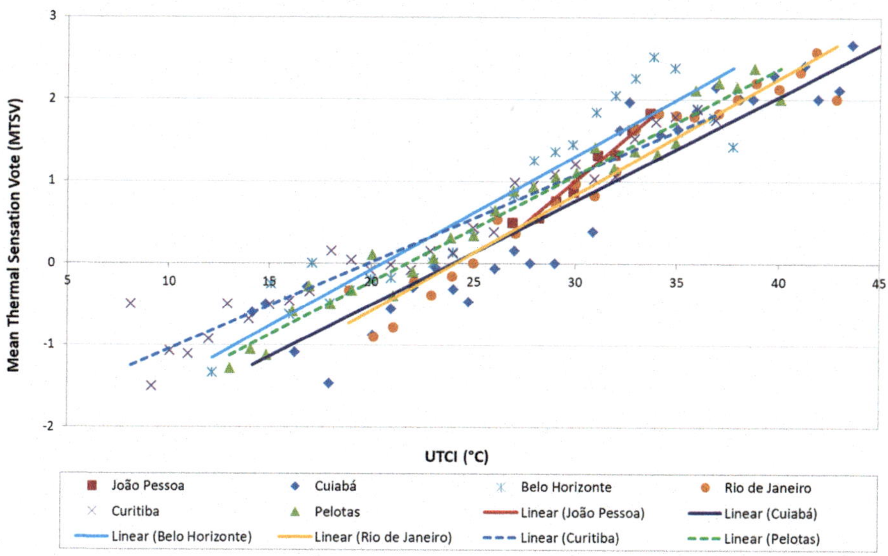


**Fig. 3** Mean thermal sensation votes, binned for a 1 °C variation of the UTCI, versus predicted UTCI for the six locations

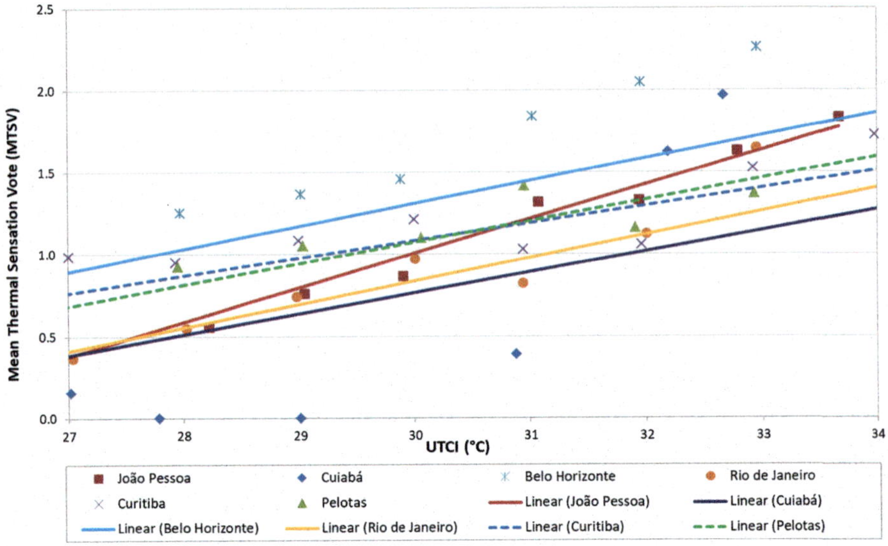


**Fig. 4** Mean thermal sensation votes, binned for a 1 °C variation of the UTCI, versus predicted UTCI for the six locations—common UTCI range

**Table 4** Number of respondents (n), neutral temperature in UTCI units, slope coefficient and intercept of the corresponding regression equation and determination coefficient (r-squared) for the mean thermal sensation votes (MTSV) compared to the predicted UTCI, as well as thermal sensitivity (°C/MTSV), by location

| Locations | n | Neutral temperatures | Slope coefficient | Intercept | r-squared | °C/MTSV |
|---|---|---|---|---|---|---|
| João Pessoa | 900 | 25.2 | 0.2095 | −5.2774 | 0.96 | 4.8 |
| Cuiabá | 683 | 24.0 | 0.1268 | −3.040 | 0.87 | 7.9 |
| Belo Horizonte | 3625 | 20.5 | 0.1385 | −2.8451 | 0.87 | 7.2 |
| Rio de Janeiro | 1325 | 24.1 | 0.1417 | −3.4138 | 0.95 | 7.1 |
| Curitiba | 1582 | 19.9 | 0.1067 | −2.1184 | 0.94 | 9.4 |
| Pelotas | 1974 | 21.7 | 0.1298 | −2.822 | 0.97 | 7.7 |

higher neutral temperatures. As for thermal sensitivity, respondents in João Pessoa were much more sensitive to changes in the UTCI, requiring a 4.8 °C variation in the UTCI for changing their thermal vote. That sensitivity is lowest in Curitiba, where population is exposed to higher seasonal and daily temperature fluctuations. It could thus be argued that to some extent thermal sensitivity is a factor of seasonal and daily temperature fluctuations local populations are exposed to.

However, as the range of outdoor exposure, expressed by UTCI units, was not equal in the six locations, a more detailed analysis of thermal responses was conducted using percentages of TS responses for the different classes of thermal comfort/stress defined for the UTCI. Figures 5, 6 and 7 present the distribution of individual TS votes over the three thermal comfort/stress categories of the UTCI where there was an overlap of data. Note that the João Pessoa sample does not appear in both the 'no thermal stress' and the 'thermal comfort' ranges (Figs. 5 and 8) and it does not figure among the other datasets in Table 5 as UTCI ranges measured there lied above the thermal comfort range.

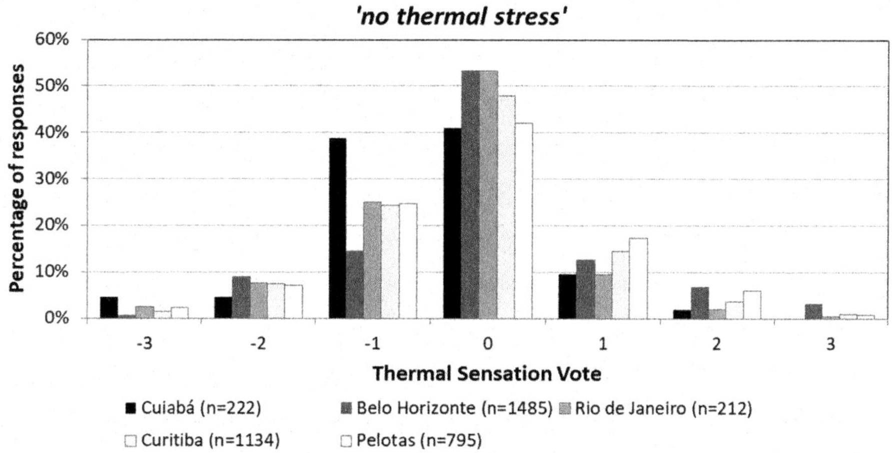


**Fig. 5** Percentages of individual thermal sensation votes for the 'no thermal stress' class of the UTCI

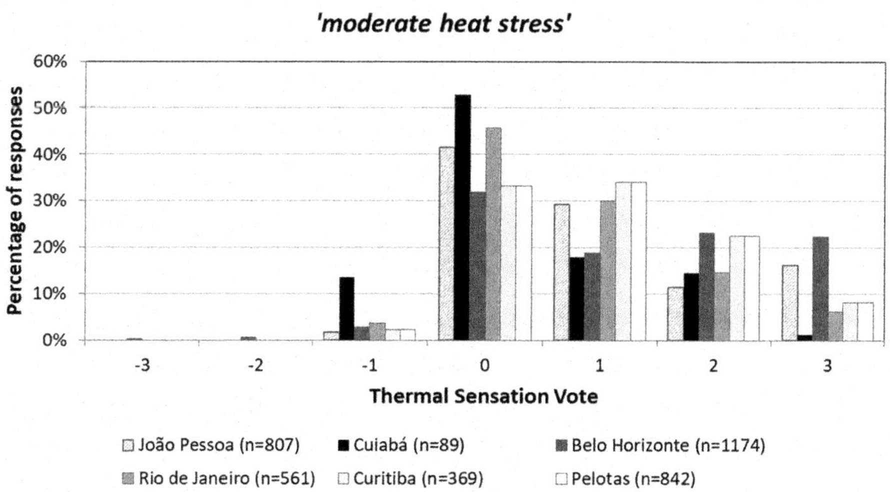


**Fig. 6** Percentages of individual thermal sensation votes for the 'moderate heat stress' class of the UTCI

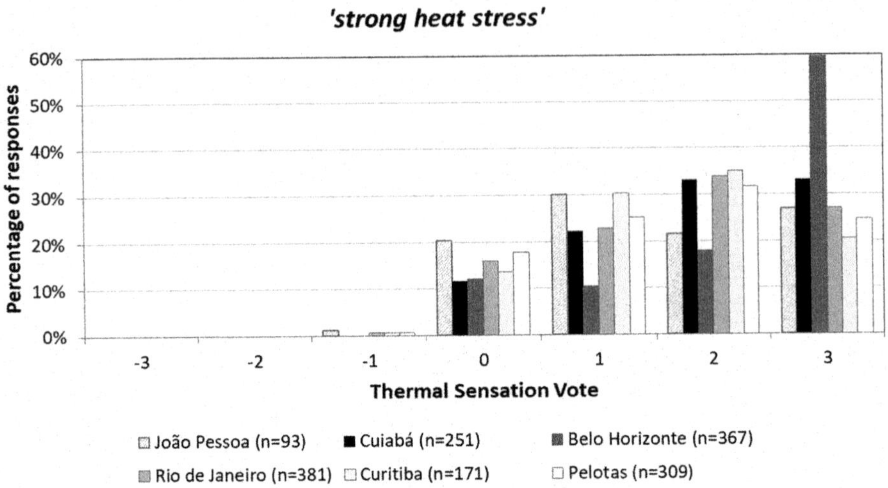


**Fig. 7** Percentages of individual thermal sensation votes for the 'strong heat stress' class of the UTC

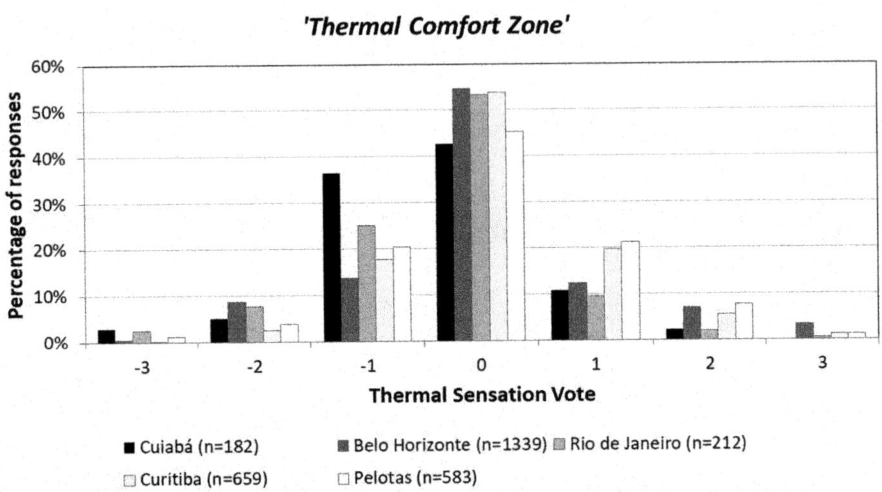


**Fig. 8** Percentages of individual thermal sensation votes for the 'thermal comfort' class of the UTCI

The overlapping of responses for the five locations with existing data within the 'no thermal stress' range shows a similar pattern for most of them, except for Cuiabá. There, responses tend to skew to the left, which suggest a cold sensation of the respondents. The two graphs for rising heat stress categories (Figs. 6 and 7) show some consistency in thermal responses, with a trend to the right side of the scale (hot sensation), which is more accentuated for Belo Horizonte in the 'strong heat stress' category. The somewhat high percentage of neutral thermal sensation observed in the

**Table 5** Descriptive statistics, skewness and kurtosis from a midpoint (TSV = 0) for the different cities—'thermal comfort' class of the UTCI

| | Cuiabá | Belo Horizonte | Rio de Janeiro | Curitiba | Pelotas |
|---|---|---|---|---|---|
| Mean | −0.06 | 0.01 | −0.05 | 0.02 | 0.01 |
| Median | 0.00 | 0.00 | 0.00 | 0.00 | 0.00 |
| Standard deviation | 0.15 | 0.12 | 0.12 | 0.12 | 0.14 |
| Variance | 0.02 | 0.02 | 0.01 | 0.01 | 0.02 |
| Kurtosis | 2.90 | −1.48 | −0.14 | 0.69 | −0.16 |
| Skewness | −1.52 | −0.45 | −0.80 | −0.08 | 0.00 |

moderate and also in the strong heat stress environmental conditions may represent the resilience and adaptability of the Brazilian population to cope with hot thermal environments.

For understanding the behavior of each population as regards the range 18 °C ≤ UTCI ≤ 26 °C that complies with the definition of the 'Thermal Comfort Zone' (TCZ) according to the Glossary of Terms for Thermal Physiology (2003, Bröde et al. 2012a), data from five cities were overlapped for comparisons (Fig. 8). Skew and kurtosis for each percentage distribution are presented in Table 5. The middle point in the thermal sensation scale (TSV = 0) was used for determining such factors by assuming a normal distribution with symmetrical pattern. For that, thermal sensation votes were multiplied by the respective percentage for each survey location.

The comparison for the range equivalent to the comfort zone between cities, taking into account their percentages in each of the 7-point thermal sensation votes (Table 5) shows the largest skew from the midpoint for Cuiabá, with a negative sign, meaning a larger concentration of responses for cold sensation. Skewness drops for the southern locations (Curitiba and Pelotas) and are higher for Rio than in Belo Horizonte, possibly due to local climate influences as shown in Table 1. Kurtosis, which shows how quickly percentage data tails off at each location, is highest for Cuiabá, which means that the bulk of thermal responses are not much distributed over the entire cold section of the TS spectrum, but in fact more concentrated in TS = −1 (cool sensation), as noticed in the graph. In contrast, for Belo Horizonte, the distribution of comfort votes is comparatively flatter.

The Dynamic Thermal Sensation (DTS) was assessed from UTCI data by assigning UTCI values to respective DTS according to operational procedures (Bröde et al. 2012a). Comparisons were drawn to reported thermal sensation for the whole dataset and for different thermal comfort/stress categories of the UTCI (Table 6).

In general, DTS fit well to thermal sensation responses in terms of the mean with a small positive bias (calculated as DTS minus reported thermal sensation) in some locations suggesting an overestimation of the thermal sensation by the DTS and also negative biases due to an underestimation of thermal sensation by the index. Biases found for the entire datasets indicate that the discrepancy is low, less than half of a thermal vote. If broken down into comfort/stress categories, biases can fluctuate between underestimation and overestimation of reported thermal sensation

**Table 6** Number of respondents (n), mean error (bias), root-mean-squared error (rmse) and Pearson correlation coefficient (r) between DTS and reported thermal sensation by location, for the entire series and for selected UTCI thermal comfort/stress categories

| | All dataset | | João Pessoa | | Cuiabá | | Belo Horizonte | | Rio de Janeiro | | Curitiba | | Pelotas | |
|---|---|---|---|---|---|---|---|---|---|---|---|---|---|---|
| | n | bias | n | bias | n | bias | n | bias | n | bias | n | bias | n | bias |
| | 10202 | | 900 | | 685 | | 3630 | | 1328 | | 1685 | | 1974 | |
| Mean bias | | *−0.05* | | 0.06 | | 0.29 | | −0.25 | | 0.30 | | −0.14 | | 0.00 |
| Rmse | | *1.08* | | 1.10 | | 1.05 | | 1.20 | | 1.00 | | 0.96 | | 1.01 |
| r | | *0.59* | | 0.24 | | 0.74 | | 0.55 | | 0.66 | | 0.62 | | 0.60 |
| UTCI comfort/stress class | | | *n* | bias | *n* | bias | *n* | bias | *n* | bias | *n* | bias | *n* | bias |
| Slight cold stress | *9* | *−0.61* | *0* | – | *0* | – | *0* | – | *0* | – | *9* | −0.61 | *0* | – |
| No thermal stress | *3812* | *−0.04* | *0* | – | *221* | 0.33 | *1475* | −0.02 | *212* | 0.36 | *1122* | −0.24 | *782* | −0.03 |
| Moderate heat stress | *4447* | *−0.15* | *798* | 0.07 | *84* | 0.60 | *1773* | −0.42 | *559* | 0.19 | *382* | −0.04 | *851* | −0.14 |
| Strong heat stress | *1599* | *0.14* | *102* | 0.06 | *253* | −0.02 | *378* | −0.33 | *381* | 0.35 | *172* | 0.29 | *313* | 0.42 |
| Very strong heat stress | *335* | *0.52* | *0* | – | *127* | 0.60 | *4* | 0.68 | *176* | 0.51 | *0* | – | *28* | 0.44 |
| Thermal comfort range | *2922* | *0.04* | *0* | – | *176* | 0.41 | *1322* | 0.03 | *212* | 0.36 | *643* | −0.16 | *569* | −0.03 |

depending on the level of thermal stress. In the highest latitudes, i.e. in the cities with colder climate, there is an underestimation of cold sensation and an overestimation of heat sensation. For the less variable conditions of João Pessoa, bias is low and close to zero whereas for the mostly warm conditions assessed in Rio with the highest average UTCI during field campaigns, there was a consistent overestimation of hot sensation, which possibly indicates a thermal adaptation of local population to heat stress conditions, as stated previously. Belo Horizonte follows a similar trend as the coldest locations except for the turning point occurring only at the 'strong heat stress' class, when bias turns positive. Cuiabá does not have a clear pattern, with subjective responses below or above predicted DTS over four classes of the UTCI.

Pearson-correlation coefficients obtained seem to be detached from the mean bias and from the different behavior of found biases over different classes of the UTCI. Highest r-value was for Cuiabá, which had one of the largest mean DTS-biases and showed an unclear behavior of the prediction error over varying classes of the UTCI, whereas the lowest correlation was for João Pessoa, which had the lowest bias values.

In terms of climate types, Rio and Cuiabá, both cities characterized by a tropical climate type (Köppen-Geiger's class 'A'—Peel et al. 2007), had a strong mean positive bias for the thermal comfort range. Such overestimation of reported thermal sensations by the DTS was most likely responsible for the upward shift of their neutral temperatures, relative to the other locations. Belo Horizonte, Curitiba and Pelotas (Köppen-Geiger's class 'C' for temperate climate types), exhibited an almost null bias, which suggests that neutral thermal sensation reported by local residents is in good agreement with estimated sensation given by the DTS. Again, observed biases may result from variations in thermal stress conditions local populations are exposed to throughout the year.

Despite the noticed effect of local climate type over a given population's thermal responses, which lead to noticeable differences in neutral temperatures, confidence intervals drawn for each survey dataset showed that such differences are not statistically significant ($p > 0.05$). There was an overlap in the confidence intervals obtained for the six locations (Fig. 9).

A more detailed analysis conducted with one-way ANOVA revels that campaigns carried out in the surveyed locations differed in terms of outdoor thermal exposure. In Rio, Cuiabá and João Pessoa, confidence intervals (CI) overlapped (with $p < 0.05$), with mean exposure conditions lying within the 'moderate heat stress' range which is representative of the climate at this locations (cf. Table 7, shown later). In Curitiba, mean outdoor thermal exposure belongs to the 'no thermal stress' class, which corresponds to the most frequent conditions at that location. In this case, the obtained confidence interval does not overlap with that of the warmer locations. Belo Horizonte and Pelotas had confidence intervals also overlapped within the 'moderate heat stress' range ($p < 0.05$). However, mean exposure during surveys in these two cities do not correspond to the most frequent conditions there (cf. Table 7).

The locations that were subjected to similar exposure conditions during surveys were evaluated for possible differences in climate adaptation. ANOVA results indicated that there are no significant differences between thermal responses assessed in Rio, Cuiabá and João Pessoa ($p > 0.05$), which might be explained by their very

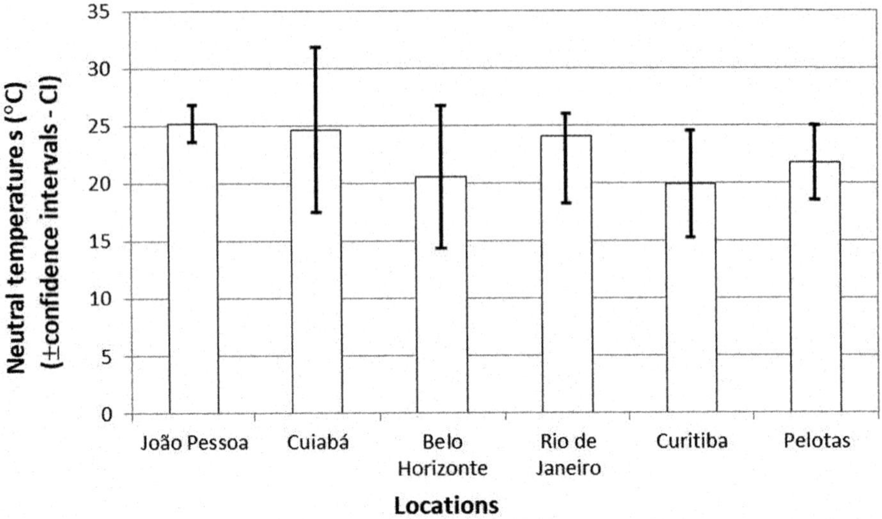


**Fig. 9** Neutral temperatures for the six locations with 95% CI

**Table 7** Predicted UTCI data with ERA5 reanalysis for the locations and thermal comfort/stress categories

| | João Pessoa | Cuiabá | Belo Horizonte | Rio de Janeiro | Curitiba | Pelotas |
|---|---|---|---|---|---|---|
| Mean UTCI | 24.8 | 28.8 | 21.9 | 26.3 | 17.5 | 16.5 |
| Mean UTCI (daytime) | 27.0 | 29.7 | 24.9 | 28.9 | 19.5 | 18.1 |
| Mean daily fluctuation | 7.0 | 15.8 | 18.0 | 13.5 | 16.0 | 15.0 |
| Absolute yearly fluctuation | 17.7 | 40.1 | 37.5 | 37.7 | 54.8 | 57.3 |
| *UTCI comfort/stress class (daytime)* | | | | | | |
| Cold (%) | 0 | 0 | 0 | 0 | 0 | 0 |
| Moderate cold stress (%) | 0 | 0 | 0 | 0 | 1 | 3 |
| Slight cold stress (%) | 0 | 0 | 3 | 0 | 9 | 12 |
| No thermal stress (%) | 35 | 31 | 45 | 32 | 66 | 64 |
| Moderate heat stress (%) | 62 | 24 | 39 | 32 | 20 | 14 |
| Strong heat stress (%) | 3 | 35 | 13 | 28 | 3 | 6 |
| Very strong heat stress (%) | 0 | 9 | 0 | 7 | 0 | 0 |

similar climate types. On the other hand, statistically significant differences were observed between thermal responses for Pelotas and Belo Horizonte ($p < 0.05$), which indicates a different level of thermal adaptation by these populations. In fact, despite the classification of these cities as Köppen-Geiger's class 'C' for temperate climate types, winter exposure conditions in Pelotas are harsher than in Belo Horizonte whereas during summer the opposite occurs, with higher UTCI values in Belo Horizonte. Therefore, different thermal adaptations are required for these populations. In turn, the climate of Pelotas is more similar to that of Curitiba. However, the average exposure conditions in Pelotas within the 'moderate heat stress' range and not within the 'no thermal stress' range did not make it possible to adequately compare the thermal responses of both populations.

## 5 Discussion

Differences found in neutral temperatures suggest an impact of local climatic features on the thermal perception of a given population. A more general description of local climate, as expressed by Köppen-Geiger's main climate classes can further confirm such effect. A negative skew of thermal sensation responses and a positive DTS bias for UTCI's thermal comfort range were noticed in the warmer locations, pointing that for comfort conditions, part of the respondents felt cold so that the predicted thermal sensation overestimated their thermal sensation. In colder locations categorized as belonging to Köppen-Geiger's 'C' climate type, skew was comparatively lower tending to zero and the same was noticed for DTS bias.

These observations can also be understood in light of UTCI data generated for a recent year (2019) using modeled weather data for the six cities. UTCI time series for the locations were obtained by extracting the UTCI from the corresponding grid cells of ERA5-HEAT climate reanalysis (Copernicus Climate Data Store 2020). Computed using the ERA5 reanalysis from the European Centre for Medium-Range Forecasts (ECMWF), ERA5-HEAT provides hourly gridded maps of the UTCI at $0.25° \times 0.25°$ spatial resolution and currently spans from 1979 to the present date (Di Napoli et al. 2020). Table 7 presents general features for each location in terms of UTCI data and the percentage of hours under various thermal comfort/stress categories.

Surveyed exposure conditions (Table 3) were warmer than estimated annual figures and closer to daytime conditions at all locations, assumed as the time period between 9 am and 3 pm, during which most of the surveys took place. It can be noticed that particularly João Pessoa and Pelotas had a large discrepancy in daytime mean UTCI during surveys when compared to predicted data. Thus, the range of thermal conditions surveyed does not exactly match the fluctuations observed over a 24-hour cycle or during the year, though the relatively lower fluctuation noticed for João Pessoa from UTCI estimates and also from onsite measurements may explain the higher sensitivity of its population to smallest changes in the UTCI. Comparatively, Curitiba, which had the highest range of conditions measured and one of the

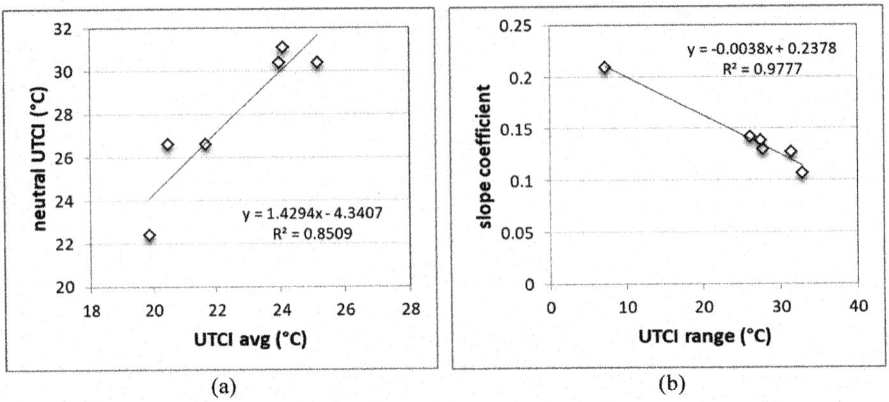


**Fig. 10** Correlations between neutral UTCI and slope coefficient to survey exposure conditions

highest fluctuations of the UTCI on annual basis, showed the lowest slope in the linear regression analysis and thus the lowest thermal sensitivity of its population.

Consequently, and generally speaking, from the six datasets, it could be argued that obtained neutral temperatures correlate well with the mean UTCI conditions surveyed and that thermal sensitivity (or the slope of the regression equation) is inversely correlated with the UTCI range, under which surveys took place (Fig. 10).

Yet, despite such relationships that might indicate a regional adaptation effect of local population to prevailing climatic features, ideally covered during the surveys, a few issues remain unexplained from a purely climatic perspective such as the correlation between subjective thermal responses and UTCI and DTS data obtained from in situ meteorological measurements.

The lack of strong correlations between subjective responses and thermal comfort index data, in our case, the UTCI (for the overall dataset $r = 0.59$, cf. Table 6) has been attributed in some studies to other components of human thermal comfort. In this respect, Höppe (2002) lists three major approaches to what is termed human thermal comfort: a psychological, a thermophysiological and another one, based on human thermal balance. Over the last two decades, various studies have found discrepancies between reported thermal sensation and a given thermal index's values. Many of those were transverse studies across populations with large samples, as the ones presented in this chapter (Nikolopoulou et al. 2001; Pantavou et al. 2013) but longitudinal studies have also been carried out with small groups (Cheng et al. 2012), more recently supported by thermal or microclimate walks (Chokhachian et al. 2018; Vasilikou and Nikolopoulou 2020) showing a similar mismatch between subjective and objective data. To tackle such shortcomings, it has been suggested that an understanding of the dynamic human parameter is required taking into consideration psychological adaptation (Nikolopoulou et al. 2001), which lead to the intersection of outdoor thermal comfort studies with environmental psychology (Lenzholzer and Nikolopoulou 2020). With the aim of presenting new research focussing on the subjectiveness involved in OTC, IJBM has launched a special issue on Subjective Approaches to Thermal Perception (Lenzholzer and Nikolopoulou 2020).

Still, in a previous study we showed that the UTCI was able to capture inter-variability within a given sample, exemplified for Curitiba, with small changes in bias, rmse and Pearson correlation presented for different subgroups (age, BMI, biological sex and urban morphology aspects) (Bröde et al. 2014). Chapter 4 of this book also looks into the sensitivity of the UTCI as regards such aspects, including historical weather information from the previous days (Sensitivity of UTCI thermal comfort prediction to personal and situational factors—residual analysis of pedestrian survey data). This way, differences in the datasets in terms of anthropometric information should not be expected to account for substantial changes in prediction bias.

In turn, thermal history and short-term acclimatization (discussed in more detail in Chap. 5 of this book—Long and short-term acclimatization effects on outdoor thermal perception versus UTCI) may have affected obtained results to a certain extent. Previously published studies for the datasets gathered in right-here-right-now surveys in Rio de Janeiro and Cuiabá showed an increase in reported thermal sensation votes and in the thermal sensitivity of subjects who reported a more frequent usage of air-conditioned environments (Krüger et al. 2015; Borges et al. 2020). Not all the six datasets contain reported information on thermal history or on air-conditioning usage, and the way thermal history was approached was not standardized, thus the consideration of such aspects in our analysis was not possible. Apart from that, the settings also varied between cities, some of them having a more urban character with a diversity of morphologies (Cuiabá, Rio de Janeiro, Curitiba and Pelotas), while others being more focused on leisure and resting areas such as public parks (João Pessoa, Belo Horizonte).

Finally, the relationship between local climate features and reported OTC data obtained from surveys with local population weakens for samples that fail to represent the full spectrum of UTCI conditions one might be exposed to locally over the year. Correlations drawn between slope and neutral temperatures to estimated UTCI data (ERA5-HEAT data) are significantly lower than those represented in Fig. 10. This last point seems to be relevant, as it could a starting guideline to be considered in standardized protocols for UTC surveys, such as the initiatives started by Johansson et al. (2014) and Lau and Krüger (2020a, b) (see also Chap. 11 of this book: Proposed framework for establishing a global database for outdoor thermal comfort research).

## 6 Conclusions

Wrapping up, it can be concluded that subjective responses tended to follow the patterns of exposure conditions during surveys, suggesting that the neutral temperature is well aligned with the mean UTCI value obtained for the monitoring campaigns and that thermal sensitivity is inversely correlated with the range of conditions surveyed. In that sense, a regional adaptation effect is verified and that effect can be linked to measured climatic conditions.

However, from looking at long-term estimates for the UTCI in each location, it was noticed that exposure conditions did not correspond in most cases to the range of conditions one might be exposed to during daytime in those locations. Moreover, anthropometric, morphological data were not similar for all samples and perhaps do not correspond to local demographic features. It can thus be concluded that standardized protocols are required for OTC surveys, samples should reflect local demographics and exposure conditions should be in tune with the range and extent of thermal comfort/stress conditions at a given location.

**Acknowledgements** The national research funding agencies CAPES (Coordenação de Aperfeiçoamento de Pessoal de Nível Superior) and CNPQ (Conselho Nacional de Desenvolvimento Científico e Tecnológico) and the regional funding agency FAPEMAT-Research Support Foundation of Mato Grosso, Brazil for financially supporting several steps of this research.

## References

Błażejczyk K, Błażejczyk M (2010) BioKlima© 2.6, Program komputerowy. http://www.igipz.pan.pl/Bioklima-zgik.html

Borges VCDAL, Callejas IJA, Durante LC (2020) Thermal sensation in outdoor urban spaces: a study in a Tropical Savannah climate, Brazil. Int J Biometeorol 64(3):533–545

Bröde P, Fiala D, Błażejczyk K, Holmér I, Jendritzky G, Kampmann B, Havenith G (2012a) Deriving the operational procedure for the Universal Thermal Climate Index (UTCI). Int J Biometeorol 56(3):481–494

Bröde P, Krüger EL, Rossi FA, Fiala D (2012b) Predicting urban outdoor thermal comfort by the Universal Thermal Climate Index UTCI—a case study in Southern Brazil. Int J Biometeorol 56(3):471–480

Bröde P, Krüger EL, Fiala D (2014) Residual analysis of UTCI predictions on outdoor thermal sensation survey data. In: 8th Windsor conference: counting the cost of comfort in a changing world, Cumberland Lodge, Windsor, UK, 10–13 April 2014

Cheng V, Ng E, Chan C, Givoni B (2012) Outdoor thermal comfort study in a sub-tropical climate: a longitudinal study based in Hong Kong. Int J Biometeorol 56(1):43–56

Chokhachian A, Lau KKL, Perini K, Auer T (2018) Sensing transient outdoor comfort: a georeferenced method to monitor and map microclimate. J Build Eng 20:94–104

Commission for Thermal Physiology of the International Union of Physiological Sciences (2003) Glossary of terms for thermal physiology. J ThermBiol 28:75–106

Copernicus Climate Data Store (2020) Thermal comfort indices derived from ERA5 reanalysis. https://doi.org/10.24381/cds.553b7518

De Onis M, Lobstein T (2010) Defining obesity risk status in the general childhood population: which cut-offs should we use? Int J Pediatr Obes 5(6):458–460

Di Napoli C, Barnard C, Prudhomme C, Cloke HL, Pappenberger F (2020) ERA5-HEAT: a global gridded historical dataset of human thermal comfort indices from climate reanalysis. Geosci Data J. https://doi.org/10.1002/gdj3.102. ISSN 2049-6060 [currently in press]

Golasi I, Salata F, de LietoVollaro E, Coppi M (2018) Complying with the demand of standardization in outdoor thermal comfort: a first approach to the Global Outdoor Comfort Index (GOCI). Build Environ 130:104–119

Hadianpour M, Mahdavinejad M, Bemanian M, Haghshenas M, Kordjamshidi M (2019) Effects of windward and leeward wind directions on outdoor thermal and wind sensation in Tehran. Build Environ 150:164–180

Höppe P (2002) Different aspects of assessing indoor and outdoor thermal comfort. Energy Build 34(6):661–665
ISO 7726 (1998) Ergonomics of the thermal environments: instruments and methods for measuring physical quantities. International Standards Organization
Jin H, Liu S, Kang J (2019) Thermal comfort range and influence factor of urban pedestrian streets in severe cold regions. Energy Build 198:197–206
Johansson E, Thorsson S, Emmanuel R, Krüger E (2014) Instruments and methods in outdoor thermal comfort studies—the need for standardization. Urbanclimate 10:346–366
Kántor N, Égerházi L, Unger J (2012) Subjective estimation of thermal environment in recreational urban spaces—Part 1: Investigations in Szeged, Hungary. Int J Biometeorol 56(6):1075–1088
Krüger E, Drach P, Bröde P (2015) Implications of air-conditioning use on thermal perception in open spaces: a field study in downtown Rio de Janeiro. Build Environ 94:417–425
Krüger EL, Silva TJV, da Silveira Hirashima SQ, da Cunha EG, Rosa LA (2020) Calibrating UTCI'S comfort assessment scale for three Brazilian cities with different climatic conditions. Int J Biometeorol 1–10 (in press)
Lau K, Krüger E (2020a) Methodological framework for establishing a global database for outdoor thermal comfort survey: progress of data acquisition and harmonization. In: Planning post carbon cities: 35th PLEA conference on passive and low energy architecture, A Coruña, Spain, September 2020
Lau K, Krüger E (2020b) Establishing a global database for outdoor thermal comfort survey: a pilot study of standardisation of methodology. In: 11th Windsor conference: resilient comfort, Windsor, UK, April 2020
Lenzholzer S, Nikolopoulou M (2020) Foreword to the special issue on subjective approaches to thermal perception. Int J Biometeorol 64(2):167–171
Mateeva Z (2011) Personal factors of human thermal perception: long-lasting climate experience. Prace i Studia Geograficzne 47:335–344
Nikolopoulou M, Lykoudis S (2006) Thermal comfort in outdoor urban spaces: analysis across different European countries. Build Environ 41(11):1455–1470
Nikolopoulou M, Baker N, Steemers K (2001) Thermal comfort in outdoor urban spaces: understanding the human parameter. Sol Energy 70(3):227–235
Pantavou K, Theoharatos G, Santamouris M, Asimakopoulos D (2013) Outdoor thermal sensation of pedestrians in a Mediterranean climate and a comparison with UTCI. Build Environ 66:82–95
Pantavou K, Lykoudis S, Nikolopoulou M, Tsiros I (2018) Thermal sensation and climate: a comparison of UTCI and PET thresholds in different climates. Int J Biometeorol 62(9):1695–1708
Peel MC, Finlayson BL, MC Mahon TA (2007) Updated world map of the Köppen-Geiger climate classification. Hydrol Earth Syst Sci Discuss Eur Geosci Union 11:1633–1644
Potchter O, Cohen P, Lin TP, Matzarakis A (2018) Outdoor human thermal perception in various climates: a comprehensive review of approaches, methods and quantification. Sci Total Environ 631:390–406
Vasilikou C, Nikolopoulou M (2020) Outdoor thermal comfort for pedestrians in movement: thermal walks in complex urban morphology. Int J Biometeorol 64(2):277–291

# Chapter 7
# Outdoor Thermal Environment and Heat-Related Symptoms of Pedestrians: An Application of the UTCI for Health Risk Assessment

**Katerina Pantavou**

**Abstract** Prolonged exposure to hot environments can cause adverse health effects ranging from thermal discomfort to heart disorders and heat stroke. Climate change and global warming increase people exposure to heat, posing a serious threat to human health. Adverse heat-related health impacts can be prevented by preparing public health systems to cope and address these threats. This study aims to examine the susceptibility of pedestrians to outdoor thermal environment. Micrometeorological conditions were monitored in urban public sites in Athens, Greece while data on individual's attributes and heat-related morbidity were collected with questionnaire-based field surveys. Participants were asked to report whether they were experiencing symptoms of heat-related illness during the field surveys. The symptoms included breathing difficulties, dizziness, headache and exhaustion. The thermal environment was assessed using the Universal Thermal Climate Index (UTCI). The relationship of experiencing or not heat-related symptoms with the UTCI was examined using logistic regression. Hierarchical clustering analysis was used to determine thresholds of the UTCI above which heat-related symptoms are expected to occur. A total of 2731 pedestrians participated in the surveys. Most of them were residents of the metropolitan area of Athens (94.2%), between 18 and 34 years old (44.8%), without a medical history of cardiovascular or respiratory disease (88.6%). Logistic regression showed that for one unit increase of the UTCI the odds of reporting heat-related symptoms increase by 4% (adjusted odds ratio (aOR) = 1.04, p-value < 0.001). The odds were increased to 6% in summer (aOR = 1.06, p-value < 0.012). The UTCI threshold for experiencing heat-related symptoms was 28.5 °C (*moderate heat stress*) for all seasons and 38.9 °C (*very strong heat stress*) for summer. These results could be used to strengthen public health capacities to protect population from heat exposure impacts.

**Keywords** UTCI · Heat · Health effects · Heat illness · Heat-related symptoms

---

K. Pantavou (✉)
Laboratory of General and Agricultural Meteorology, Agricultural University of Athens, Iera Odos 75, 118 55 Athens, Greece
e-mail: kpantavou@aua.gr



E. L. Krüger (ed.), *Applications of the Universal Thermal Climate Index UTCI in Biometeorology*, Biometeorology 4,
https://doi.org/10.1007/978-3-030-76716-7_7

# 1 Introduction

Heat illness is a medical condition caused by exposure to ambient heat. It includes minor conditions such as heat rash, cramps, syncope and exhaustion as well as the serious condition of heat stroke. Warning signs and symptoms include red cluster of pimples or small blisters, muscle pain, headache, dizziness, heavy sweating, rapid breathing, fainting, and rise of body temperature (CDC 2020).

Heat illness or heat-related morbidity occurs when the human body is unable to regulate- maintain within certain boundaries (i.e., 36.5–37.5 °C)—its internal temperature (Guyton and Hall 2006). The balance of body temperature is preserved by heat production and loss by means of thermoregulation. The body generates heat mainly in the brain, heart, liver, and with the contraction of skeletal muscles. Moreover, through mechanisms of hormonal thermogenesis (thyroid gland releases hormones to increase metabolism) and vasoconstriction (blood vessels become narrower to decrease blood flow to the skin and keep heat near the inner body) the produced heat is increased or retained. The body loses heat by sweating (sweat glands release sweat that cools the skin when evaporating) and vasodilation (blood vessels become wider to increase blood flow to the skin away from the warm inner body). In both cases, i.e. heat production or heat loss, the mechanisms behind thermoregulation are controlled by the brain (hypothalamus). When internal temperature increases, the brain sends signals to organs, glands, muscles, and the nervous system responding to environmental changes and trying to keep the body temperature in a state of equilibrium, called homeostasis (Guyton and Hall 2006).

Environmental conditions may lead to failure of thermoregulatory mechanisms by preventing heat dissipation, which results in the body producing heat faster than it can lose it. This may activate dangerous physiological responses or pathways potentially leading to organ failure apart from causing heat illness. Heat exposure may trigger ischemia, heat cytotoxicity, inflammatory response, disseminated intravascular coagulation and rhabdomyolysis (Mora et al. 2017). Evidence of 27 different pathways were found to be triggered by heat, leading to failure of brain, heart, lungs, liver, kidneys, pancreas and intestines (Mora et al. 2017). For example, thermoregulatory responses in skin blood flow and sweating are accompanied by cardiovascular adjustments, which have to be effective in order for the body to withstand exposure (Cui and Sinoway 2014). Thus, individuals with chronic cardiovascular diseases may be particularly vulnerable to heat (Basu and Samet 2002; Cui and Sinoway 2014; Sun et al. 2018). Factors such as physiology, activity, clothing, and medication intake may negatively or positively affect the thermoregulation process (Lipman et al. 2013).

The effect of hot environments on human health is commonly recognized as a J-shaped relationship between ambient temperature and heat-related health outcomes (Basu and Samet 2002). This relationship was suggested using data of all-cause and cause-specific mortality (Son et al. 2019), as well as morbidity data based on hospitalizations, emergency hospital admissions and visits, or ambulance transports (Ye et al. 2012) during periods of heatwaves or extreme temperatures (Song et al. 2017). The examined heat-related health outcomes mainly focus on cardiovascular and respiratory diseases (Son et al. 2019). The negative effect of hot environments on heat illness (Murakami et al. 2012; Wang et al. 2016), preterm deliveries (Avalos et al. 2017), intestinal infectious diseases, endocrine, nutritional, and metabolic diseases in infants (0–4 years old) (Xu et al. 2014), and mental health have also been documented (Sherbakov et al. 2018). A decline of the heat effect has been noticed when a peak temperature is observed (Sugg et al. 2016) as well as negative association with health outcomes has been also found (Pantavou et al. 2008) attributed to the effectiveness of and compliance to preventive measures enforced during these peak events.

Vulnerable populations comprise elderly populations (Son et al. 2019), women (Son et al. 2019), infants, children and pregnant women (WHO 2020). Populations at high risk are also those exposed to ambient environment for work purposes as well as low-income populations residing in old building structures (WHO 2020), in urban areas with lack of green spaces and usually without air-conditioning (Xu et al. 2013). Thresholds of air temperature above which heat-related mortality or morbidity increases vary by location, which point to a potential adaptation of the population to local climates (Hajat and Kosatky 2010; Ye et al. 2012), in addition to regional adaptation trends in terms of optimal thermal comfort ranges as shown in the previous chapter (Chap. 6: Regional adaptation of the UTCI: comparisons between different datasets in Brazil). These thresholds were found to decrease with latitude (Hajat and Kosatky 2010).

The prevailing indicator of heat stress effects in the published literature is air temperature (Ye et al. 2012; Song et al. 2017; Son et al. 2019). Few studies combine meteorological variables, i.e., air temperature, relative humidity, wind speed, solar radiation, in term of thermal indices to assess heat-related health risks. Such indices include Wet-Bulb Globe Temperature (WBGT), Apparent Temperature, Predicted Mean Vote (PMV), Physiologically Equivalent Temperature (PET) and Universal Thermal Climate Index (UTCI) (Pantavou et al. 2008, 2016, 2020; Vaneckova et al. 2011; Nastos and Matzarakis 2012; Di Napoli et al. 2018). The UTCI is increasingly used to evaluate thermal conditions and heat-related mortality risk in different climates (Nastos and Matzarakis 2012; Urban and Kyselý 2014; Di Napoli et al. 2018; Kuchcik 2020).

In light of global warming trends (IPCC 2018), which bring about an increased exposure to heat, the focus of environmental health includes heat-related disease prevention. With this in mind, heat-related morbidity is a serious public health concern especially in typical thermal conditions and in the absence of preventive measures. Moreover, the need for well-grounded, universal and integrated tools for the assessment of thermal environment becomes apparent. Towards that aim, this chapter examines the relationship between the UTCI and milder symptoms of heat-related illness experienced by pedestrians in an urban environment with the aim to better establish its usefulness. Milder symptoms of heat-related illness such as headache, dizziness, breathing difficulties and exhaustion are common, easily self-valuated. Such symptoms are usually resolved without hospitalization, and are thus not recorded in health registries. However, although mild, these symptoms degrade the quality of life while they could be precursors of a more serious heat-related health condition, consequently deserving careful investigation.

# 2 Materials and Methods

## *2.1 Study Design and Study Area*

This study is part of the Urban Biometeorology and Planning project (UBiPlan 2020). The UBiPlan project integrates data from three major projects on thermal sensation carried out independently in Athens, Greece and enriched such database with new data. The aim was to develop a database to be used for a spatial assessment of thermal conditions. Overall, the UBiPlan database includes data from field surveys conducted in 13 different urban sites covering ten different municipalities of the greater area of Athens. All four projects (Nikolopoulou et al. 2004; Pantavou et al. 2013; Tseliou et al. 2016; UBiPlan 2020) included field surveys employing questionnaire-based interviews and concomitant in-situ meteorological measurements. The surveys were conducted following similar protocols, so that data integration is considered meaningful.

Data used in this study were collected in field surveys carried out in the context of the two projects that collected data on heat-related effects on human health (Pantavou et al. 2013, 2016; UBiPlan 2020). Monitoring campaigns alongside with questionnaire surveys with pedestrians took place in warm, cold and transient seasons, i.e., summer, autumn and winter, spanning from around 9:00 a.m., to sunset. The monitoring locations were outdoor public sites (squares, pedestrian streets, promenades) in the metropolitan area of Athens, Greece.

Athens is the capital of Greece with a Csa type-Mediterranean climate (Kottek et al. 2006). The warm and dry period is from April to October and the cold and rainy period from mid-October until the end of March. Monthly average air temperature ranges between 9.3 °C in January and 27.0 °C in July while monthly maxima reach 32.6 °C in July. Monthly average relative humidity ranges between 48% in July and 73% in November and December (NOA 2020). The metropolitan area of Athens has a total population of 3,828,434 citizens which is about the one third of the country's population (ELSTAT 2020).

## *2.2 Microclimate Monitoring*

Micrometeorological conditions were monitored on-site and at the height of 1.1 m above the ground using a mobile weather station. Measured meteorological variables included air temperature ($T_a$, °C), relative humidity (RH, %), average wind speed ($v_a$, m s$^{-1}$), and globe temperature ($T_{gl}$, °C; PVC sphere with 40 mm in diameter painted grey with RAL 7001). Data were stored at 1 min intervals with a data logger. Features of the monitoring equipment used in this study are presented in Table 1.

**Table 1** Type of instruments used in the field surveys

| Variable | Abb | Type | Instrument | |
|---|---|---|---|---|
| | | | Project 1 | Project 2 |
| Air temperature | $T_a$ | Thermo-hygrometer | Rotronic (HC2-S3C03-PT15) | Rotronic (MP101A-T7-W4W) |
| Relative humidity | RH | | | |
| Average wind speed | $v_a$ | Anemometer | Second Wind C3 cup anemometer | Atmos 22 |
| Globe temperature | $T_{gl}$ | Pt100 | Tailor made (PVC sphere 40 mm diameter painted with RAL 7001) | Tailor made (PVC sphere 40 mm diameter painted with RAL 7001) |
| Data logger | Sampling time: 30 s<br>Storing time: 1 min | | Campbell (CR10X) | Delta-T (GP2) |

*Project 1* Pantavou et al. (2013, 2016); *Project 2* UBiPlan (2020)

## 2.3 *Questionnaire Description*

The questionnaire was designed to capture heat-related morbidity of pedestrians and information related to demographics and subjective attributes. Questionnaire items (Table 2) were selected based on studies of similar design for examining thermal sensation (Nikolopoulou et al. 2004; Tseliou et al. 2016) and in accordance to international standards (ISO 10551 1995; ISO 8996 2004; ISO 9920 2007). Heat-related morbidity was reported by participants; i.e., whether they were experiencing symptoms of heat-related illness such as breathing difficulties, headache, dizziness and exhaustion (CDC 2020) at the moment of the interview. These are common symptoms that are feasible to be self-evaluated and quantified, normally do not result in hospitalization and are thus usually not recorded in health registries, despite being potential precursors for a more serious underlying heat-related health condition. Demographics included biological sex, age, height, weight and city of residence. Subjective attributes involved clothing description, main activity for the last half hour before the interview (sleeping, sitting, standing, walking, sports), recent exposure history (outdoors, indoors/vehicle with/without air-conditioning), outdoor exposure duration (i.e., less than 5 min, 5–15 min, 15–30 min, 30 min to 1 h, more than 1 h), purpose of visit (rest, work, passing by, entertainment/meeting friends/shopping/other), medical history (respiratory or cardiovascular), thermal sensation based on a seven-point bipolar scale from cold (−3) to hot (+3), thermal comfort (comfortable, uncomfortable) and thermal preference (cooler, neither warmer nor cooler/no change, warmer).

**Table 2** Questionnaire items and response categories

| Variable | Questionnaire items | Response categories |
|---|---|---|
| Biological sex | Sex: | Male/female |
| Age | Age (years) | <12/13–17/18–24/25–34/35–44/45–54/55–64/>65 |
| Place of residence | Where do you live? | (City/municipality) |
| Height | Please state your height | … (cm) |
| Weight | Please state your weight | … (kg) |
| Clothing | The clothes that best describe participant's outfit are | Estimated according to ISO 9920 (2007) |
| Activity | Over the last half hour, you have been mainly | Sleeping/sitting/standing/walking/doing sports |
| Exposure duration | How long have you been in this place? | <5 min/5 to 15 min/15 to 30 min/30 min to 1 h/>1 h |
| Exposure history | Where were you during the last half hour? | Indoors-car-public transport, with/without air conditioning/outdoors |
| Visit purpose | Reason to be in this place | Rest/work/passing by/entertainment-meeting friends-shopping-other |
| Medical history | Do you have any chronic condition like | Cardiovascular-respiratory/none |
| Health symptoms | At this moment, do you have[a] | Breathing difficulties/headache/dizziness/exhaustion/no symptoms |
| Thermal sensation | How do you feel at this precise moment? | Cold/cool/slightly cool/neither cold nor hot/slightly warm/warm/hot |
| Thermal comfort | Do you find this | Comfortable-neutral/uncomfortable |
| Thermal preference | Please state how you would prefer to be now | Cooler/neither cooler nor warmer-no change/warmer |

[a]Participants could select as many as apply

## 2.4 *Data Processing and the UTCI*

Clothing insulation ($I_{cl}$, in clo units) was estimated using the clothing description of participants (ISO 9920 2007). Body mass index was derived from height and weight (BMI, kg m$^{-2}$). Mean radiant temperature ($T_{mrt}$, °C) was estimated based on $T_a$, $T_{gl}$, and $v_a$ (ISO 7726 2001). Questionnaire responses used in the two projects (Pantavou et al. 2013, 2016; UBiPlan 2020) were harmonised to allow integration of the two datasets. Furthermore, pedestrians were classified as local (residing in the same municipality of the monitoring site) and non-local residents, based on the responses regarding the place of residence. Also, heat-related symptoms were collapsed into

**Table 3** UTCI assessment scale in terms of thermal stress

| UTCI (°C) range | Stress category |
|---|---|
| Above +46 | Extreme heat stress |
| +38 to +46 | Very strong heat stress |
| +32 to +38 | Strong heat stress |
| +26 to +32 | Moderate heat stress |
| +9 to +26 | No thermal stress |
| +9 to 0 | Slight cold stress |
| 0 to −13 | Moderate cold stress |
| −13 to −27 | Strong cold stress |
| −27 to −40 | Very strong cold stress |
| below −40 | Extreme cold stress |

a dichotomous variable, i.e., reporting and non-reporting heat-related symptoms at the moment of the interview.

The UTCI was estimated to assess thermal environment using the BioKlima software (Błażejczyk and Błażejczyk 2010). This approach uses a sixth-order polynomial equation utilizing $T_a$ (°C), RH (%), $v_a$ (m s$^{-1}$), and $T_{mrt}$ (°C) (Bröde et al. 2012).The UTCI assumes a self-adapting clothing insulation, a permanent walking speed of 4 km h$^{-1}$ (1.11 m s$^{-1}$) and an internal heat production of 135 W m$^{-2}$. Its original assessment scale (Błazejczyk et al. 2013) is presented in Table 3.

## 2.5 Statistical Analysis

The statistical analysis involved frequency distributions for categorical variables and mean, median and standard deviation values for continuous variables. The differences between experiencing and not heat-related symptoms during the interview was examined with chi-squared and t-tests. Binary logistic regression was used to examine the association of heat-related symptoms with the UTCI and pedestrians' attributes treating heat-related symptoms as a dichotomous variable (i.e., experiencing or not heat-related symptoms during the interview). UTCI data and other questionnaire items were treated as independent covariates and factors. Odds ratios (OR) and p-values were produced from logistic regression models. Multivariate logistic analysis was performed adjusting for variables that were significant in the univariate analysis. Hosmer–Lemeshow goodness-of-fit test was used to assess the goodness of fit of each model. A p-value higher than significance level indicates a good logistic regression model fit.

Hierarchical cluster analysis was applied to examine whether pedestrians can be classified in distinct groups of people experiencing or not heat-related symptoms, based on the UTCI. The UTCI transition point between the two groups was defined as the average of the UTCI for the two closest cases belonging to different groups. Statistical significance was considered at a significance level $p \leq 0.05$.

# 3 Results

## 3.1 Sample Description

The dataset consists of 2731 (about 60% of those belonging to the sample of the UBiPlan project) responses. Most of the participants were males (54.7%). About 44.8% were between 18 and 34 years old. Approximately 94% of the respondents were residents of the metropolitan area of Athens and 31% local residents (residents in the municipality of the monitoring site). About 88.6% of the participants reported no medical history of cardiovascular or respiratory disease. In the last half hour before the interview, 62.8% of the participants reported that they had been indoors, in a vehicle or had used public transport, and 66.2% had been in a non-air conditioning environment. 96.3% of the participants reported that their activity during the last half hour before the interview was low, i.e., they were mainly sitting (37.5%), standing (31.6%) or walking (27.1%). More than half of the respondents (57.9%) were visiting the monitoring site for less than 15 min, 37.9% were visiting the site for work purposes and 24.9% were passing through. Mean BMI of the participants was estimated at 24.5 ± 4.5 kg $m^{-2}$.

The interviews were almost evenly distributed between summer (July, 39.8%) and winter (January–February 33.5%) while autumn (October-early November) interviews were slightly less (26.6%). Participants' characteristics were similar across seasons except for exposure history and purpose of visit. Pedestrians were more likely ($p < 0.001$) to have spent the last half hour before the interview outdoors in autumn (41.9%) or winter (38.8%) than in summer (32.3%) or in a place without air-conditioning in autumn (89.3%) than in winter (60%) and summer (56.4%; $p < 0.001$). More participants reported that they visited the monitoring site for work in summer (41.6%) than in autumn (35.8%) or winter (35.4%). Mean $I_{clo}$ (±standard deviation) was estimated 0.4 ± 0.1 clo for summer, 0.7 ± 0.2 clo for autumn and 1.2 ± 0.3 clo for winter.

## 3.2 Exposure Conditions

Field surveys were conducted under a range of meteorological conditions. Air temperature was recorded between 6.9 and 39.3 °C, and the relative humidity ranged between 22 and 79% (Table 4). Grey globe temperature ranged from 7.9 to 43.8 °C while mean air speed was estimated 1.0 ± 0.6 m $s^{-1}$. UTCI ranged from 4.9 to 43.8 °C corresponding to the *slight cold stress* and *very strong heat stress* categories.

About 58.7% of the participants reported thermal sensation between *slightly cool* and *slightly warm*, 69.7% felt *comfortable* and 39.7% reported that they would prefer *neither cooler nor warmer/ no change* in their thermal sensation.

Most interviews (53.5%) were conducted in heat stress conditions (UTCI above 26 °C, corresponding to *moderate/strong/very strong heat* stress categories; Fig. 1a). For UTCI values between 32 and 38 °C (*strong heat stress*), 50.2% of the respondents reported thermal sensation *warm* or *hot*. This percentage increased to 69.1% for UTCI values above 38 °C (*very strong heat stress*) (Fig. 1b). Over 35.4% of the participants reported that they felt uncomfortable when the UTCI ranged between 32 and 38 °C (*strong heat stress*) (Fig. 1c). Over 49.1% would prefer to experience a *cooler* thermal sensation when the UTCI stress category was *moderate, strong or very strong heat stress* (Fig. 1d). Moreover, participants seemed to be more sensitive to low temperatures. About 76.6% of the participants reported *cold* or *cool* thermal sensation while 52.5% reported that they felt uncomfortable in *slight cold stress* conditions (0 °C < UTCI < 9 °C).

**Table 4** Meteorological conditions during the field surveys

| Season | Meteorological variable | Minimum | Maximum | Median | Mean | Standard deviation |
|---|---|---|---|---|---|---|
| Summer | $T_a$ (°C) | 27.9 | 39.3 | 32.4 | 32.3 | 1.6 |
| | RH (%) | 22 | 50 | 33 | 34 | 7 |
| | $v_a$ (ms$^{-1}$) | 0.4 | 9.3 | 1.0 | 1.1 | 0.8 |
| | $T_{gl}$ (°C) | 29.4 | 43.8 | 34.8 | 35.5 | 3.3 |
| | $T_{mrt}$ (°C) | 30.1 | 77.2 | 43.2 | 46.0 | 10.2 |
| | UTCI (°C) | 28.9 | 43.8 | 34.3 | 34.9 | 3.4 |
| Autumn | $T_a$ (°C) | 16.8 | 27.9 | 22.5 | 22.0 | 2.6 |
| | RH (%) | 36 | 73 | 56 | 55 | 7 |
| | $v_a$ (ms$^{-1}$) | 0.4 | 2.6 | 0.7 | 0.8 | 0.4 |
| | $T_{gl}$ (°C) | 17.5 | 35.7 | 25.3 | 26.7 | 5.4 |
| | $T_{mrt}$ (°C) | 17.8 | 63.3 | 38.4 | 39.6 | 13.4 |
| | UTCI (°C) | 15.4 | 34.2 | 25.6 | 26.5 | 5.2 |
| Winter | $T_a$ (°C) | 6.9 | 20.4 | 12.9 | 12.1 | 3.7 |
| | RH (%) | 23 | 79 | 49 | 51 | 14 |
| | $v_a$ (ms$^{-1}$) | 0.4 | 3.5 | 0.8 | 1.0 | 0.5 |
| | $T_{gl}$ (°C) | 7.9 | 27.7 | 16.6 | 16.0 | 5.5 |
| | $T_{mrt}$ (°C) | 10.3 | 69.5 | 23.2 | 28.7 | 13.5 |
| | UTCI (°C) | 4.9 | 27.3 | 16.8 | 16.1 | 6.1 |

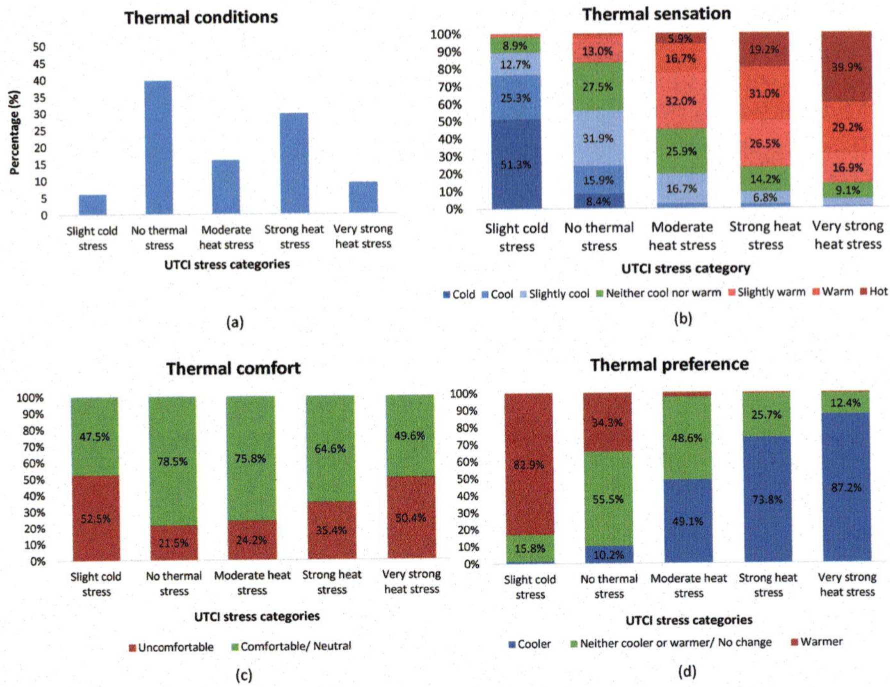


**Fig. 1** **a** Assessment of the thermal environment during the interviews according to UTCI thermal stress categories, **b** thermal sensation ($p < 0.001$), **c** thermal comfort ($p < 0.001$) and **d** thermal preference ($p < 0.001$) reported by pedestrians for each UTCI stress category

## *3.3 Heat-Related Symptoms*

Heat-related symptoms were reported by 405 (14.8%) participants. Most of them reported that they were experiencing *exhaustion* (45.4%) at the time of the interview. About 31.9% reported experiencing *headache*, 19.5% *dizziness* and 17.8% *breathing difficulties*. Pedestrians were more likely to report heat-related symptoms in summer (20.4%) than autumn (13.8%) or winter (9.1%) (Fig. 2a). *Exhaustion* was more frequently reported in summer (49.6%) and *headache* in autumn (36.4%) and winter (36.9%) (Fig. 2b).

Mean UTCI was higher for pedestrians reporting heat-related symptoms than for those who did not (29.3 ± 8.8 °C compared to 25.9 ± 9.4 °C, $p < 0.003$). The percentage of pedestrians reporting symptoms increased with UTCI heat stress categories (Fig. 3a), i.e., from *moderate stress* to *very strong stress*. Indications of increased reporting in UTCI *moderate* (35.1%), *strong* (50%), *very strong heat stress* (44.3%) categories was found for *exhaustion* (Fig. 3b).

Females reported more often heat-related symptoms than males (18.4% versus 11.6%; $p < 0.001$) as did participants between 25 and 35 years old (20.3%; $p < 0.001$) compared to the remaining age groups. Similarly, non-local residents (16.3%

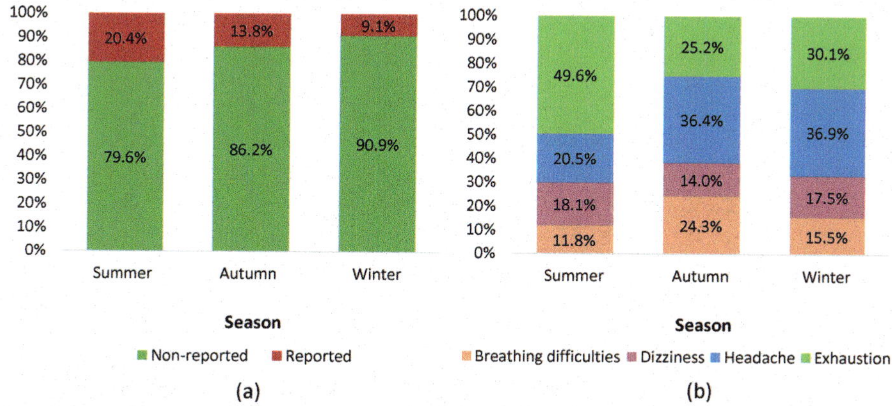


**Fig. 2** Seasonal frequency distribution of heat-related health symptoms **a** collapsed into two categories (reporting and non-reporting heat-related symptoms; $p < 0.001$) and **b** for each reported symptom ($p < 0.001$)

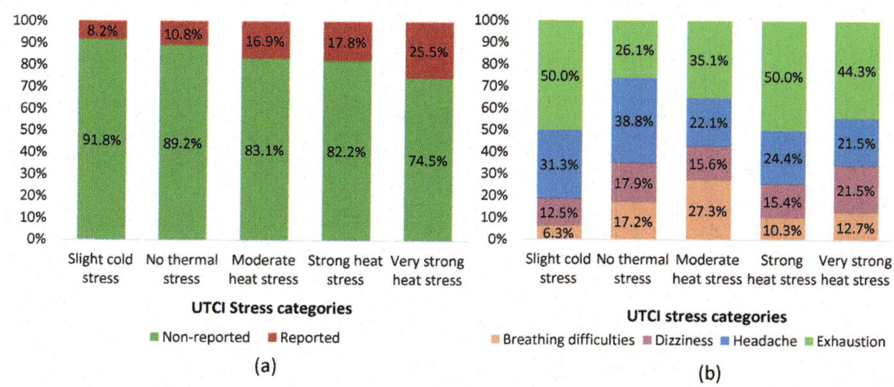


**Fig. 3** Frequency distribution of heat-related health symptoms against UTCI stress categories **a** collapsed into two categories (reporting and non-reporting heat-related symptoms; $p < 0.001$) and **b** for each reported symptom ($p < 0.001$)

versus 11.8%; $p = 0.001$) and participants who had been indoors, in a car or public transport the last half hour before the interview (16.1% versus 12.5% outdoors; $p = 0.06$) or with medical history of cardiovascular or respiratory disease (24.8% versus 13.5%; $p < 0.001$) were more likely to report symptoms. Moreover, the prevalence of heat-related symptoms was higher among participants visiting the monitoring site for over 1 h (19.1%; $p = 0.001$), reporting thermal sensation *hot* (28.7% $p < 0.001$), thermal comfort *uncomfortable* (22.9% versus 11.3%; $p < 0.001$) or thermal preference *cooler* (22.2%; $p < 0.001$).

Binary logistic regression (Table 5) showed that for 1 °C increase of the UTCI the odds of reporting heat-related symptoms increases by 4% (adjusted OR (aOR) = 1.04, p-value < 0.001). The odds were increased to 6% for the pedestrians participated

**Table 5** Multivariate logistic regression for the association of heat-related symptoms (breathing difficulties, dizziness, headache, exhaustion) with UTCI and pedestrians' attributes. Only variables found to be statistically significant were considered

| Variable | | All seasons | | | | Summer | | | |
|---|---|---|---|---|---|---|---|---|---|
| | | aOR | Lower confidence interval | Upper confidence interval | P-value | aOR | Lower confidence interval | Upper confidence interval | P-value |
| UTCI | | 1.04 | 1.03 | 1.05 | <0.001 | 1.06 | 1.01 | 1.11 | 0.012 |
| Biological sex[a] | Males | 0.60 | 0.48 | 0.75 | <0.001 | 0.72 | 0.53 | 0.98 | 0.037 |
| Place of residence[b] | Non-local | 1.35 | 1.05 | 1.74 | 0.020 | 1.34 | 0.95 | 1.90 | 0.097 |
| Medical history[c] | None | 0.46 | 0.34 | 0.62 | < 0.001 | 0.51 | 0.32 | 0.82 | 0.005 |
| Exposure duration[d] | <15 min | 0.66 | 0.51 | 0.87 | 0.003 | 0.56 | 0.39 | 0.80 | 0.001 |
| | 15 min to 1 h | 0.67 | 0.46 | 0.93 | 0.023 | 0.61 | 0.39 | 0.97 | 0.031 |
| Exposure history[e] | Indoors-car-public transport | 1.33 | 1.04 | 1.70 | 0.025 | 1.32 | 0.93 | 1.86 | 0.117 |

Reference category [a]Females, [b]local residence, [c]medical history of cardiovascular or respiratory disease, [d]more than 1 h, [e]outdoors in summer field surveys (aOR = 1.06, p-value < 0.012). Biological sex and subjective attributes were also associated with reporting heat-related symptoms. Females were 40% more likely to report heat-related symptoms than males (aOR = 0.60, p-value < 0.001). Moreover, non-local residents (35%, aOR = 1.35, p-value = 0.020), pedestrians with a medical history of cardiovascular or respiratory (54%, aOR = 0.46, p-value < 0.001) or who had been indoors, in a car or public transport the last half hour before the interview (33%, aOR = 1.33, p-value = 0.025), and who had been at the monitoring site over 1 h (about 33%, aOR: 0.66/0.67; p-value ≤ 0.023) were more likely to report heat-related symptoms. The Hosmer–Lemeshow goodness-of-fit test indicated that regression models adequately fitted the data (p-value = 0.189 for all season and p-value = 0.113 for summer).

Hierarchical cluster analysis showed that the respondents can be reasonably divided into two groups, as follows: those who were experiencing and those who were not experiencing heat-related symptoms, in terms of the UTCI. The UTCI cut-off point was found 28.5 °C (*moderate heat stress*) for all seasons and 38.9 °C (*very strong heat stress*) for summer.

## 4 Discussion

This study examined the association of mild symptoms of heat-related illness with the thermal environment, assessed in terms of the UTCI while controlling for subjective attributes. *Breathing difficulties, dizziness, headache* or *exhaustion* were reported by 14.8% of the pedestrians who participated in the field surveys, which were conducted under non-extreme thermal conditions. Heat-related symptoms were associated with UTCI values. Females, non-local residents, people with history of a cardiovascular/respiratory disease, those recently exposed to an indoor environment or being in a vehicle, participants visiting the monitoring site for more than one hour or reporting thermal sensation as *hot,* experienced more often heat-related symptoms.

The findings of this chapter confirmed the relationship of UTCI and *hot* thermal sensation with health effects prevalence. Results on the confounding subjective attributes were in agreement with earlier findings in the literature. A similar increase of the likelihood of manifestation of heat-related symptoms (4%) for increases in the PET, another widely used thermal stress index (Potchter et al. 2018) was found in a different Mediterranean setting (Pantavou et al. 2020). The category *moderate heat stress* of the UTCI was also identified as the transition level for experiencing heat-related symptoms in a previous study in Athens, Greece (Pantavou et al. 2016). Females vulnerability to heat has already been suggested in both heat-related mortality and morbidity studies (Basu 2009; Ye et al. 2012; Pantavou et al. 2016, 2020; Näyhä et al. 2017). Differences in health-effects between males and females could be attributed to differences in physical characteristics, physiological variables of temperature regulation, aerobic fitness (Gagnon and Kenny 2012) as well as to occupational exposure (Fouillet et al. 2006). Nevertheless, females' sensitivity to heat was also identified among farmworkers in Florida (Mutic et al. 2018). Whereas ratio of body surface to body mass, subcutaneous fat thickness, sudomotor activity and sweat production rate have been found to vary between males and females (Seidell et al. 1988; Kaciuba-Uscilko and Grucza 2001; Gagnon and Kenny 2012; Dehghan et al. 2013). Individuals with cardiovascular health problems demonstrate reduced thermal tolerance to heat stress (Kenney et al. 2014) since thermoregulatory responses in skin blood flow and sweating are accompanied by cardiovascular adjustments.

In this study, the sample of vulnerable population, i.e., children, elderly people and pregnant women, was relatively small while a response bias could have potentially interfered as participants self-reported heat-related symptoms. Still, the analysis involved a relatively large number of responses, representative of the active part of the population involved in daily activities. Moreover, the analysis focused on heat-related symptoms that were feasible to self-evaluate. This study is one of a few examining the heat-related symptoms of pedestrians. The findings suggest that the UTCI is consistent with changes in the reported thermal sensation, comfort and preference enhancing its reliability and usefulness as a thermal environment assessment tool. The UTCI *moderate heat stress* category could be considered as a threshold of experiencing heat-related symptoms for pedestrians.

## 5 Conclusions

Heat is a major health risk of increasing research interest due to climate change and the rising global ambient temperatures. Public health guidance and recommendations should be enhanced with thorough assessments of the most current evidence and approaches. In this study, the UTCI was examined as a risk indicator of experiencing heat-related symptoms under non-extreme thermal conditions, and was found to be a credible candidate for the assessment of heat-related health effects. A threshold for experiencing heat-related symptoms could be considered the category of *moderate heat stress.*

Future research including extreme thermal conditions and concurrent air quality assessment could be beneficial for extending the results of this study. Furthermore, it would be interesting to examine the possible impact of heat exposure on the health condition of vulnerable populations. The results of this study are part of the UBiPlan project that aims to examine the spatial distribution of thermal burden and public health impact in order to improve design guidelines and to identify spatial patterns associated with optimized thermal conditions in outdoor urban spaces.

**Acknowledgements** I would like to thank Dr. Spyridon Lykoudis for his helpful comments and Emmanouil Melas for his support in the field surveys.

**Funding** This project has received funding from the Hellenic Foundation for Research and Innovation (HFRI) and the General Secretariat for Research and Technology (GSRT), under grant agreement No. 146.

## References

Avalos LA, Chen H, Li DK, Basu R (2017) The impact of high apparent temperature on spontaneous preterm delivery: a case-crossover study. Environ Health 16(1):1–13

Basu R, Samet JM (2002) Relation between elevated ambient temperature and mortality: a review of the epidemiologic evidence. Epidemiol Rev 24(2):190–202

Basu R (2009) High ambient temperature and mortality: a review of epidemiologic studies from 2001 to 2008. Environ Health 8(1):1–13

Błażejczyk K, Błażejczyk M (2010) BioKlima© 2.6, Program komputerowy. http://www.igipz.pan.pl/Bioklima-zgik.html

Błażejczyk K, Jendritzky G, Bröde P, Fiala D, Havenith G, Epstein Y, Kampmann B (2013) An introduction to the universal thermal climate index (UTCI). Geogr Pol 86(1):5–10

Bröde P, Fiala D, Błażejczyk K, Holmér I, Jendritzky G, Kampmann B, Havenith G (2012) Deriving the operational procedure for the Universal Thermal Climate Index (UTCI). Int J Biometeorol 56(3):481–494

CDC (2020) Heat stress—heat related illness. https://www.cdc.gov/niosh/topics/heatstress/heatrelillness.html. Accessed 13 Mar 2020

Cui J, Sinoway LI (2014) Cardiovascular responses to heat stress in chronic heart failure. Curr Heart Fail Rep 11(2):139–145

Dehghan H, Mortazavi SB, Jafari MJ, Maracy MR (2013) Cardiac strain between normal weight and overweight workers in hot/humid weather in the Persian Gulf. Int J Prev Med 4(10):1147

Di Napoli C, Pappenberger F, Cloke HL (2018) Assessing heat-related health risk in Europe via the Universal Thermal Climate Index (UTCI). Int J Biometeorol 62(7):1155–1165

ELSTAT (2020) Demographic characteristics/2011. https://www.statistics.gr/en/statistics/pop. Accessed 28 Sep 2020

Fouillet A, Rey G, Laurent F, Pavillon G, Bellec S, Guihenneuc-Jouyaux C, Hémon D (2006) Excess mortality related to the August 2003 heat wave in France. Int Arch Occup Environ Health 80(1):16–24

Gagnon D, Kenny GP (2012) Does sex have an independent effect on thermoeffector responses during exercise in the heat? J Physiol 590(23):5963–5973

Guyton AC, Hall JE (2006) Textbook of medical physiology, 11th edn. WB Sauders Co., Philadelphia

Hajat S, Kosatky T (2010) Heat-related mortality: a review and exploration of heterogeneity. J Epidemiol Community Health 64(9):753–760

IPCC (2018) Global Warming of 1.5°C.https://www.ipcc.ch/sr15/. Accessed 15 Oct 2020

ISO 10551 (1995) Ergonomics of the physical environment—subjective judgement scales for assessing physical environments. Switzerland, Geneva

ISO 7726 (2001) Ergonomics of the thermal environment—instruments for measuring physical quantities. Switzerland, Geneva

ISO 8996 (2004) Ergonomics of the thermal environment—determination of metabolic rate. Switzerland, Geneva

ISO 9920 (2007) Ergonomics—estimation of the thermal characteristics of a clothing ensemble. Switzerland, Geneva

Kaciuba-Uscilko H, Grucza R (2001) Gender differences in thermoregulation. Curr Opin Clin Nutr Metab Care 4(6):533–536

Kenney WL, Craighead DH, Alexander LM (2014) Heat waves, aging, and human cardiovascular health. Med Sci Sports Exerc 46(10):1891

Kottek M, Grieser J, Beck C et al (2006) World Map of the Köppen-Geiger climate classification updated. Meteorol Z 15:259–263

Kuchcik M (2020) Mortality and thermal environment (UTCI) in Poland—long-term, multi-city study. Int J Biometeorol 1–13

Lipman GS, Eifling KP, Ellis MA, Gaudio FG, Otten EM, Grissom CK (2013) Wilderness Medical Society practice guidelines for the prevention and treatment of heat-related illness. Wilderness Environ Med 24(4):351–361
Mora C, Counsell CW, Bielecki CR, Louis LV (2017) Twenty-seven ways a heat wave can kill you: deadly heat in the era of climate change. Circ Cardiovasc Qual Outcomes 10(11):e004233
Murakami S, Miyatake N, Sakano N (2012) Changes in air temperature and its relation to ambulance transports due to heat stroke in all 47 prefectures of Japan. J Prev Med Public Health 45(5):309
Mutic AD, Mix JM, Elon L, Mutic NJ, Economos J, Flocks J, McCauley LA (2018) Classification of heat-related illness symptoms among Florida farmworkers. J Nurs Scholarsh 50(1):74–82
Nastos PT, Matzarakis A (2012) The effect of air temperature and human thermal indices on mortality in Athens. Greece Theor Appl Climatol 108(3):591–599
Näyhä S, Rintamäki H, Donaldson G, Hassi J, Jousilahti P, Laatikainen T, Ikäheimo TM (2017) The prevalence of heat-related cardiorespiratory symptoms: the vulnerable groups identified from the National FINRISK 2007 Study. Int J Biometeorol 61(4):657–668
Nikolopoulou M, Lykoudis S, Kikira M (2004) Thermal comfort models for open urban spaces. In: Nikolopoulou M (ed) Designing open spaces in the urban environment: a bioclimatic approach. Centre Renewable Energy Sources (C.R.E.S). http://alpha.cres.gr/ruros/desgui.htm. Accessed 15 Oct 2020
NOA-National Observatory of Athens (2020) Climatological bulletin. http://www.meteo.noa.gr/ENG/iersd_climatological.htm. Accessed 27 Sep 2020
Pantavou K, Theoharatos G, Nikolopoulos G, Katavoutas G, Asimakopoulos D (2008) Evaluation of thermal discomfort in Athens territory and its effect on the daily number of recorded patients at hospitals' emergency rooms. Int J Biometeorol 52(8):773–778
Pantavou K, Theoharatos G, Santamouris M, Asimakopoulos D (2013) Outdoor thermal sensation of pedestrians in a Mediterranean climate and a comparison with UTCI. Build Environ 66:82–95
Pantavou KG, Lykoudis SP, Nikolopoulos GK (2016) Milder form of heat-related symptoms and thermal sensation: a study in a Mediterranean climate. Int J Biometeorol 60(6):917–929
Pantavou K, Giallouros G, Lykoudis S, Markozannes G, Constantinou E, Panagi A, Nikolopoulos GK (2020) Impact of heat exposure on health during a warm period in Cyprus. Euro-Mediterr J Environ Integr 5:1–12
Potchter O, Cohen P, Lin TP, Matzarakis A (2018) Outdoor human thermal perception in various climates: a comprehensive review of approaches, methods and quantification. Sci Total Environ 631:390–406
Seidell JC, Oosterlee A, Deurenberg P, Hautvast JG, Ruijs JH (1988) Abdominal fat depots measured with computed tomography: effects of degree of obesity, sex, and age. Eur J Clin Nutr 42(9):805–815
Sherbakov T, Malig B, Guirguis K, Gershunov A, Basu R (2018) Ambient temperature and added heat wave effects on hospitalizations in California from 1999 to 2009. Environ Res 160:83–90
Son JY, Liu JC, Bell ML (2019) Temperature-related mortality: a systematic review and investigation of effect modifiers. Environ Res Lett 14(7):073004
Song X, Wang S, Hu Y, Yue M, Zhang T, Liu Y, Shang K (2017) Impact of ambient temperature on morbidity and mortality: an overview of reviews. Sci Total Environ 586:241–254
Sugg MM, Konrad CE, Fuhrmann CM (2016) Relationships between maximum temperature and heat-related illness across North Carolina, USA. Int J Biometeorol 60(5):663–675
Sun Z, Chen C, Xu D, Li T (2018) Effects of ambient temperature on myocardial infarction: a systematic review and meta-analysis. Environ Pollut 241:1106–1114
Tseliou A, Tsiros IX, Nikolopoulou M, Papadopoulos G (2016) Outdoor thermal sensation in a Mediterranean climate (Athens): the effect of selected microclimatic parameters. Archit Sci Rev 59(3):190–202
UBiPlan (2020) Urban biometeorology and planning. https://ubiplanproject.wordpress.com/. Accessed 23 Feb 2020

Urban A, Kyselý J (2014) Comparison of UTCI with other thermal indices in the assessment of heat and cold effects on cardiovascular mortality in the Czech Republic. Int J Environ Res Public Health 11(1):952–967

Vaneckova P, Neville G, Tippett V, Aitken P, FitzGerald G, Tong S (2011) Do biometeorological indices improve modeling outcomes of heat-related mortality? J Appl Meteorol Climatol 50(6):1165–1176

Wang Y, Bobb JF, Papi B, Wang Y, Kosheleva A, Di Q, Dominici F (2016) Heat stroke admissions during heat waves in 1,916 US counties for the period from 1999 to 2010 and their effect modifiers. Environ Health 15(1):1–9

WHO-World Health Organization (2020) Information and public health advice: heat and health. https://www.who.int/globalchange/publications/heat-and-health/en/. Accessed 21 Sep 2020

Xu Y, Dadvand P, Barrera-Gómez J, Sartini C, Marí-Dell'Olmo M, Borrell C, Basagaña X (2013) Differences on the effect of heat waves on mortality by sociodemographic and urban landscape characteristics. J Epidemiol Commun Health 67(6):519–525

Xu Z, Hu W, Su H, Turner LR, Ye X, Wang J, Tong S (2014) Extreme temperatures and paediatric emergency department admissions. J Epidemiol Commun Health 68(4):304–311

Ye X, Wolff R, Yu W, Vaneckova P, Pan X, Tong S (2012) Ambient temperature and morbidity: a review of epidemiological evidence. Environ Health Perspect 120(1):19–28

# Chapter 8
# Mapping the Universal Thermal Climate Index (In Different Scales)

**Krzysztof Błażejczyk and Anna Błażejczyk**

**Abstract** One of the important aims of climatic and bioclimatic studies is to define spatial differentiation of particular climate elements and/or bioclimatic indicators. Several approaches are used for this purpose. Isoline maps aim to create spatial variability of any climate variable based on data points from a network of meteorological stations. The second commonly used visualisation method is based on typological maps, which assume that values observed in particular locations represent wider areas with similar properties of natural and socio-economic environment. The combination of isoline and typological approach are raster maps. In these maps, the space is divided into several regular geometrical fields (raster). For each raster, meteorological variables and bioclimatic indices are defined. The aim of the chapter is to present methods of cartographical representation of the Universal Thermal Climate Index in different spatial scales (global, regional, local).

**Keywords** Isoline maps · Typological maps · Raster maps · Spatial distribution of the UTCI · Cartographical representation of the UTCI

## 1 Introduction

One of the main aims of climatic and bioclimatic studies is to define geographical patterns for particular climate elements (e.g. air temperature, precipitation, solar radiation) and/or bioclimatic indicators. In the past, the spatial variability of these elements was mainly based on location-specific research using data from a network of meteorological stations. Observed weather station data were commonly used to create *isoline* maps, i.e. maps where lines connect points with identical values inside

K. Błażejczyk (✉)
Institute of Geography and Spatial Organization of the Polish Academy of Sciences, Climate Impacts Research Laboratory, Twarda 51/55, 00-818, Warsaw, Poland
e-mail: k.blaz@twarda.pan.pl

A. Błażejczyk
Laboratory of Bioclimatology and Environmental Ergonomics, Warsaw, Poland
e-mail: anna@blazejczyk.eu



E. L. Krüger (ed.), *Applications of the Universal Thermal Climate Index UTCI in Biometeorology*, Biometeorology 4,
https://doi.org/10.1007/978-3-030-76716-7_8

a continuum and show any spatial gradient of the presented variable. This method is still widely used in synoptic research and for presenting climate/bioclimate maps over an extended region, e.g. countries and continents (an overview of warning systems used in various national weather services is shown by Di Napoli and colleagues in this chapter: UTCI as an operational forecasting tool of human biometeorological conditions in Europe). The isolines can be generated manually (as done in the past) or automatically (more recent years) as an effect of mathematical interpolation of observed values.

The second commonly used visualisation method of spatial distribution for climate data are *typological* maps. This method assumes that values observed at specific points are representative of a wider area with similar natural and socio-economic properties. As an example, we can mention maps with climate types (e.g. Koeppen–Geiger classification, Peel et al. 2007) or bioclimate maps (e.g. Mieczkowski's TCI maps, Mieczkowski 1985). Typological maps are applied in different spatial scales, from global to local (e.g. topoclimatic maps, Oke 1987; Geiger 1969).

The combination between the isoline and the typological approach gives us raster maps. In raster maps, the space is divided into several regular geometrical fields (usually a matrix of grid cells). For each raster, meteorological variables are presented, which can be later used for the calculation of bioclimatic indices. There are different ways to define climate elements over a raster, e.g. kriging interpolation (Ustrnul and Czekierda 2003) or by dividing typological pictures into raster modes (Kozłowska-Szczęsna and Błażejczyk 1996).

These methods of cartographical representation for climate and bioclimate have been recently used to represent the spatial variability of human thermal stress as defined by the Universal Thermal Climate Index (UTCI). Some examples of different approaches of UTCI mapping are presented in more detail in this chapter.

## 2 Global UTCI Maps

The first attempt to present a global distribution of UTCI characteristics (namely annual frequency cold and heat stress conditions) was prepared for the final workshop of COST 730 action in Geneva in 2009 by G. Jendritzky, B. Tinz and O. Baddour. The authors have applied in this purpose ECMWF-reanalysis (ERA15/ERA40) data with spatial resolution of about 1.1°, i.e. ca. 120 km.

A few years later, Pappenberger et al. (2015) have published maps of cold stress (UTCI < −13 °C) and heat stress (UTCI > 32 °C) frequency in January and July for the period 2009–2012. The UTCI was calculated globally using meteorological input from the European Centre for Medium-Range Weather Forecasts (ECMWF). The horizontal resolution of the model was approximately 80 × 80 km and the time step was 3 h.

This research was continued within the framework of the latest global climate reanalysis dataset created by the ECMWF. Meteorological variables were used to calculate a worldwide historical gridded UTCI dataset, called ERA5-HEAT, thereby drawing UTCI maps. The automated routine implemented to generate ERA5-HEAT is described in detail by Di Napoli et al. (2020a). It generates UTCI time series as a sequence of spatial grids at 0.25° resolution (approximately 28 km at the equator) and 1 h intervals (Di Napoli et al. 2020b). These are freely available from the Climate Data Store, CDS, developed as part of the Copernicus Climate Change Service C3S implemented by ECMWF (CDS 2020).

## 3 Regional UTCI Maps

Over the last decade, several regional maps of UTCI distribution were done for various regions in Europe, Asia and Africa. From the cartographical point of view, the maps represent three different approaches: raster, isoline or cartogram (Nusrat and Kobourov 2016). The decision on which method is most effective and informative depends mostly on the data sources available and on the purpose of the study.

### *3.1 Raster Maps*

The global ECMWF database of meteorological variables is applied in regional UTCI research. Figure 1 is an example of that. It shows a map of heat stress during the heatwave of June 2019 in Europe. The map presents a distribution of particular thermal stress categories based on averaged UTCI values for the period between June 24 and July 2, 2019. It can be underlined that in the Mediterranean Basin, moderate and even strong heat stress conditions were observed not only over land, but also over sea surface. Analytic maps of June 2019 heatwave can also be found at the Copernicus webpage (ESOTC 2019).

UTCI records from ECMWF reanalysis data were recently applied in several regional bioclimate studies. Di Napoli et al. (2018) have used them in studies of regional patterns of heat-related mortality in Europe. The research covers the period 1979–2016 with a time resolution of 3 h. The spatial pictures identify Mediterranean and Black Sea regions as the most exposed to strong heat stress during midday hours.

Ge et al. (2017) have applied reanalysis data to calculate UTCI over China. Their study refers to tourism. The generated map showed the south-eastern part of the country with the greatest exposure to high UTCI values and increased frequency of heat stress conditions. On the other hand, the Tibetan Plateau is the region with the lowest UTCI and the greatest occurrence of cold stress. Similar results were also obtained for specific Chinese regions by Kong et al. (2017) and Wu et al. (2019).

ECMWF reanalysis data were also applied in the analysis of spatial and temporal variations of UTCI over the China-Pakistan Economic Corridor (CPEC) from 1979 to 2018 by Zeng et al. (2020). The authors found inter-regional differences in UTCI time trends and indicated a potential applicability of UTCI in regional planning within CPEC.

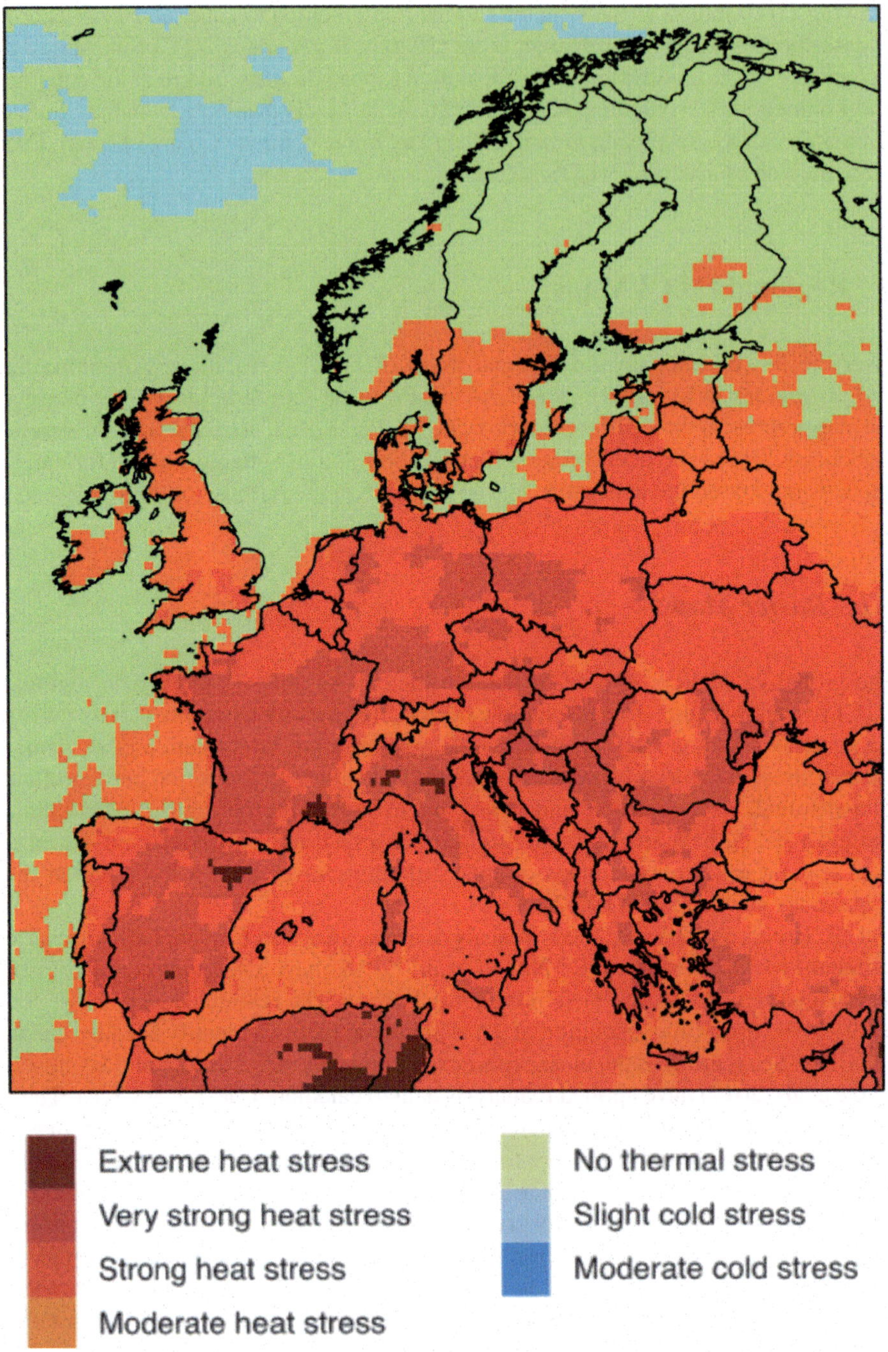


**Fig. 1** Distribution of particular thermal stress categories over Europe during the June 2019 heat-wave occurred between June 24 and July 2, 2019 (*Source* https://www.ecmwf.int/en/newsletter/162/news/heat-health-hazard-index-based-ecmwf-data)

In Europe, Brecht (2019) has applied a set of Global Climate Models to analyze the spatial variability of UTCI over Germany. In his representation, a gradual increase of UTCI was observed from the northern to the southern part of the country.

## *3.2 Isoline Maps*

Regional UTCI studies use data not only from reanalysis and climate models but also from networks of meteorological stations. Cartographical representations of the spatial distribution of UTCI by the isoline method were published for Russia, Benin and Poland.

Vinogradova (2020) has calculated the UTCI for 500 meteorological stations in Russia for the period 2001–2015. Daily meteorological data were obtained from the archive of the All-Russian Research Institute of Hydrometeorological Information – World Data Center. UTCI values were calculated for daytime (at 3 pm) and night time (at 3 am), as representative of the warmest and coolest hours of the day, respectively. Because of its great geographical extent, Russia's bioclimate presents a wide range of UTCI values both in winter and summer.

Boko et al. (2014) have produced maps of mean monthly distribution of UTCI values over the territory of Benin (equatorial Africa). They have used meteorological data from 11 stations for the period 1971–2010. They noticed the greatest heat stress in the northernmost stations (at the southern edge of the Sahel region) during summer months.

A spatial distribution of multiannual UTCI values was also accomplished for Poland by Tomczyk and Owczarek (2020), and by Kuchcik et al. (2021). Tomczyk and Owczarek (2020) used daily meteorological data from 40 stations for the period 1966–2015. The UTCI was calculated at 12 UTC to consider the warmest part of the day. Their analysis shows a spatial distribution of the number of days per annum with strong and very strong heat stress in summer and in July, in particular. Kuchcik et al. (2021) used meteorological data at 12 UTC from the period 1951–2018 for 22 synoptic stations. The analysis concentrates both on the spatial distribution of the UTCI as well as on its time trends. Figure 2 shows the spatial distribution of the highest, mean, and lowest annual UTCI values. The UTCI distribution represents well the spatial climatic variability of Poland.

A novel approach was recently proposed by Roshan et al. (2018). The authors have applied daily meteorological data for 155 Iranian synoptic stations spanning the period 1995–2014. The purpose of their research was to cluster stations in terms of their bioclimatic conditions' similarity as embodied by UTCI. Daily UTCI outputs were used for clustering using the Euclidean Distance by the Ward method (Anderberg 1973). Their study revealed that the division of the stations into 7 clusters with different UTCI patterns is the most appropriate for the territory of Iran.

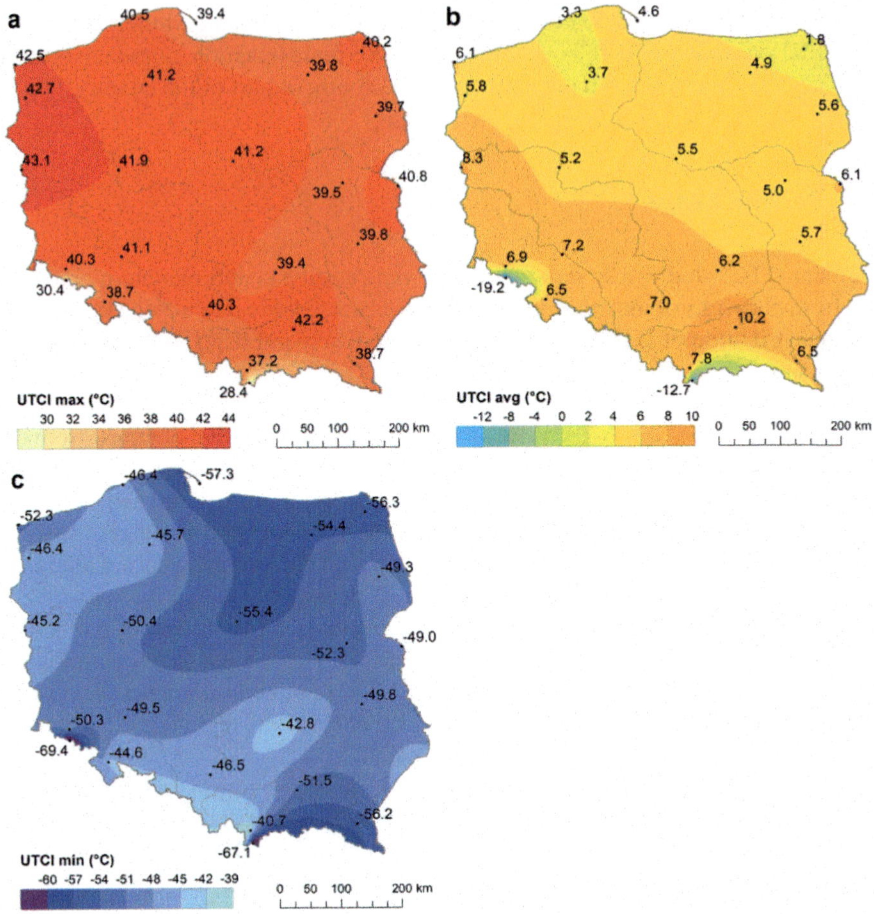


**Fig. 2** The highest **a**, yearly average **b**, and the lowest **c** values of UTCI in Poland, 1951–2018

## *3.3 Cartogram Maps*

Another cartographical approach for the representation of UTCI spatial variability was introduced by Błażejczyk et al. (2020a). In their work, the influence of air circulation on UTCI spatial differences over the northern Carpathians (border area between Poland, Slovakia and Ukraine) was studied. Daily meteorological data for 12 UTC were calculated from 21 stations for the period 1986–2015. In order to represent the spatial distribution of UTCI, cartogram maps were employed. Results show that air circulation strongly impact UTCI characteristics (obtained values and the frequency of days with cold and heat stress). Distinct differences between southward and northward exposed stations and between western and eastern parts of the studied region were also found (Fig. 3).

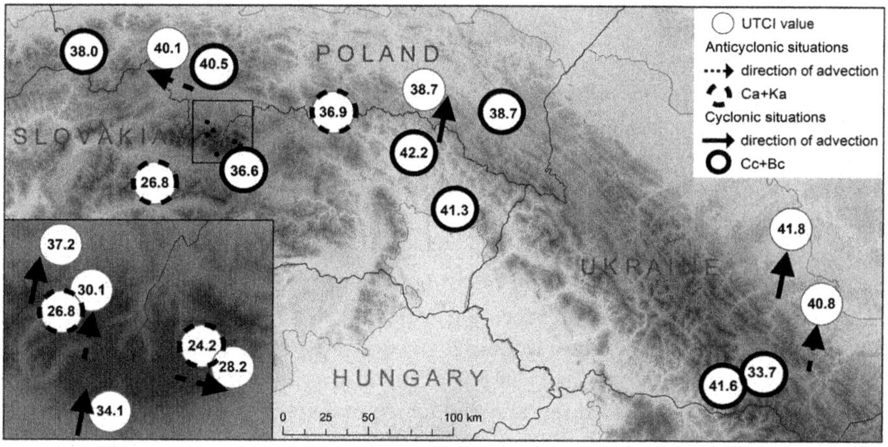


**Fig. 3** Diagram map of highest UTCI values at particular stations and atmospheric circulation registered in northern Carpathians region (Poland-Slovakia-Ukraine), 1986–2015

# 4 Local-Scale UTCI Maps

Global and regional databases are very useful when addressing the distribution of UTCI characteristics (mean, maximum, minimum values, frequency of various UTCI thermal stress categories) over large territories. However, climatic and bioclimatic analyses generally refer to small areas such as cities and neighborhoods. In regional climate research we pay attention to general features of radiation, air circulation and geographical factors responsible for large-scale bioclimatic differentiation. However, in local-scale studies we concentrate on specific features of the environment (e.g. relief, land use, physical properties of surface, plant cover) which can potentially modify overall climatic conditions. In local climate research, deviations of particular meteorological variables from standard weather conditions observed at the nearest meteorological station are usually the focus of attention (Kunert and Błażejczyk 2011).

Several topoclimatic field studies carried out in different geographical regions provide information of the general impacts of particular geographical elements on microclimatic conditions (Geiger 1969; Oke 1987; Słowińska et al. 2010; Erell et al. 2011; Matuschek and Matzarakis 2011; Błażejczyk and Błażejczyk 2014; Provençal et al. 2016; Trimmel et al. 2019). Results lead to the general conclusion that the patterns of change in microclimate are the same all over the world but their values can differ from region to region. The following scheme of climate modifications in local scale, caused by various environmental factors, can be drawn:

*Relief*: incoming solar radiation is significantly greater at equator-facing slopes (south in the northern and north in the southern hemisphere) than at other slopes,—sun-facing slopes are warmer than others,—air temperature decreases and incoming solar radiation rises according to increases in altitude,—bottoms of valleys and basins are areas of so called "cold lakes air",—tops and elevated ridges are exposed to stronger winds than valleys.

*Forests and parks*: incoming solar radiation under canopies can be reduced up to 90%,—maximum air temperatures and their amplitudes are significantly lower during daytime than in open areas,—air humidity is higher by 5–10%,-wind speed can be reduced by 40–90%,—on sunny days, heat stress can be reduced but in days with low temperature cold stress is likely to be enhanced.

*Ground moisture*: of sandy and dry areas warms rapidly during daytime and cools fast in night time hours generating large daily temperature fluctuations (a similar effect is produced by artificial surfaces, e.g. asphalt, concrete pavements, in urbanized areas). An opposite temperature rhythm is observed over wet areas.

*Land use*: coastal areas of lakes, large rivers and seas are cooler, windier and more humid (this refers to water bodies and their surroundings up to dozens of meters),—rural areas modify meteorological variables depending on their structure (e.g. wet meadows versus dry cultivated fields).

*Urban areas*: great microclimatic modifications are observed in *urbanized areas*. Such areas strongly influence wind flows (direction and velocity), radiation field (shaded and exposed to solar radiation clusters) and ambient temperature (urban heat island, intra-urban differences due to morphology, parks and water bodies). These modifications generate considerable spatial variability in microclimatic and micro-bioclimatic conditions.

The general relations mentioned above can reach different ranges in various climatic zones. Results from urban climate studies have been applied to develop models of spatial differentiation of urban areas. Park et al. (2014) have used a combination of two models: HURES (**H**uman-**U**rban **R**adiation **E**xchange **S**imulation) and ENVI-met 3.1 (http://www.envi-met.com) to simulate the UTCI within downtown areas in Nanaimo (BC, Canada) and Changwon (Republic of Korea). Results from those simulations show distinct differences for the UTCI in studied areas and over consecutive hours. The ENVI-met model was also applied for the calculations of the UTCI in Kashgar city, China (Wu et al. 2019). The representation of other models using high-resolution urban UTCI modeling is shown by Geletič and colleagues in the following chapter of this book (In this chapter: Application of the UTCI in high-resolution urban climate modeling techniques).

The most complex topoclimatic database refers to the mid-latitude climates of the northern hemisphere. Several experimental, topoclimatic studies were carried out in different types of landscapes, both natural (forests, meadows, fields, wetlands, slopes, ridges etc.) and man-made (urban, rural, industrial) in Poland, France, southern Scandinavia, western Ukraine, Bulgaria, European Russia, United States, Mongolia, Turkmenistan, among others. Reviews of their results are presented and discussed in Błażejczyk (1990), Błażejczyk et al. (1992), Grzybowski (1990), Kuchcik (2001), Paszyński et al. (1999).

The obtained results from a wide set of topoclimatic research can serve as a basis for defining coefficients of changes in solar radiation (zr), wind speed (zv), air temperature (zt) and relative humidity (zf) in different types of local landscapes (Table 1). While the experimental research was carried out during warm seasons,

**Table 1** Coefficients of change in global solar radiation (zr), air temperature (zt), wind speed (zv) and relative humidity (zf) in selected types of landscape

| Type of landscape | | Coefficients of change | | | |
|---|---|---|---|---|---|
| | | zr | zt | zv | zf |
| Relief features | Plains | 1.00 | 1.00 | 1.00 | 1.00 |
| | Tops and upper parts of hills | 1.00 | 1.00 | 1.40 | 1.00 |
| | Valley bottoms (H—is valley depth, W—is valley width): | | | | |
| | >20 m H and <200 m W | 0.95 | 0.85 | 0.70 | 1.10 |
| | <20 m H and <200 m W | 1.05 | 0.90 | 0.80 | 1.00 |
| | >20 m H and >200 m W | 1.05 | 0.95 | 0.90 | 1.05 |
| | <20 m H and >200 m W | 1.00 | 1.00 | 1.00 | 1.00 |
| | Slopes (EL—elevation above slope foot): | | | | |
| | Equator facing, EL 20–50 m | 1.20 | 1.20 | 1.00 | 0.95 |
| | Equator facing, EL >50 m | 1.20 | 1.20 | 1.00 | 0.95 |
| | Pole facing, EL 20–50 m | 0.80 | 0.85 | 1.00 | 1.10 |
| | Pole facing, EL >50 m | 0.80 | 0.85 | 1.00 | 1.10 |
| | East/west, EL 20–50 m | 1.00 | 0.95 | 1.00 | 1.00 |
| | East/west, EL >50 m | 1.00 | 0.95 | 1.00 | 1.00 |
| Land use | Fields and wetlands | 1.00 | 1.00 | 1.00 | 1.00 |
| | Meadows | 1.00 | 0.95 | 1.00 | 1.00 |
| | Forests | 0.30 | 0.90 | 0.20 | 1.10 |
| | Ground transportation belts | 1.00 | 1.05 | 0.95 | 0.90 |
| | Rural settlement | 1.00 | 1.10 | 0.80 | 1.00 |
| | Intra-forest-settlement | 0.60 | 0.95 | 0.60 | 1.00 |
| | Downtown | 0.80 | 1.25 | 0.60 | 0.90 |
| | Industrial areas | 0.80 | 1.30 | 0.60 | 0.90 |
| | Water banks and water bodies | 1.00 | 0.85 | 1.10 | 1.20 |
| Ground moisture | Dry | 1.00 | 1.00 | 1.00 | 1.00 |
| | Humid | 1.00 | 0.95 | 1.00 | 1.10 |
| | Wet | 1.00 | 0.90 | 1.00 | 1.20 |

Caution: zt coefficients are valid within the air temperature range of 5–35 °C (*Source* Błażejczyk (2002), modified by Kunert (2010))

the coefficients of changes in air temperature are valid within the range of 5-35 °C (Błażejczyk 2002).

For defining landscape units, several environmental layers must be available: land use and land cover, ground moisture, hypsometry, plant cover. Values of particular meteorological components necessary for UTCI analysis are calculated by means of the reclassification of environmental layers using coefficients listed in Table 1. As basic values of meteorological elements, we can use variables monitored at the nearest meteorological station and/or they can be obtained by defining specific weather scenarios typical for particular regions and seasons. In the next step, the modelled meteorological variables are used for the calculation of UTCI.

In Poland, the first attempt to present a spatial distribution of UTCI was done by Błażejczyk (2011, 2013) for Warsaw. Environmental raster layers (relief, land use, vegetation cover) were developed by digitizing paper maps that had a spatial resolution of 120 × 120 m (Kozłowska-Szczęsna et al. 1996). For each raster, different types of landscape were defined and consequently relative values of particular climate elements were adopted. UTCI was calculated with the use of IDRISI Tajga (2009) software package. Several weather scenarios were applied to define expected values of meteorological variables (Table 2).

To assess thermal stress, the following simplified UTCI equation was applied:

$$\text{UTCI} = 0.84 \times T_a + 0.246 \times T_{mrt} - 2.45 \times v_a + 0.204 \times \text{vp} - 0.01 \quad (1)$$

**Table 2** Weather scenarios applied in UTCI modelling for Warsaw

| Air temperature ($T_a$, °C) | Total cloud cover (N, %) | Wind speed ($v_a$, m/s) |
|---|---|---|
| Cool (10) | Cloudy (80) | Calm (2)<br>Windy (4)<br>Very windy (8) |
| Cool (10) | Sunny (20) | Calm (2)<br>Windy (4)<br>Very windy (8) |
| Warm (20) | Cloudy (80) | Calm (2)<br>Windy (4)<br>Very windy (8) |
| Warm (20) | Sunny (20) | Calm (2)<br>Windy (4)<br>Very windy (8) |
| Hot (30) | Cloudy (80) | Calm (2)<br>Windy (4)<br>Very windy (8) |
| Hot (30) | Sunny (20) | Calm (2)<br>Windy (4)<br>Very windy (8) |

Relative humidity was set at 50% as constant value

where: $T_a$ is air temperature (°C), $T_{mrt}$ is the mean radiant temperature (°C), v is wind speed at 10 m above ground (m/s), vp is water vapour pressure (hPa) calculated based on relative humidity (RH) and air temperature values, as follows:

$$\mathrm{vp} = 6.112 \times 10^{[7.5\times \mathrm{Ta}/(237.7+\mathrm{Ta})]} \times 0.01 \times \mathrm{RH}) \quad (2)$$

To calculate $T_{mrt}$, the following procedure was applied:

$$T_{mrt} = \left[\left(R_{prim} + 0.5 \times L_g + 0.5 \times L_a\right) / \left(5.384 \times 10^{-8}\right)\right]^{0.25} - 273 \quad (3)$$

where:

$$L_a = 5.5 \times 10^{-8} \times (273 + T_a)^4 \cdot [0.82 - 0.25 \times 10^{(-0.094\times 0.75\cdot \mathrm{vp})]}, \quad (4)$$

$$L_g = 5.5 \times 10^{-8} \times \left(273 + T_{\mathrm{gr}}\right)^4, \quad (5)$$

Ground surface temperature ($T_{\mathrm{gr}}$) is assessed as follows:

$$T_{\mathrm{gr}} = \mathrm{ta} - (\text{for cloudiness, N} > 80\%) \quad (6)$$

$$T_{\mathrm{gr}} = 1.25 \times T_a - (\text{for N} < 80\% \text{ and } T_a > 0°\mathrm{C}) \quad (7)$$

$$T_{\mathrm{gr}} = 0.9 \times T_a - (\text{for N} < 80\% \text{ and } T_a < 0°\mathrm{C})] \quad (8)$$

To calculate solar radiation absorbed a by nude man ($R_{prim}$) the SolAlt model (Błażejczyk 2001, 2004; Błażejczyk and Kunert 2011) was applied.

$$R_{prim} = 0.7 \times [73.98 \times \ln(\mathrm{h}) - 100.4] \times \mathrm{zr} - \text{ for N} <= 20\% \text{ (sunny weather)} \quad (9)$$

$$R_{prim} = 0.665 \times \mathrm{h}^{1{,}039} \times \mathrm{zr} - \text{for N} > 80\% \text{ (cloudy weather).} \quad (10)$$

where h is solar angle taken for midday hours as 30° for Ta = 10 °C, 50° for Ta = 20 °C and 60° for Ta = 30 °C; zr is the coefficients of change in global solar radiation a shown in Table 1.

In Warsaw, the smallest differentiation of heat stress is observed at moderate temperature (20 ºC) and wind speed (4 m/s), both by cloudy and sunny weather. Irrespective of cloudiness, in most of the urban area UTCI is within the range of "no thermal stress" (9-26 ºC). In sunny hot summer days, moderate heat stress (26-32 ºC) can be found only in industrial and very dense urban areas. During extreme weather conditions, a diversified urban morphology causes a great spatial differentiation of biothermal conditions. On cloudy, cool and windy days, UTCI in the downtown area can mean moderate cold stress range. On the other hand, during sunny, hot, humid and calm days, several hot spots with extreme heat stress are found in the city center (Fig. 4).

The Błażejczyk's method was used by Dobek et al. (2013) for the city of Lublin (eastern Poland). The authors applied environmental layers (relief, land use) at a resolution of 5 × 5 m. Those layers were created based on Digital Elevation Model

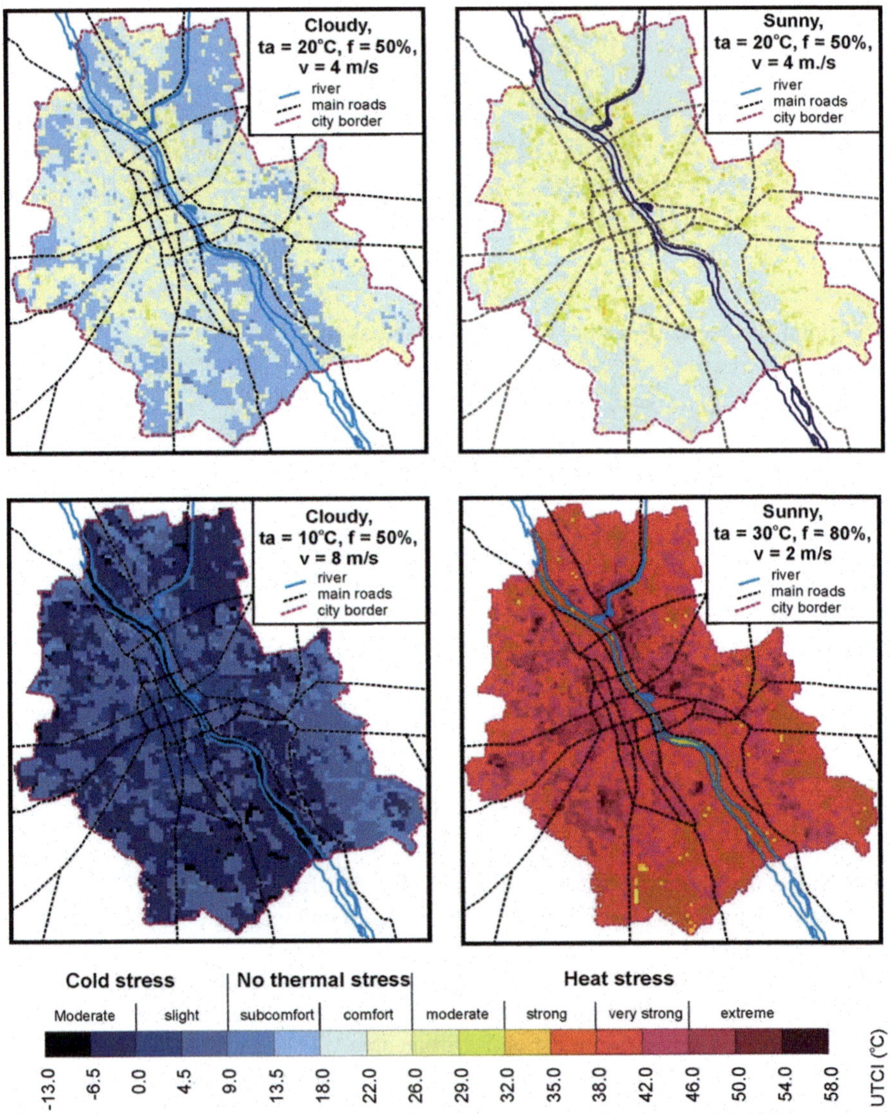


**Fig. 4** Spatial distribution of UTCI in Warsaw (Poland) in selected weather scenarios

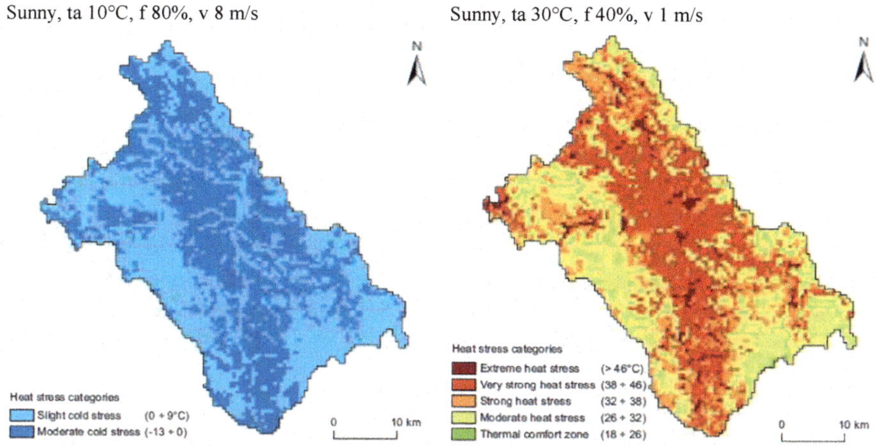


**Fig. 5** Distribution of different UTCI categories over Kłodzka Land basin at different weather scenarios (*Source* Milewski 2013)

and Topographic Database derived from topographic maps at a scale 1:10,000. The UTCI maps were generated for several weather scenarios. Similarly to Warsaw, the greatest heat stress is expected in central parts of the city.

The Błażejczyk's method was also used by Milewski (2013) to present local differentiation of UTCI in the mountain region of Kłodzka Land basin (eastern Sudetes, Poland). Raster mode environmental data with a resolution of 0.5 × 0.5 km were derived from various sources. The land-use layer is based on Corine Land Cover 2006 data. The layer with types of relief was developed using the Digital Terrain Elevation Data (DTED). Automatic reclassification of environmental layers allowed the definition of landscape units, and then the calculation of meteorological variables. UTCI maps were calculated for different weather scenarios of the warm season. The example shows a distinct spatial diversity of UTCI, especially during hot, sunny and not-windy summer weather (Fig. 5).

In 2010, Kunert has modified the Błażejczyk's method (Kunert 2010). She applied the typological approach to establish types of landscape. Based on environmental information, spatial units of homogeneous landscape features (e.g. flat grassland with moderate ground moisture, valleys bottoms with wet meadows) were defined as a combination of information listed in Table 1. For every type of landscape zr, zt, zf and zf coefficients (cf. Table 1) were classified. Based on such coefficients, projected values of meteorological variables were defined for particular weather scenarios. For the calculation of UTCI, a full algorithm is used as available in the BioKlima 2.6 software package (https://www.igipz.pan.pl/bioklima.html).

First, the modified method was applied in UTCI maps of Mazovia Lowland in central Poland (Błażejczyk and Błażejczyk 2014). Landscape units were defined for 1 × 1 km raster based on environmental layers of land use, hypsometry and ground moisture. While the study area is very flat, the main factors differentiating spatial diversity of UTCI were land use and ground moisture. The maps present UTCI values in two weather scenarios. In both images, there is the well-defined Warsaw agglomeration, which generates higher UTCI than surrounding areas (Fig. 6).

Recently, the method has been applied to study the spatial variability of UTCI in the sub-mountain health resort of Przerzeczyn (Błażejczyk et al. 2020b). In the first step, environmental maps were created. The land-use layer was developed based on Corine Land Cover 12 data. While the relief is not very differentiated, the principal relief units were defined using a hypsometric layer at topographic map. The same map and information from the map with soil types were applied to identify ground moisture (Fig. 7). Finally, for every part of the area with individual combinations of environmental factors (type of land use, relief and ground moisture) the zr, zt,

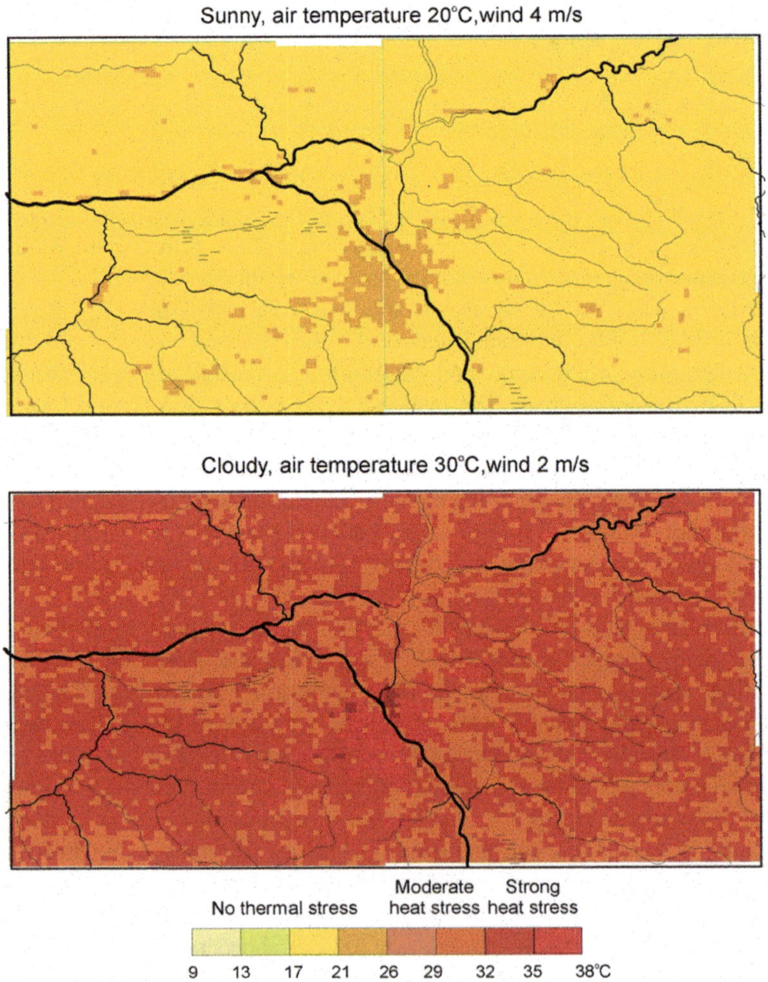


**Fig. 6** Distribution of UTCI over Mazovia Lowland (central Poland) during two summer weather scenarios

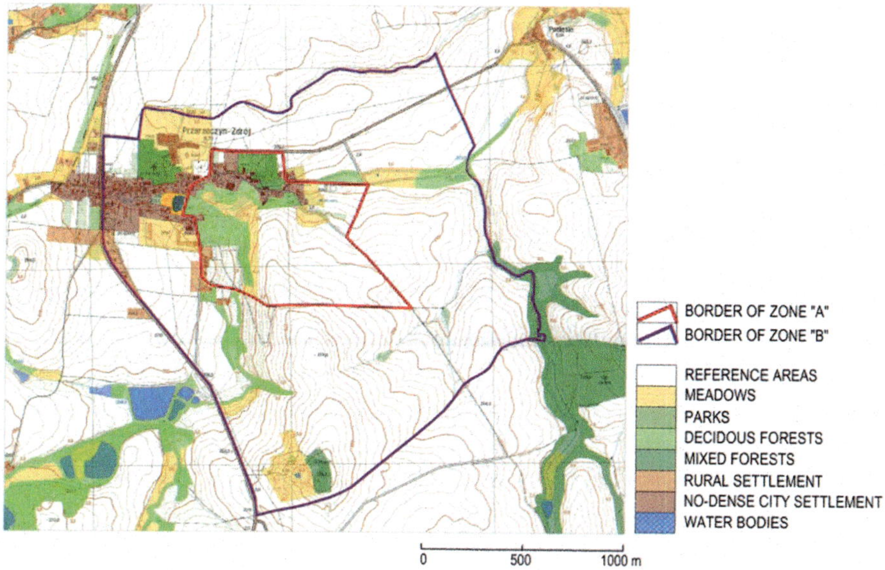


**Fig. 7** Hypsometry and land use of Przerzeczyn health resort (health resort areas are divided into core zone "A" and buffer zone "B")

zv and zf coefficients were defined. In the next step, these coefficients were applied to calculate values of particular climate elements: global solar radiation ($K_{glob}$), air temperature ($T_a$), wind speed ($v_a$) and relative humidity of air (RH).

UTCI values were calculated in BioKlima 2.6 software using the full UTCI formula (Błażejczyk and Błażejczyk 2010) for different weather scenarios typical for warm seasons in Poland: clear sky and overcast, four air temperatures (1, 10, 20 and 30 °C), two wind speeds (2 and 8 m/s). The relative humidity was set constant at 70%. The combination of weather scenarios is shown in Table 3. The table also indicates the solar angle (typical for midday hours in particular seasons in Poland) used in particular weather scenarios to calculate the mean radiant temperature. The procedure applies for this purpose the SolGlob model, which estimates absorbed solar radiation by a pedestrian based on global solar radiation ($K_{glob}$) data (Błażejczyk 2001, 2004; Błażejczyk and Kunert 2011).

$$R_{prim} = 0.17269 \times K_{glob}^{0.9763} - \text{ for } K_t <= 0.8 \tag{11}$$

$$R_{prim} = 2.58454 \times K_{glob}^{0.5842} - \text{ for } K_t \text{ between 0.8 and 1.05} \tag{12}$$

$$R_{prim} = 30.3982 \times K_{glob}^{0.2326} - \text{ for } K_t \text{ between 1.05 and 1.2} \tag{13}$$

$$R_{prim} = 6.24967 \times K_{glob}^{0.4861} - \text{ for } K_t \text{ greater than 1.2} \tag{14}$$

**Table 3** Definition of weather scenarios used in UTCI simulation

| Season | Global solar radiation (W/m$^2$) | Air temperature (°C) | Wind speed (m/s) | Solar angle (°) |
|---|---|---|---|---|
| Spring and autumn | 200 (sunny) | 1 (cold) | 2 (calm) | 15 |
| | 200 (sunny) | 1 (cold) | 8 (windy) | 15 |
| | 100 (cloudy) | 1 (cold) | 2 (calm) | 15 |
| | 100 (cloudy) | 1 (cold) | 8 (windy) | 15 |
| | 400 (sunny) | 10 (cool) | 2 (calm) | 30 |
| | 400 (sunny) | 10 (cool) | 8 (windy) | 30 |
| | 100 (cloudy) | 10 (cool) | 2 (calm) | 30 |
| | 100 (cloudy) | 10 (cool) | 8 (windy) | 30 |
| Summer | 800 (sunny) | 20 (warm) | 2 (calm) | 50 |
| | 800 (sunny) | 20 (warm) | 8 (windy) | 50 |
| | 200 (cloudy) | 20 (warm) | 2 (calm) | 50 |
| | 200 (cloudy) | 20 (warm) | 8 (windy) | 50 |
| | 800 (sunny) | 30 (hot) | 2 (calm) | 50 |
| | 800 (sunny) | 30 (hot) | 8 (windy) | 50 |
| | 200 (cloudy) | 30 (hot) | 2 (calm) | 50 |
| | 200 (cloudy) | 30 (hot) | 8 (windy) | 50 |

where $K_t$ represents theoretically possible global radiation at cloudless sky:

$$K_t = K_{glob}/\left(-0.0015 \times \mathrm{h}^3 + 0.1796 \times \mathrm{h}^2 + 9.6375 \times \mathrm{h} - 11.9\right) \quad (15)$$

The $K_t$ values are applied also to assess ground surface temperature ($T_{\mathrm{gr}}$).

$$T_{\mathrm{gr}} = T_{\mathrm{a}} - \text{ for } K_t <= 0.8 \quad (16)$$

$$T_{\mathrm{gr}} = 1.25 \times T_{\mathrm{a}} - \text{ for } K_t > 0.8 \text{ and } T_{\mathrm{a}} > 0\,°\mathrm{C} \quad (17)$$

$$T_{\mathrm{gr}} = 0.9 \times T_{\mathrm{a}} - \text{ for } K_t > 0.8 \text{ and } T_{\mathrm{a}} < 0\,°\mathrm{C} \quad (18)$$

In the analysis of the spatial variability of UTCI at local scale not only absolute index values but also deviations from reference conditions can be taken into account. Table 4 shows examples of UTCI deviations from reference conditions (i.e. observed at a location typical for setting up a standard meteorological station, i.e. flat, elevated grass area with free air flow) in different types of landscape during extreme weather scenarios.

During cold, cloudy and windy days, UTCI at reference conditions may drop to −21.6 °C. Over water bodies, UTCI deviates from reference conditions by −1.4 °C. In parks and urban areas, UTCI is 3.5–7.6 °C higher than at the reference area. In

**Table 4** Deviation of UTCI from reference conditions at different types of landscape during selected weather scenarios of warm season in Poland

| Type of landscape | Weather scenarios | | | |
|---|---|---|---|---|
| | Cloudy Cold Windy | Sunny Cold Calm | Cloudy Hot Windy | Sunny Hot Calm |
| Reference conditions (flat, elevated grass field) | UTCI at reference area (°C) | | | |
| | −21.6 | 2.3 | 24.6 | 35.4 |
| | Deviations of UTCI from reference area (K) | | | |
| Bottoms of valleys | 0.9 | −1.1 | −0.5 | −1.3 |
| Coniferous/mixed forests | 17.5 | −2.6 | −6.1 | −7.7 |
| Decideous forests | 12.2 | −3.2 | −3.7 | −3.7 |
| Fresh meadows | −0.1 | −0.1 | −1.0 | −1.8 |
| Parks | 7.1 | −4.1 | −3.3 | −3.7 |
| Rural settlements | 3.5 | 1.6 | 2.6 | 4.4 |
| No-dense urban settlements | 7.6 | 2.7 | 1.2 | 1.9 |
| Water bodies | −1.4 | −0.4 | −2.7 | −4.5 |

*Note* negative UTCI deviations are marked in grey

densely vegetated areas, UTCI is by 12.2–17.5 °C higher. During cold, sunny and calm weather, UTCI in reference conditions can reach 2.3 °C. The greatest negative UTCI deviations are observed inside forests and parks (from −1.1 to −4.1 °C). Only in urban areas UTCI is higher than in reference conditions, by 1.6–2.7 °C. During hot cloudy and windy days, UTCI at reference conditions reaches 24.6 °C. In urban areas UTCI is higher by 1.2–2.6 °C. However, in forests, parks and over water bodies thermal stress is reduced by 2.7–6.1 °C. A similar spatial pattern of UTCI deviations is noticed during hot, sunny and calm days (Table 4).

The aforedescribed research dealt with recommendations for health resorts from the point of view of its bioclimatic potential, to be considered in therapeutic and health prophylactic purposes. Spatial images of UTCI deviations under various weather scenarios indicated parts of the resort where patients can experience softening or hardening of thermal stress (in relation to reference conditions). The maps (Fig. 8) show that a given area can reduce or intensify cold or heat stress. For example, during cold, cloudy and windy weather, positive UTCI deviations (softening of cold stress occurring in open, flat areas) are experienced inside forests, parks and urbanized areas. However, in hot, sunny and calm weather, urbanized areas intensify heat stress whereas forests and parks can significantly reduce it. Detailed analysis of UTCI distribution in various weather scenarios can be a base for planning therapeutic and recreational activities in particular parts of resorts. Depending on actual weather

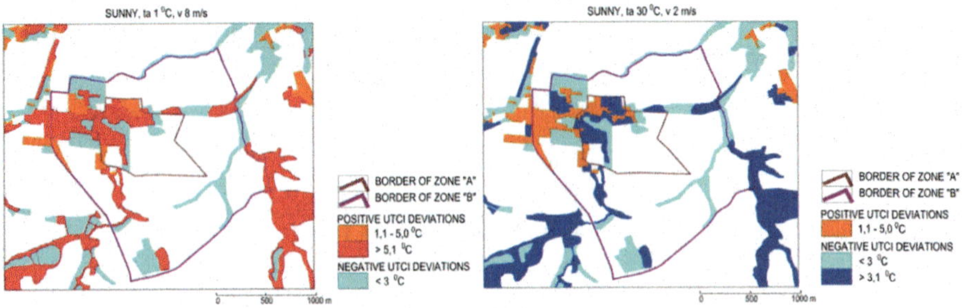


**Fig. 8** Deviations of UTCI from reference conditions (marked in white) over Przerzeczyn health resort during two weather scenarios

conditions, particular forms of climate therapy[1] can be organized in various types of landscape (Table 5).

## 5 Conclusions

This chapter presented an overview of mapping approaches of the UTCI developed for different spatial scales, from global to local. In global and regional mapping, relevant advances were achieved thanks to open global databases of meteorological variables (ECMWF). Both, global and regional mapping of UTCI provide a great opportunity to analyze spatial patterns and peculiarities of thermal stress variability. The overview showed recent studies with great potential for identifying regions with specific UTCI characteristics which can be applied for many practical purposes such as human health and tourism. They are also applied in the analysis of regional differences in climate change scenarios.

It should be stressed that the mapping of UTCI at a local scale does not follow a single, specific methodology. The most complex approaches were done for urban areas as well as for landscapes of mid-latitude geographical regions. In the present chapter, we presented a method for UTCI mapping at a local scale based on topoclimatic field research carried out in the mid-latitude climate belt. In research carried out in climates of high and low latitudes, it is necessary to validate the modification coefficients used in the presented method in order to make them globally applicable. The first attempts of UTCI mapping at a local scale show possible applications in urban climate analyses, local planning as well as for health resort treatment, recreational and health prophylactic purposes.

[1] According to the "Encyclopedia of balneology, physical medicine, bioclimatology and health resort geology" (Ponikowska 2015) there are three principal forms of climate therapy: heliotherapy (using in treatment the exposure to direct solar radiation during sun bathing), aerotherapy (exposure to complex weather stimuli during air bathing) and terrain kinesitherapy (exposure to complex weather stimuli during outdoor physical activity, e.g. walking, jogging, biking etc.).

**Table 5** Recommendations for different forms of climate therapy in various types of landscape in Przerzeczyn health resort

| Weather scenario | Recommendations | |
|---|---|---|
| | Preferences | Limitations |
| Sunny/cold/calm | Terrain kinesitherapy (TK) in any type of landscape | Heliotherapy (HT) not recommended<br>Aerotherapy (AT) only in parks and forests |
| Sunny/cold/wind | TK in parks and forests | HT and AT impossible in any type of landscape |
| Cloudy/cold/calm | TK in any type of landscape | HT impossible, AT only in parks and forests |
| Cloudy/cold/windy | TK in parks and forests | HT impossible, AT not recommended |
| Sunny/cool/calm | TK and AT in any type of landscape | Limited HT in parks |
| Sunny/cool/wind | TK and AT in parks and forests | HT not recommended anywhere |
| Cloudy/cool/calm | TK in any type of landscape | HT impossible, limited AT |
| Cloudy/cool/windy | TK in parks and forests | HT impossible, AT not recommended |
| Sunny/warm/calm | TK and AT in any type of landscape | HT only in open areas (fields, meadows) |
| Sunny/warm/wind | TK in any type of landscape,AT in parks and forests | Limited HT |
| Cloudy/warm/calm | TK and AT in any type of landscape | HT impossible |
| Cloudy/warm/windy | TK in any type of landscape,AT in parks, forests s and rural settlements | HT impossible |
| Sunny/hot/calm | HT in open areas (fields, meadows)<br>AT in any type of landscape | TK very limited (overheating risk) especially in bottoms of valleys, decideuous forests and urban area |
| Sunny/hot/wind | HT and AT in bottom of valleys and rural settlements | TK limited to parks and coniferous/mixed forests |
| Cloudy/hot/calm | AT in any type of landscape | TK very limited (overheating risk) especially in bottoms of valleys, decideuous forests and urban area<br>HT impossible |
| Cloudy/hot/windy | AT in any type of landscape | TK limited to parks and coniferous/mixed forests<br>HT impossible |

## References

Anderberg MR (1973) Cluster analysis for applications. Academic Press

Błażejczyk K (1990) Biotopoclimatic differentiation of selected types of landscape. IGiPZ PAN Conf Pap 4:175–187

Błażejczyk K, Krawczyk B, Skoczek J (eds) (1992) Topoclimatic and microclimatic research in various climate zones. Zeszyty IGiPZ PAN 5
Błażejczyk K (2001) Human heat balance as a tool in bioclimatic research. Przegląd Geogr 73(4):535–554
Błażejczyk, K. (2002). Influence of air circulation and local factors on climate and bioclimate of Warsaw agglomeration. Dokumentacja Geograficzna 26.
Błażejczyk K (2004) Radiation balance in man in various meteorological and geographical conditions. Geogr Pol 77(1):63–76
Błażejczyk K, Błażejczyk M (2010) BioKlima© 2.6, Program komputerowy. http://www.igipz.pan.pl/Bioklima-zgik.html
Błażejczyk K (2011) Mapping of UTCI in local scale (the case of Warsaw). Prace i Studia Geograficzne WGSR UW 47:275–283
Błażejczyk K (2013) Distribution of universal thermal climate index (UTCI) in Warsaw. Geogr Pol 86(1):79–80
Błażejczyk K, Błażejczyk A (2014) Assessment of bioclimatic variability on regional and local scales in Central Europe using UTCI. Scientific Annals of "Alexandru Ioan Cuza" University of Iasi, 60, 1, s. IIc, Geography series:67–82
Błażejczyk K, Kunert A (2011) Bioclimatic principles of recreation and tourism in Poland, 2nd edn. Monografie IGiPZ PAN, vol 13, Warszawa
Błażejczyk K, Nejedlik P, Skrynyk O, Halaś A, Skrynyk O, Baranowski J, Mikulova K (2020a) Thermal stress in northern carpathians and air circulation. Misc Geogr. https://doi.org/10.2478/mgrsd-2020-0022.9
Błażejczyk, A., Kuchcik, M., Baranowski, J. & Błażejczyk, K. (2020b) Assessment of therapeutic properties of climate in Przerzeczyn health resort. Manuscript, available by request at bioklimatologia@wp.pl.
Boko NPM, Medeou KF, Vissin EW, Błażejczyk K, Houssou CS (2014) Caracterisation des ambiances bioclimatiques dans les villes littorales du Benin (Afrique de l'ouest). In: Proceedings XXVIIe colloque de l'association internationale de climatologie 2–5 juillet 2014—Dijon (France), pp 605–611
Brecht BM (2019) Die urbane wärmebelastung unter einfluss lokaler faktoren und zukünftiger klimaänderungen. PhD thesis, KIT-Fakultät für Physik des Karlsruher Instituts fürTechnologie (KIT)
CDS (2020) Thermal comfort indices derived from ERA5 reanalysis. https://doi.org/10.24381/cds.553b7518
Di Napoli C, Barnard Ch, Prudhomme Ch, Cloke HL, Pappenberger F (2020) ERA5-HEAT: A global gridded historical dataset of human thermal comfort indices from climate reanalysis. Geosci Data J. https://doi.org/10.1002/gdj3.102
Di Napoli C, Barnard C, Prudhomme C, Pappenberger F (2020b) A heat health hazard index based on ECMWF data. https://www.ecmwf.int/en/newsletter/162/news/heat-health-hazard-index-based-ecmwf-data
Di Napoli C, Pappenberger F, Cloke HL (2018) Assessing heat-related health risk in Europe via the universal thermal climate index (UTCI). Int J Biometeorol 62:1155–1165
Dobek M, Demczuk P, Nowosad M (2013) Spatial variation of the universal thermal climate index in Lublin in specified weather scenarios. Ann Univ Mariae Curie-Skłodowska, Lub-Pol, Sect B 68(1):21–37
Erell E, Pearlmutter D, Williamson T (2011) Urban microclimate: designing the spaces between buildings. Routledge, London
ESOTC (2019) Climate in 2019, Spotlight on heat and cold stress. https://climate.copernicus.eu/ESOTC/2019/heat-and-cold-stress
Ge Q, Kong Q, Xi J, Zheng J (2017) Application of UTCI in China from tourism perspective. Theoret Appl Climatol 128:551–561
Geiger R (1969) Topoclimates. In: Landsberg HE, Flohn H (eds) World survey of climatology, vol 2. General Climatology, 2. Elsevier, Amsterdam-London-New York, pp 105–138

Grzybowski J (ed) (1990) The questions of contemporary topoclimatology. In: IGiPZ PAN conference papers, vol 4
IDRISI Taiga: Guide to GIS and Image Processing (2009) http://web.pdx.edu/~nauna/resources/TaigaManual.pdf
Kong Q, Ge Q, Zheng J (2017) Spatio-temporal changes in extreme UTCI indices in China. Geogr Res 36(6):1171–1182
Kozłowska-Szczęsna T, Błażejczyk K, Krawczyk B (1996) Atlas of Warsaw, No 4, Natural geographical environment. IGiPZ PAN, Warszawa
Kuchcik M (ed) (2001) Contemporary topoclimatic research. In: Dokumentacja geograficzna IGiPZ PAN, vol 23
Kuchcik M, Błażejczyk K, Halaś A (2021) Long-term changes in hazardous heat and cold stress in humans: multi-city study in Poland. Int J Biometeorol. https://doi.org/10.1007/s00484-020-02069-7
Kunert A (2010) Modeling of UTCI index in various types of landscape. In: Matzarakis A, Mayer H, Chmielewski F-M, (eds), Proceedings 7th conference of biometeorology, berichte des meteorologischen instituts der albert-ludwigs-universität freiburg, vol 20, pp 302–307
Kunert A, Błażejczyk K (2011) Local-scale air-temperature differences in various types of Poland's landscape. Przegląd Geogr 83(1):69–90
Matuschek O, Matzarakis A (2011) A mapping tool for climatological applications. Meteorol Appl 18(2):230–237
Mieczkowski Z (1985) The tourism climatic index: a method of evaluating world climates for tourism. Can Geogr/le Géographe Can 29(3):220–233
Milewski P (2013) Application of the UTCI to the local bioclimate of Poland's Ziemia Kłodzka region. Geogr Pol 86(1):47–54
Nusrat S, Kobourov S (2016) The state of the art in cartograms. In: Maciejewski R, Ropinski T, Vilanova A (eds), EuroVis 2016, STAR—state of the art report, vol 35 (3)
Oke TR (1987) Boundary layer climates, 2nd edn. Methuen, London—New York
Pappenberger F, Jendritzky G, Staiger H, Dutra E, Di Giuseppe F, Richardson DS, Cloke HL (2015) Global forecasting of thermal health hazards: the skill of probabilistic predictions of the universal thermal climate index (UTCI). Int J Biometeorol 59(3):311–323
Park S, Tuller SE, Jo M (2014) Application of universal thermal climate index (UTCI) for microclimatic analysis in urban thermal environments. Landsc Urban Plan 125:146–155
Paszyński J, Miara K, Skoczek J (1999) The energy exchange at the earth−atmosphere bound−ary as a base for topoclimatological mapping. Dokumentacja Geograficzna nr, 14
Peel MC, Finlayson BL, McMahon TA (2007) Updated world map of the Köppen-Geiger climate classification. Hydrol Earth Syst Sci 11(5):1633–1644
Ponikowska I (ed) (2015) Encyclopedia of balneology, physical medicine, bioclimatology and health resort geology. Aluna, Konstancin-Jeziorna. (REFERENCE NOT IN THE TEXT)
Provençal S, Bergeron O, Leduc R, Barrette N (2016) Thermal comfort in Quebec City, Canada: sensitivity analysis of the UTCI and other popular thermal comfort indices in a mid-latitude continental city. Int J Biometeorol 60(4):591–603
Roshan G, Yousefi R, Błażejczyk K (2018) Assessment of the climatic potential for tourism in Iran through biometeorology clustering. Int J Biometeorol 62(4):525–542
Słowińska S, Słowiński M, Lamentowicz M (2010) Relationships between Local Climate and Hydrology in Sphagnum Mire: Implications for Palaeohydrological Studies and Ecosystem Management. Pol J Environ Stud 19(4)
Tomczyk AM, Owczarek M (2020) Occurrence of strong and very strong heat stress in Poland and its circulation conditions. Theor Appl Climatol 139(3):893–905
Trimmel H, Weihs P, Faroux S, Formayer H, Hamer P, Hasel K, Laimighofer J, Leidinger D, Masson V, Nadeem I, Oswald SM, Revesz M, Schoetter R (2019) Thermal conditions during heat waves of a mid-European metropolis under consideration of climate change, urban development scenarios and resilience measures for the mid-21st century. Meteorol Z. (Contrib. Atm. Sci.), PrePub https://doi.org/10.1127/metz/2019/0966

Ustrnul Z, Czekierda D (2003) Construstion of the air temperature maps for Poland using GIS. Archiwum Fotogrametrii, Kartografii i Teledetekcji 13ªA:243–254

Vinogradova V (2020) Using the universal thermal climate index (UTCI) for the assessment of bioclimatic conditions in Russia. Int J Biometeorol. https://doi.org/10.1007/s00484-020-01901-4

Wu F, Yang X, Shen Z (2019) Regional and seasonal variations of outdoor thermal comfort in China from 1966 to 2016. Sci Total Environ 665:1003–1016

Zeng D, Wu J, Mu Y, Deng M, Wei Y, Sun W (2020) Spatial-temporal pattern changes of UTCI in the China-Pakistan economic corridor in recent 40 years. Atmosphere 11:858. https://doi.org/10.3390/atmos11080858

# Chapter 9
# Application of the UTCI in High-Resolution Urban Climate Modeling Techniques

**Jan Geletič, Michal Lehnert, Jaroslav Resler, and Pavel Krč**

**Abstract** Urban population is affected by many processes typical for urban landscapes. Spatial patterns of physical, social, behavioral and environmental elements of urban environments and their interactions with human beings have been therefore gaining interest of scholars as well as politicians and local actors for decades. The combination of climate change, growing urban population and increasing computing capabilities opens up space for new areas of spatial analyses in urban environments—among them analyses of human thermal comfort. Consequently, modeling of thermal exposure as a cardinal factor of thermal comfort in real outdoor urban environments represents a pending and challenging task for urban climate research.

**Keywords** PALM · Urban climate modeling · Spatiotemporal variability of the UTCI · Heat stress

## 1 Introduction

Two of the most important processes, characteristically modeled in urban environments, are radiative and heat exchange processes. These processes are crucial for the energy budget of the surfaces, which strongly affect boundary-layer dynamics, as well as spatiotemporal distribution of (bio)meteorological variables—in other words, human thermal exposure. Demand for spatially more detailed and complex information about climate in very specific urban areas has led to development of

J. Geletič · M. Lehnert · J. Resler · P. Krč
Department of Complex Systems, Institute of Computer Science of the Czech Academy of Sciences, Pod Vodárenskou Věží 271/2, 182 07 Prague, Czech Republic

J. Geletič (✉)
Global Change Research Institute of the Czech Academy of Sciences, Bělidla 986/4a, 603 00 Brno, Czech Republic
e-mail: geletic@cs.cas.cz

M. Lehnert
Department of Geography, Faculty of Science, Palacký University in Olomouc, 17. Listopadu 12, 771 46 Olomouc, Czech Republic

 
E. L. Krüger (ed.), *Applications of the Universal Thermal Climate Index UTCI in Biometeorology*, Biometeorology 4,
https://doi.org/10.1007/978-3-030-76716-7_9

numerical models capable of simulating such processes at fine scale (micro and local scales).

Fine-scale modeling of processes in the urban canopy layer is a challenging task where the uncertainty is an intrinsic part of the modeling process. Nevertheless, substantial progress made in last decades led to the development of computational fluid dynamics (CFD) models, e.g. MITRAS (Schlünzen et al. 2003), ENVI-met (Bruse 2004), MUKLIMO_3 (Sievers 2012, 2016), UrbClim (de Ridder et al. 2015) or most recently PALM (Maronga et al. 2020). The progress and development of fine-scale models, however, brings many new problems, such as the level of detail of input data (and their accuracy, collection and/or mapping), the effect of neighborhood on microclimate and the complex model validation.

Difficulties and uncertainties are connected particularly to the heterogeneity of surface characteristics (e.g., albedo, emissivity, heat capacity, plant canopy, roughness etc.), their accuracy and level of detail (e.g., parameterized trees or detail information about each trunk, branch and crown) and diverse human activities (e.g., anthropogenic heat, air pollution). This results in a high variability of related quantities of urban atmosphere (primarily air temperature ($T_a$), relative humidity (RH), mean radiant temperature ($T_{mrt}$) and wind components). Characteristic length scales of surface heterogeneities in urban areas are 1–100 m, thus models with input grid sizes far below the kilometer scale are needed to resolve the characteristic atmospheric features triggered by a specific urban environment. In this regard, atmospheric modeling of urban areas usually considers a limited number of relevant processes; in many cases, it is performed with coarser resolution which cannot properly describe the conditions inside street canyons (Resler et al. 2017). One of the most challenging questions is the approach to the turbulence simulation; is it sufficient to use a parameterized model of turbulence (e.g., Reynolds-averaged Navier-Stokes; RANS) or is it more efficient to utilize a large-eddy simulation (LES) model with explicitly resolved turbulence (Fröhlich and von Terzi 2008).

LES models solve the three-dimensional prognostic equations for momentum, temperature, humidity, and other scalar quantities (such as chemical species). The PALM modeling system with its PALM-4U components for detailed modeling of urban-canopy-related processes which is used in the following case study is the first open-source meteorological model based on the LES principles with the implementation of most urban-canopy-related processes.

### *1.1 Modeling of Thermal Exposure in Urban Environments with the PALM-4U Tool*

For a calculation of thermal exposure in urban environment, several variables are needed: air temperature ($T_a$), relative humidity (RH), wind speed ($v_a$; more precisely air velocity) and radiation (r; frequently summarized in the integrating variable of mean radiant temperature—$T_{mrt}$) (Höppe 1999; Geletič et al. 2018). However, the thermal environment is particularly diverse, which leads to an enormous variability and complex mutual interactions between those variables. For instance, heat and moisture exchanges, aerodynamic, and radiation conditions are markedly different inside the street canyon and on the city-scale level. Enhanced heat storage in urban areas results from intensive absorption of shortwave (*SW*) and longwave (*LW*) radiation and its multiple reflections due to the prevalence of horizontally and vertically oriented impervious urban surfaces (Voogt and Oke 2003). In addition, the enclosed spaces between the buildings in the city lead to the absorption of emitted longwave radiation, which can also be referred to as the reduction of total heat losses by radiation. Human activities in urban environments produce another extra source of heat. The combination and interactions of these factors result in a complex spatiotemporal pattern of $T_a$, RH, $v_a$ and $T_{mrt}$ indicating that fine-scale modeling of the climate in street-level needs to be analyzed as a complex system, which can be provided by PALM-4U (Belda et al. 2020; Gehrke et al. 2020; Resler et al. 2017, 2020a).

PALM-4U enables us to utilize several sophisticated modules embedded in PALM, namely the land surface model (LSM; Gehrke et al. 2020), the building surface model (BSM, formerly USM; see Resler et al. 2017), the plant canopy model (PCM), and the radiative transfer model (RTM; Krč et al. 2021), which uses external radiation forcing from the Weather Research and Forecasting (WRF) model in our setting. Furthermore, the following components are applied in PALM-4U: the Cartesian topography, the human biometeorology module (BIO; Fröhlich and Matzarakis 2020), online chemistry modules (Khan et al. 2020), and online and offline domain nesting features (Hellsten et al. 2020; Kadasch et al. 2020).

## *1.2 The Mean Radiant Temperature*

Of particular importance for the thermal assessment of a given thermal environment is the mean radiant temperature ($T_{mrt}$). $T_{mrt}$ at an arbitrary point in space is defined as the temperature of an imaginary object, at which that object would be in radiative equilibrium with its surroundings, i.e. the absorbed irradiance would be equal to the emitted radiance. An overview and limitations of in situ measurements approaches for assessing the $T_{mrt}$ are described by Kántor and Unger (2011).

For $T_{mrt}$ modeling, three models are typically used: RayMan (Pro), SOLWEIG and ENVI-met. RayMan is developed on the basis of the Association of German Engineers' environmental meteorology standards (*Verein Deutscher Ingenieure*) VDI 3787 and 3789 (Verein Deutscher Ingenieure (VDI) 1994, 1998, 2001). RayMan was validated by many authors, most thorough studies were published by Krüger et al. (2014) and Lee and Mayer (2016). RayMan is a closed source model, where relevant information regarding radiation is missing. Without detailed knowledge of input variables and parameters it is very hard to truthfully describe all small discrepancies of $T_{mrt}$ in the urban environment. SOLWEIG model's treatment of *SW* and *LW* radiation fluxes is extensively documented by Lindberg et al. (2008, 2016) and validated in many papers (e.g., Lindberg et al. 2016; Kántor et al. 2018). ENVI-met is the only CFD model used for spatial analysis of outdoor microclimate (based on RANS principle, more specifically). ENVI-met follows the same VDI standard as RayMan, but no up-to-date description of the model is available. There exists only one source with a description of $T_{mrt}$ calculation (Huttner 2012). Currently, ENVI-met is probably the most used model for a spatiotemporal calculation of $T_{mrt}$ (same as the following calculation of biometeorological indices, e.g. UTCI and PET). The most comprehensive review of available ENVI-met validation studies is given by Tsoka et al. (2018). ENVI-met model is also closed source, as RayMan. Practically, this means that a lack of transparency is similar as in the case of RayMan. All three models were analyzed in detail in a comparative numerical simulation and validation study by Gál and Kántor (2020).

One of the most recent possibilities of $T_{mrt}$ calculation is the PALM model. $T_{mrt}$, together with radiation fluxes, are obtained with the RTM model (Krč et al. 2021), which provides *SW* and *LW* radiation for the BIO module (Fröhlich and Matzarakis 2020). The importance of radiative transfer processes in the PALM model system 6.0 is described in detail by Salim et al. (2020). A new approach for the calculations of $T_{mrt}$ in RTM allows the BIO module to perform more precise calculations of derived biometeorological indices; this brings a new perspective for the spatiotemporal differentiation of physical level of thermal comfort at pedestrian level with the use of validated model results (Resler et al. 2020a).

## 1.3 Application of the Universal Thermal Climate Index (UTCI)

The calculation of the UTCI is based on the Fiala-UTCI multi-node dynamic of human heat transfer and regulation model (Fiala et al. 2012) and includes a clothing model that reflects behavioral adaptation of clothing insulation by the general urban population in response to momentary outdoor air temperature (Havenith et al. 2012). The UTCI is then defined as the isothermal air temperature of the reference condition that would elicit the same dynamic response of the physiological model for any combinations of $T_a$, RH, $v_a$ and r (Jendritzky et al. 2012). In the case study described in this chapter, the PALM model SVN revision 4508 was employed (Resler et al. 2020b). More specifically, for $T_{mrt}$ calculations, the ellipsoidal parameterization set in RTM was used for calculations using the BIO module implemented in the PALM-4U. The analyzed output represents 10-min averages of the calculated UTCI. In the model, these values are calculated at a given time step, which can vary in range from 0.1 to 1.0 s, depending on current meteorological conditions.

## 2 Case Study Prague-Dejvice

Dejvice, a characteristic neighborhood of a Central European city located in the Czech capital Prague, was chosen for testing PALM. This area was indicated as a "hot-spot" and vulnerable area with elevated temperatures during heat waves (Geletič et al. 2020). The area is located in Köppen's climate type Cfb (temperate oceanic climate). The average annual temperature recorded in a nearby meteorological station, at Praha-Klementinum (WMO 11514) is 10.8 °C (climate normal for 1980–2010); the mean temperature in July is 20.8 °C. Statistical data for Prague (CZE_Prague, WMO 11518, obtained from EnergyPlus Weather Converter v. 7.1.0.010; EPW, 2021) show that the hottest month is August, with a daily average of 17.6 °C and a temperature range of 10.9 K, with maxima reaching 32 °C. For a baseline of 23.3 °C, August has 320 cold degree-hours (CDH) and July has a somewhat higher value of 360 CDH. In this study, we consider days around summer solstice, hence days in late July, which represents the peak monthly global solar radiation in this location (4,880 W $m^{-2}$).

The domain includes a complex terrain; the mean elevation of the domain is 225 m a.s.l. and elevation spread is ±15 m. Substantial parts of the domain represent typical historical residential areas with a combination of old and new buildings (Local Climate Zone 'LCZ' 5—'open midrise', according to the classification scheme of urban areas aimed at the standardization of methods of observation and documentation in urban heat island studies, as proposed by Stewart and Ok 2012) and a variety of other urban components (such as gardens, parks or parking lots). However, the south-west and north-east parts of the domain are mostly sparsely built-up (Fig. 1). The building heights alongside the streets range approximately from 20 to 30 m, the highest building in the domain is 60 m high. Both main boulevards are approximately 40 m wide, with the boulevards (with a few exceptions) not exhibiting much greenery. The majority of trees are located in the intra-blocks and parks.

A typical summer period, from 18 to 21 July, was simulated. Daily maximum temperature exceeded 30 °C only on the last day simulated (30.1 °C); the previous days' maxima were 27.6 °C and 28.1 °C, respectively. The period between the afternoon of 19 July and the late afternoon of 21 July was mostly clear, with some high-altitude cirrus. The late afternoon and evening of 21 July was cloudy, mostly with low-level cumulus. The highest values of relative humidity occurred at night (93%), dropping to 27% during the day. The wind above roof level was north-westerly, light, mostly below 3.2 m $s^{-1}$, and often as low as 1.6 m $s^{-1}$ (for more details see Resler et al. 2020a).

**Fig. 1** Aerial image of study domain with selected locations in Prague-Dejvice, Czech Republic (*data source* Czech Office for Surveying, Mapping and Cadastre)

# 3 Results

## 3.1 Spatiotemporal Pattern of UTCI in a Hot Sunny Day

Results of the spatiotemporal pattern of the UTCI in the pilot domain Prague-Dejvice enable us to illustrate and comprehensively assess the spatiotemporal pattern of thermal exposure in a case study for the chosen urban area. Six characteristic spatial patterns of the UTCI related to the time of day can be drawn from model results (Geletič et al. 2021). Following a similar procedure to the one presented by Vanos et al. (2020), we present averages for different periods, termed 'phases', in order to express spatial patterns of the UTCI/thermal exposure with respect to UTCI-related heat stress categories (Bröde et al. 2012).

In the "*late-night phase*" (00:00–03:00 UTC), during the late-night hours before sunrise highest values of the UTCI are typically close to the walls of buildings, due to sensible heat fluxes and *LW* radiation emitted from façades. Thermal radiation emitted from façades is most intense in narrow alleys and in small courtyards with dense tree canopies resulting in higher UTCI values (Fig. 2a). Conversely, open spaces with few trees or bare soil (e.g., parking lots, lawns) show characteristically lower UTCI values during the late-night phase.

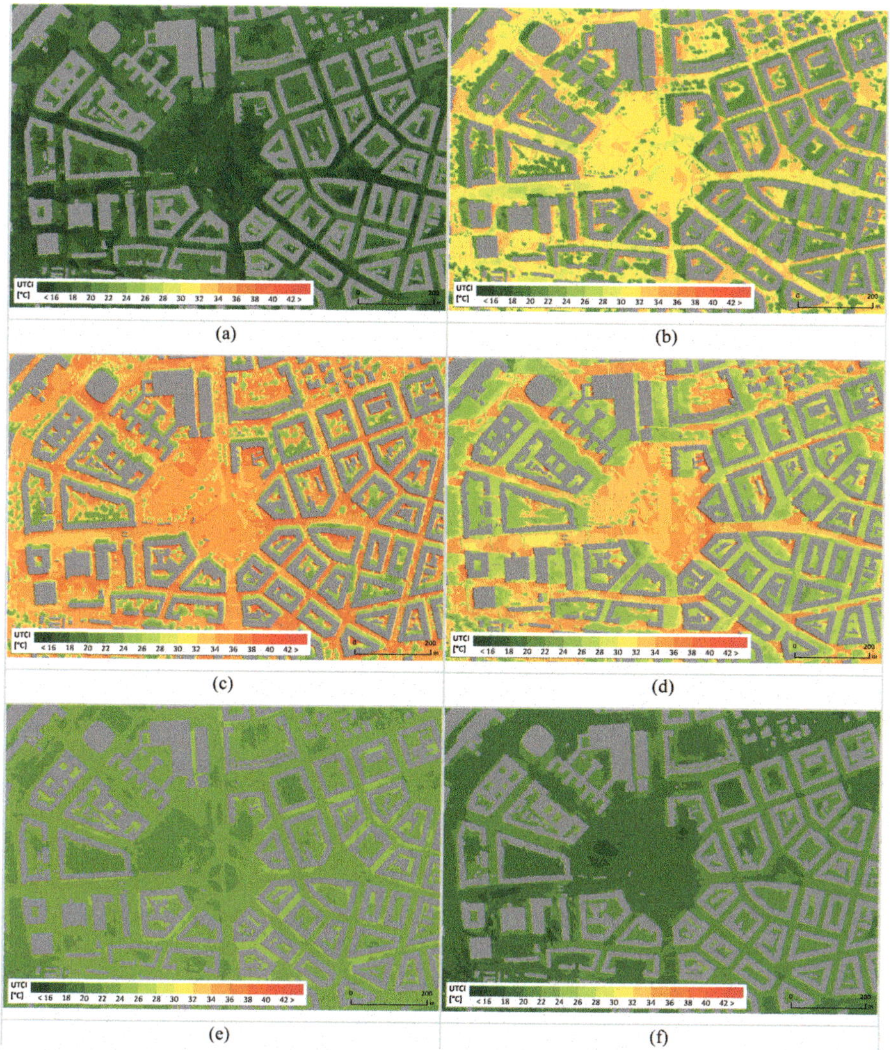


**Fig. 2** Spatial variability of the average UTCI in Prague-Dejvice on 20 July 2018: late-night phase (**a**; 00:00–03:00 UTC), sunrise transitional phase (**b**; 04:00–08:00 UTC), morning phase (**c**; 09:00–12:00 UTC), afternoon phase (**d**; 13:00–16:00 UTC), evening transition phase (**e**; 17:00–20:00 UTC) and early night phase (**f**; 21:00–00:00 UTC)

Following the "*sunrise transitional phase*" (04:00–08:00 UTC; Fig. 2b), the morning pattern of thermal exposure ("*morning phase*") is formed (09:00–12:00 UTC; Fig. 2c). During this phase, the highest UTCI values are seen in the vicinity of the sunlit south and east sides of existing buildings. In addition, closed courtyards with low vegetation coverage start to heat up rapidly, leading to a steep rise in UTCI values. In contrast, lower UTCI values are noticed in the shade of trees and buildings.

The comparatively lower values of the UTCI due to shade are even more pronounced during the "*afternoon phase*" (13:00–16:00 UTC, Fig. 2c), when the overall highest daytime values of the UTCI are reached. The maximum UTCI values shift from south-facing to south-west-facing façades in sunlit areas (Fig. 2d). Nevertheless, a developed turbulent flow results in generally increased spatiotemporal variability of UTCI patterns.

After the "*evening transition phase*" (17:00–20:00 UTC; Fig. 2e), the "*early-night phase*" (21:00–00:00; Fig. 2f) is formed with its unique spatial thermal exposure pattern. During this phase, higher UTCI values are clearly observed in narrow streets, and especially in courtyards with a high proportion of impervious surfaces, typically asphalt or concrete. In contrast, $T_{mrt}$ values in open spaces covered by lawn/grass surfaces and a few trees become substantially lower; UTCI values in these spaces resemble those expected for parks and gardens at any given time. The spatial UTCI distribution pattern during the early night phase moves towards the characteristic pattern of the late-night phase, as described, and the daily cycle of thermal exposure repeats itself (in the case of stable weather conditions).

## 3.2 *Microscale Diurnal Course of the UTCI in an Urban Area*

The daily course of the UTCI for the same period (19–21 of July) is in micro-scale largely dictated by street orientation and by locational features of the analyzed point. The key drivers are in the closest vicinity of each point of interest for analysis, as well as its properties. Daily maxima of the UTCI were observed in the east-to-west (or west-to-east) oriented streets without vegetation, typically on the north side of the street, close to south-facing walls (cf. green continuous line for the E3 point in Fig. 3). These locations are usually sunlit during most of the day with the maxima reaching up to 40 °C. The UTCI also increases in streets provided with buildings with a large windows fraction, and/or where high surface albedo predominates on both sides of the street (at both sunlit and shaded façades). It can be hypothesized that this effect is attributed to the strong reflection of short-wave radiation (already discussed in Belda et al. 2020 and Geletič et al. 2021). This effect would then be followed by the absorption of reflected radiation by pavement surfaces, which usually have a lower albedo and higher roughness (Lobaccaro et al. 2019), though these findings are still theoretical assumptions, which need to be verified by specially-targeted validation campaigns (Geletič et al. 2021). In contrast, areas located on the shaded side of such streets present the lowest maxima of the UTCI (around 33 °C), as south-facing walls are typically irradiated only in the early morning or in the late afternoon (cf. orange continuous line for the E4 point on Fig. 3).

A north-to-south street is characterized by a single daily maximum UTCI value. In the morning to noon phase, UTCI maxima for the west side of streets (near east-facing walls; cf. red continuous line for the N1 point in Fig. 3) appear before noon

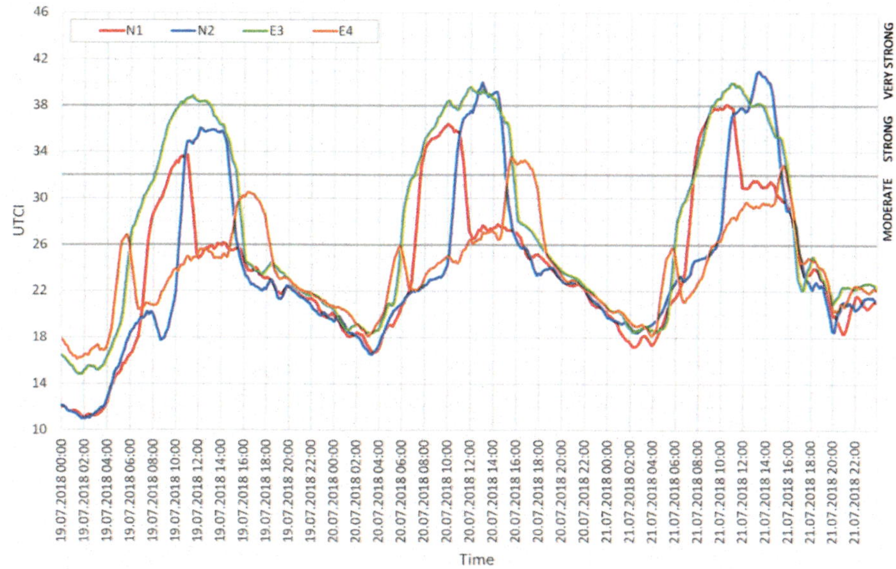


**Fig. 3** Moving averages ($n = 5$) of the UTCI index with heat stress intensity (right x-axis) in selected street-canyon locations (points N1, N2, located in a north-to-south street, points E3, E4 located in streets running east-to-west, cf. Fig. 1)

and can reach 36–38 °C. The maxima for the east side of the streets (adjacent to west-facing walls; cf. blue continuous line for the N2 point in Fig. 3) shifts to the afternoon and can be slightly higher as in east-west oriented streets; they can reach 41 °C. Finally, both sides of the streets show similar UTCI values at noon (Fig. 3). However, the results herein suggest that UTCI maxima on the east side of streets (adjacent to west-facing walls) reach values of up to 5 °C higher than those on the west side of streets (adjacent to east-facing walls); thus the effect of the momentary solar gain by the walls is the same as that of the already-heated surfaces (warmed by short-wave reflection and long-wave emission during morning hours).

The ability of trees in moderating heat stress in street canyons is demonstrated here and in previous studies. Model simulations estimate the cooling effect of trees within a range from 4 to 9 °C UTCI during the day and only minor UTCI increases are estimated for the night time period. Such increase is the effect of tree crowns, limiting radiative cooling of the heated surfaces. Preliminary results show increases within the range 0.6–1.0 K for coniferous trees and 0.2–0.5 for broad-leaved trees (in summer period, with considerable soil moisture).

A decrease of the UTCI is mainly determined by the parameters of the street canyon; street width, building height, trees placement and their properties (tree crown canopy and leaf area density, primarily). Densely planted streets with small distances between trees (cf. Fig. 4a and related UTCI in Fig. 4b), forming a "green tunnel" effect, yield lower values of the UTCI during daytime than streets with scattered trees (Resler et al. 2017). Leaf area density represents the second important factor

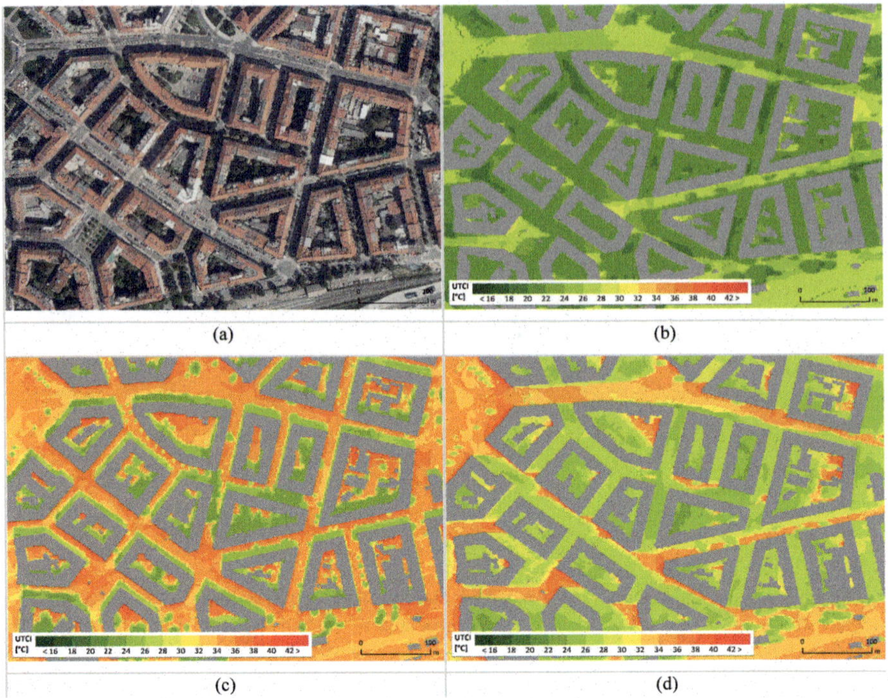


**Fig. 4** Orthophoto map (**a**) and examples of spatial variability of 10-min UTCI averages in street canyons and courtyards in the densely built-up quarter of Prague-Dejvice on 20 July 2018 at 06:00 UTC (**b**), 11:00 UTC (**c**) and 15:00 UTC (**d**)

for changes in the UTCI; trees without or with scattered leaves will provide less shading than densely-leaved crowns, with diminished possibilities to reduce heat stress. On the other hand, leaves and branches reduce the radiative cooling effect of the sky during night-time. Similar findings on the effect of vegetation on heat stress mitigation were obtained from measurements by Middel and Krayenhoff (2019) using the MaRTy device, modeled using ENVI-met and measured by Gatto et al. (2020), and measured by Park et al. (2012). A review of the cooling effect of urban green spaces is presented by Aram et al. (2019).

The results for daytime show that UTCI values in the closed courtyards with trees were about 5–10 °C lower than those in the courtyards with impervious surfaces (Fig. 4c). Trees are able to mitigate spells of moderate heat stress substantially, thus avoiding the occurrence of strong heat stress. Maximum UTCI values occurred in the afternoon, but this pattern might change due to level of shading in the courtyard. Interestingly, during the first part of the night, UTCI values remained slightly higher in paved courtyards (1–2 °C), a response to higher surface temperatures arising from heat storage in the urban fabric (Fig. 4d). Higher levels of long-wave radiation emitted by the tree crowns in comparison with the radiation from the sky contribute to higher $T_{mrt}$ and consequently to higher UTCI values in the course of the night, in a similar fashion to the situation in street canyons. During the second part of the night, UTCI differences between paved courtyards and green courtyards with trees tend to vanish. Still, in all cases, UTCI values during the night remain substantially below the level of thermal stress.

The presence of grass rather than paved surfaces in open public places or courtyards leads to less pronounced cooling effects than the effect of trees, in a range between 2 and 5 °C UTCI. These numbers probably are somewhat overestimated when comparisons are made with empirical studies performed in this region (e.g., Müller et al. 2014; Lehnert et al. 2020). The cooling effect of the grass in the urban environment depends heavily on soil moisture and other conditions (e.g. the structure of the soil layers and the leaf area density, LAD). A decrease in soil moisture, in particular, leads to an increase in surface temperature. This effect has been observed (Lehnert 2013) and also modeled (Resler et al. 2020a). Furthermore, on hot days, grass on high-quality soil and subsoil layers typically exhibits higher soil moisture than surfaces with only a shallow soil layer separated from the subsurface (e.g. non-irrigated green tram-lines) (Resler et al. 2020a).

### *3.3 Potential Application in Urban Planning*

Concerning urban adaptation and mitigation strategies, an adequate management of vegetation coverage in urban areas is relevant for mitigating heat stress and the impact of heat waves in cities. Results for summer time are in this case representative, since during that season the urban population is more strongly affected by heat-wave events (Geletič et al. 2018). UTCI output as well as $T_{mrt}$ can thus provide reliable information, since both variables are affected by tree-shading effects and radiative fluxes (especially direct short-wave radiation), as recently observed by several authors reporting the heat-mitigating effect of plants (e.g., Morakinyo et al. 2017; Resler et al. 2017; Rui et al. 2018; Gatto et al. 2020). However, several authors point out that due to the complexity of urban environments, an inconsequent planting of trees in streets with heavy traffic can have adverse effects, decreasing thermal discomfort, but at the same time increasing air pollution. Those effects were described by Resler et al. (2017, 2020a) in realistic domains during a validation campaign in two locations in Prague, including sensitivity testing of adaptation measures on synthetic (multiple copies of realistic) domain (Belda et al. 2020).

Moreover, potential municipality stakeholders may have unique ideas and requests regarding the presentation of relevant findings. The format of results shown to stakeholders represent an important point for discussion; for a basic analysis, maps or animations can be used, for a detailed analysis of selected locations, linear graphs can be more useful. In that case, PALM brings new complexity to the microscale urban climate research. At any rate, findings must be presented clearly, showing advantages or disadvantages of evaluated strategies, in a comprehensible form to urban planners and for the layman in modeling research.

# 4 Summary and Further Perspectives

The spatiotemporal thermal exposure patterns based on innovative modeling approach follow well previous research results. It was illustrated that PALM and other LES models enable spatially and temporally exhaustive, exact, and reputable tools for thermal exposure analyses which are crucial for further applications. Utilization of PALM in thermal comfort research brings several new opportunities and perspectives for practical applications of this research. First, components of blue-green infrastructure (Lehnert et al. 2020) in urban areas and various versions of their placements might be assessed in real urban environments not only from the point of view of thermal exposure, as illustrated in this chapter. Further calculations, for instance the assessment of thermal exposure of a pedestrian walking on a selected route in time (Middel and Krayenhoff 2019), might also be substantially improved. Undoubtedly, the model itself has some limitations so far and these limitations need to be emphasized and further minimized (Belda et al. 2020; Resler et al. 2020a, b). It should be stressed that the PALM model was not validated against measurements focused primarily on thermal comfort; the current validations concentrated primarily on urban meteorology (Resler et al. 2020a), wind dynamics (Gronemeier et al. 2020) and effects of land-surface interactions (Gehrke et al. 2020). Apart from that, it cannot be forgotten that thermal comfort should be considered as complex phenomena reflecting not only physical, but also social/behavioral and psycho-physiological factors. However, we trust that further developments in thermal exposure modeling, which we aimed to illustrate in this chapter, will subsequently accelerate investigation in other areas of thermal comfort research.

**Acknowledgements** The terrain-mapping campaign of building properties was co-financed by the Strategy AV21 project "Energy interactions of buildings and the outdoor urban environment" and by the Czech Academy of Sciences. We would like to thank Prof. Jiří Cajthaml and the students of the Faculty of Civil Engineering of the Czech Technical University, Prague, for their help with the terrain-mapping campaign. The PALM simulations, and pre- and postprocessing were performed with the HPC infrastructure of the Institute of Computer Science of the Czech Academy of Sciences (ICS), supported by the long-term strategic development financing of the ICS (RVO:67985807). Parts of the simulations were performed on the supercomputer of the IT4I Czech supercomputing center, supported by the Ministry of Education, Youth, and Sports from the Large Infrastructures for Research, Experimental Development, and Innovations project "IT4Innovations National Supercomputing Center—LM2015070". The WRF and CAMx simulations were performed on the HPC infrastructure of the Department of Atmospheric Physics of the Faculty of Mathematics and Physics of Charles University, Prague, supported by the Operational Program Prague—Growth Pole of the Czech Republic project "Urbanization of weather forecast, air-quality prediction, and climate scenarios for Prague" (CZ.07.1.02/0.0/0.0/16_040/0000383), which is co-financed by the EU. Author Jan Geletič was supported by the Czech Academy of Sciences under the program for research and mobility support of starting researchers (MSM100302001).

## References

Aram F, García EH, Solgi E, Mansournia S (2019) Urban green space cooling effect in cities. Heliyon 5(4):

Belda M, Resler J, Geletič J, Krč P, Maronga B, Sühring M, Auvinen M (2020) Sensitivity analysis of the PALM model system 6.0 in the urban environment. Geosci Model Dev Discuss 1–32

Bröde P, Fiala D, Błażejczyk K, Holmér I, Jendritzky G, Kampmann B, Havenith G (2012) Deriving the operational procedure for the Universal Thermal Climate Index (UTCI). Int J Biometeorol 56(3):481–494

Bruse M (2004) ENVI-met 3.0: updated model overview. University of Bochum. Retrieved from: www.envi-met.com

De Ridder K, Lauwaet D, Maiheu B (2015) UrbClim–A fast urban boundary layer climate model. Urban Climate 12:21–48

EPW—EnergyPlus Weather. Available online: https://energyplus.net/weather-location/europe_wmo_region_6/CZE//CZE_Prague.115180_IWEC. Accessed on 13 Jan 2021

Fiala D, Havenith G, Bröde P, Kampmann B, Jendritzky G (2012) UTCI-Fiala multi-node model of human heat transfer and temperature regulation. Int J Biometeorol 56(3):429–441

Fröhlich J, Von Terzi D (2008) Hybrid LES/RANS methods for the simulation of turbulent flows. Prog Aerosp Sci 44(5):349–377

Fröhlich D, Matzarakis A (2020) Calculating human thermal comfort and thermal stress in the PALM model system 6.0. Geosci Model Dev 13(7):3055–3065

Gál CV, Kántor N (2020) Modeling mean radiant temperature in outdoor spaces, A comparative numerical simulation and validation study. Urban Climate 32:

Gatto E, Buccolieri R, Aarrevaara E, Ippolito F, Emmanuel R, Perronace L, Santiago JL (2020) Impact of urban vegetation on outdoor thermal comfort: comparison between a Mediterranean city (Lecce, Italy) and a Northern European city (Lahti, Finland). Forests 11(2):228

Gehrke KF, Sühring M, Maronga B (2020) Modeling of land-surface interactions in the PALM model system 6.0: land surface model description, first evaluation, and sensitivity to model parameters. Geosci Model Dev Discuss 1–34

Geletič J, Lehnert M, Savić S, Milošević D (2018) Modelled spatiotemporal variability of outdoor thermal comfort in local climate zones of the city of Brno, Czech Republic. Sci Total Environ 624:385–395

Geletič J, Lehnert M, Jurek M (2020) Spatiotemporal variability of air temperature during a heat wave in real and modified landcover conditions: Prague and Brno (Czech Republic). Urban Climate 31:

Geletič J, Lehnert M, Krč P, Resler J, Krayenhoff ES (2021) High-resolution modelling of thermal exposure during a hot spell: a case study using PALM-4U in Prague, Czech Republic. Atmosphere 12(2):175

Gronemeier T, Surm K, Harms F, Leitl B, Maronga B, Raasch S (2020) Validation of the dynamic core of the PALM model system 6.0 in urban environments: LES and wind-tunnel experiments. Geosci Model Develop Discuss 1–26

Havenith G, Fiala D, Błazejczyk K, Richards M, Bröde P, Holmér I, Jendritzky G (2012) The UTCI-clothing model. Int J Biometeorol 56(3):461–470

Hellsten A, Ketelsen K, Sühring M, Auvinen M, Maronga B, Knigge C, Raasch S (2020) A nested multi-scale system implemented in the large-eddy simulation model PALM model system 6.0. Geosci Model Develop Discuss 1–45

Höppe P (1999) The physiological equivalent temperature–a universal index for the biometeorological assessment of the thermal environment. Int J Biometeorol 43(2):71–75

Huttner S (2012) Further development and application of the 3D microclimate simulation ENVI-met (Doctoral dissertation, Universitätsbibliothek Mainz)

Jendritzky G, de Dear R, Havenith G (2012) UTCI—why another thermal index? Int J Biometeorol 56(3):421–428

Kadasch E, Sühring M, Gronemeier T, Raasch S (2020) Mesoscale nesting interface of the PALM model system 6.0. Geosci Model Develop Discuss 1–48
Kántor N, Unger J (2011) The most problematic variable in the course of human-biometeorological comfort assessment—the mean radiant temperature. Central Eur J Geosci 3(1):90–100
Kántor N, Gál CV, Gulyás Á, Unger J (2018) The impact of façade orientation and woody vegetation on summertime heat stress patterns in a central European square: comparison of radiation measurements and simulations. Adv Meteorol
Khan B, Banzhaf S, Chan EC, Forkel R, Kanani-Sühring F, Ketelsen K, Sühring M (2020) Development of an atmospheric chemistry model coupled to the PALM model system 6.0: implementation and first applications. Geosci Model Develop Discuss 1–34
Krč P, Resler J, Sühring M, Schubert S, Salim MH, Fuka V (2021) Radiative Transfer Model 3.0 integrated into the PALM model system 6.0. Geosci Model Develop 14(5):3095–3120
Krüger EL, Minella FO, Matzarakis A (2014) Comparison of different methods of estimating the mean radiant temperature in outdoor thermal comfort studies. Int J Biometeorol 58(8):1727–1737
Lee H, Mayer H (2016) Validation of the mean radiant temperature simulated by the RayMan software in urban environments. Int J Biometeorol 60(11):1775–1785
Lehnert M (2013) The soil temperature regime in the urban and suburban landscapes of Olomouc, Czech Republic. Moravian Geograph Rep 21(3):27–36
Lehnert M, Tokar V, Jurek M, Geletič J (2020) Summer thermal comfort in Czech cities: measured effects of blue and green features in city centres. Int J Biometeorol 1–13
Lindberg F, Holmer B, Thorsson S (2008) SOLWEIG 1.0–Modelling spatial variations of 3D radiant fluxes and mean radiant temperature in complex urban settings. Int J Biometeorol 52(7):697–713
Lindberg F, Onomura S, Grimmond CSB (2016) Influence of ground surface characteristics on the mean radiant temperature in urban areas. Int J Biometeorol 60(9):1439–1452
Lobaccaro G, Acero JA, Sanchez Martinez G, Padro A, Laburu T, Fernandez G (2019) Effects of orientations, aspect ratios, pavement materials and vegetation elements on thermal stress inside typical urban canyons. Int J Environ Res Public Health 16(19):3574
Maronga B, Banzhaf S, Burmeister C, Esch T, Forkel R, Fröhlich D, Raasch S (2020) Overview of the PALM model system 6.0. Geosci Model Develop 13(3):1335–1372
Middel A, Krayenhoff ES (2019) Micrometeorological determinants of pedestrian thermal exposure during record-breaking heat in Tempe, Arizona: introducing the MaRTy observational platform. Sci Total Environ 687:137–151
Morakinyo TE, Kong L, Lau KKL, Yuan C, Ng E (2017) A study on the impact of shadow-cast and tree species on in-canyon and neighborhood's thermal comfort. Build Environ 115:1–17
Müller N, Kuttler W, Barlag AB (2014) Counteracting urban climate change: adaptation measures and their effect on thermal comfort. Theoret Appl Climatol 115(1):243–257
Park M, Hagishima A, Tanimoto J, Narita KI (2012) Effect of urban vegetation on outdoor thermal environment: field measurement at a scale model site. Build Environ 56:38–46
Resler J, Krč P, Belda M, Juruš P, Benešová N, Lopata J, Kanani-Sühring F (2017) PALM-USM v1. 0: a new urban surface model integrated into the PALM large-eddy simulation model. Geosci Model Develop 10(10):3635–3659
Resler J, Eben K, Geletič J, Krč P, Rosecký M, Sühring M, Vlček O (2020a). Validation of the PALM model system 6.0 in real urban environment; case study of Prague-Dejvice, Czech Republic. Geosci Model Develop Discuss 1–57
Resler J, Eben K, Geletič J, Krč P, Rosecký M, Sühring M, Belda M, Fuka V, Halenka T, Huszár P, Karlický J, Benešová N, Ďoubalová J, Honzáková K, Keder J, Nápravníková Š, Vlček O (2020b) PALM 6.0 revision 4508. [dataset]
Rui L, Buccolieri R, Gao Z, Ding W, Shen J (2018) The impact of green space layouts on microclimate and air quality in residential districts of Nanjing, China. Forests 9(4):224
Salim MH, Schubert S, Resler J, Krč P, Maronga B, Kanani-Sühring F, Schneider C (2020) Importance of radiative transfer processes in urban climate models: a study based on the PALM model system 6.0. Geosci Model Develop Discuss 1–55

Sievers U (2012) Das kleinskalige Strömungsmodell MUKLIMO_3 Teil 1: Theoretische Grundlagen, PC-Basisversion und Validierung. Berichte des Deutschen Wetterdienstes, Band 240, Offenbach am Main, Germany (German)

Sievers U (2016) Das kleinskalige Strömungsmodell MUKLIMO_3. Teil 2: Thermodynamische Erweiterungen. Berichte des Deutschen Wetterdienstes, Band 248, Offenbach am Main, Germany (German)

Schlünzen KH, Hinneburg D, Knoth O, Lambrecht M, Leitl B, Lopez S, Wolke R (2003) Flow and transport in the obstacle layer: first results of the micro-scale model MITRAS. J Atmos Chem 44(2):113–130

Stewart ID, Oke TR (2012) Local climate zones for urban temperature studies. Bull Am Meteor Soc 93(12):1879–1900

Tsoka S, Tsikaloudaki A, Theodosiou T (2018) Analyzing the ENVI-met microclimate model's performance and assessing cool materials and urban vegetation applications–A review. Sustain Cities Soc 43:55–76

Vanos JK, Baldwin JW, Jay O, Ebi KL (2020) Simplicity lacks robustness when projecting heat-health outcomes in a changing climate. Nat Commun 11(1):1–5

Verein Deutscher Ingenieure (VDI) (1994) Umweltmeteorologie–Wechselwirkung zwischen Atmosphäre und Oberflächen—Berechnung der kurz-und langwelligen Strahlung (Environmental meteorology–interaction between atmosphere and surfaces—Computation of short-and longwave radiation). (VDI 3789, Part 2). Beuth, Berlin

Verein Deutscher Ingenieure (VDI) (1998) Umweltmeteorologie: Methoden zur human-biometeorologischen Bewertung von Klima und Lufthygiene für die Stadt- und Regionalplanung, Teil 1: Klima [Environmental meteorology: methods for the human biometeorological evaluation of climate and air quality for urban and regional planning at regional level. Part 1: Climate (VDI 3787, Part I). Beuth, Berlin

Verein Deutscher Ingenieure (VDI) (2001) Umweltmeteorologie: Wechselwirkungen zwischen Atmosphäre und Oberflächen, Berechnung der spektralen Bestrahlungsstärken im solaren Wellenlängenbereich [Environmental Meteorology: Interactions between atmosphere and surfaces, calculation of spectral irradiances in the solar wavelength range] (VDI 3789, Part 3). Beuth, Berlin

Voogt JA, Oke TR (2003) Thermal remote sensing of urban climates. Remote Sens Environ 86(3):370–384

# Chapter 10
# The Universal Thermal Climate Index as an Operational Forecasting Tool of Human Biometeorological Conditions in Europe

**Claudia Di Napoli, Alessandro Messeri, Martin Novák, João Rio, Joanna Wieczorek, Marco Morabito, Pedro Silva, Alfonso Crisci, and Florian Pappenberger**

**Abstract** In operational weather forecasting standard environmental parameters, such as air temperature and humidity, are traditionally used to predict thermal conditions in the future. These parameters, however, are not enough to describe the thermal stress induced by the outdoor environment to the human body as they neglect the human heat budget and personal characteristics (e.g. clothing). The Universal Thermal Climate Index (UTCI) overcomes these limitations by using an advanced

---

C. Di Napoli (✉) · F. Pappenberger
European Centre for Medium-Range Weather Forecasts, Shinfield Road, Reading RG2 9AX, UK
e-mail: claudia.dinapoli@ecmwf.int

F. Pappenberger
e-mail: Florian.Pappenberger@ecmwf.in

A. Messeri · M. Morabito
Interdepartmental Centre of Bioclimatology, University of Florence, Florence, Italy
e-mail: alessandro.messeri@unifi.it

M. Morabito
e-mail: marco.morabito@ibe.cnr.it

A. Messeri
Department of Agrifood Production and Environmental Sciences, University of Florence, Florence, Italy

M. Novák
Czech Hydrometeorological Institute, Regional Office in Usti n.L., Usti n. L., Czech Republic

Faculty of Environment, Jan Evangelista Purkyne University, Usti n. L., Czech Republic

M. Novák
e-mail: martin.novak@chmi.cz

J. Rio · P. Silva
Portuguese Institute for Sea and Atmosphere, Lisbon, Portugal
e-mail: joao.rio@ipma.pt

P. Silva
e-mail: pedro.silva@ipma.pt


E. L. Krüger (ed.), *Applications of the Universal Thermal Climate Index UTCI in Biometeorology*, Biometeorology 4,
https://doi.org/10.1007/978-3-030-76716-7_10

thermo-physiological model coupled with a state-of-the-art clothing model. Several systems have been recently developed to operationally forecast human biometeorological conditions via the UTCI, i.e. by computing UTCI from the forecasts of air temperature, humidity, wind speed and radiation as provided by numerical weather prediction models. Here we describe the UTCI-based forecasting systems developed in Czech Republic, Italy, Poland, Portugal and at the pan-European scale. Their characteristics are illustrated and their potential as warning systems for thermal hazards discussed.

**Keywords** Thermal health hazards · Heat stress · Cold stress · Forecasting · Universal Thermal Climate Index · Biometeorology · Preparedness

## 1 Introduction

Air temperature is the core figure of any weather forecast. It seldom is, however, the only atmospheric parameter affecting the well-being of an individual. Wind may cool down the human body by taking away its heat. Sunlight and humidity may warm it up instead by radiation exposure and by limiting sweating, respectively. Knowing how the outdoor environment influences the heat exchange between the human body and its surroundings is mandatory to make the forecasts of air temperature, wind, radiation and humidity meaningful from a physiological point of view. Being based on an advanced heat budget model, the UTCI is able to provide this knowledge as a one-dimensional biometeorological-sound measure quantifying the thermal comfort/discomfort perceived by the human body when exposed to outdoors conditions. As the World Meteorological Organisation underlines the need for national meteorological services to incorporate biometeorological forecasts into their suite of products and services offered to the public (WMO 2004), the production and dissemination of UTCI forecasts in operational mode, i.e. as a 24/7 automated process, would represent a significant advancement in this sense.

The UTCI has been extensively applied in climatological studies to investigate heat- and cold-related hazards and their negative impacts on human health, also within vulnerable groups (Nastos and Matzarakis 2012; Morabito et al. 2014; Urban and Kyselý 2014; Krzyżewska et al. 2017; Romaszko et al. 2017; Błażejczyk et al. 2018; Di Napoli et al. 2018; Skutecki et al. 2019; Urban et al. 2021 for a more detailed review, refer to Chap. 2 of this book). Less research has however focused on assessing the UTCI as a predictor tool of thermal health hazards for public weather

J. Wieczorek
The Institute of Meteorology and Water Management – National Research Institute, Warsaw, Poland
e-mail: joanna.wieczorek@imgw.pl

M. Morabito · A. Crisci
Institute of BioEconomy National Research Council, Florence, Italy
e-mail: alfonso.crisci@ibe.cnr.it

services (Novák 2013). Computational power and data accessibility as well as the skill of numerical weather prediction (NWP) models have for years favoured the use of simpler standard parameters, such as air temperature and humidity, in the forecast of environmental conditions detrimental to human health. Recently, though, UTCI forecasts computed from a state-of-the-art NWP model have been shown to successfully predict thermal health hazards at different spatial and temporal scales across the globe thus demonstrating that technical capabilities are now mature for the implementation and use of the UTCI for operational purposes (Pappenberger et al. 2015).

To date, five operational forecasting systems based on the UTCI exist in Europe. They have been developed by the Czech Hydrometeorological Institute (CHMI), the Polish Institute of Meteorology and Water Management – National Research Institute (IMGW-PIB), the Portuguese Institute of the Sea and the Atmosphere (IPMA), the European Centre for Medium-Range Forecasts (ECMWF) and an Italian multi-institute consortium composed of the Italian National Research Council - Institute of BioEconomy (IBE-CNR), the University of Florence - Centre of Bioclimatology (CIBIC) and the Laboratory of Monitoring and Environmental Modelling for Sustainable Development (LaMMA). Each system is based on the *offline coupling* between a NWP model and the UTCI model, i.e. forecast outputs of air temperature, humidity, wind speed and radiation from a NWP model are passed to the UTCI model for the computation of UTCI forecasts. UTCI forecasts are either location specific (*point forecasts*) or mapped (*gridded forecasts*) and extend into the future at distinct time steps (*lead time*). These steps are embedded within a fully automated procedure, called operational chain, that develops over three main stages. First, the forecasts generated by a NWP model for air temperature, humidity, wind speed and radiation are retrieved. Second, these forecasts are input as forcings into the UTCI model (mostly used in its operational form, Bröde et al. 2012) which produces UTCI forecasts as outputs. Third, UTCI forecasts are disseminated to end-users. This chapter describes for the first time the five UTCI-based forecasting systems that, to the authors' knowledge, are currently operational in Europe. Their characteristics are illustrated and their role as thermal hazards warning systems discussed.

## 2 UTCI Forecasting Systems in Europe

The UTCI-based forecasting systems currently operational in Czech Republic, Italy, Poland, Portugal and at the pan-European scale are here presented and described. Their main characteristics are summarised in Table 1.

**Table 1** Summary table of UTCI forecasting systems currently operational in Europe

| | Czech Republic | Italy | Poland | Portugal | Europe |
|---|---|---|---|---|---|
| NWP model | ALADIN | GFS | GFS | ECMWF | ECMWF |
| Spatial resolution | 2.3 km | 12 km | City | City | 9–18 km |
| Temporal resolution | 1 h | 1 h | 6 h | 1–3 h | 6 h |
| Outputs | Maps, charts | Maps | Maps | Maps, charts | Maps |
| Maximum lead time | 3 days | 5 days | 2 days | 6 days | 10 days |
| End-users | National weather service | General public | General public, National weather service | General public, Civil protection agencies, Public health authorities | Civil protection agencies |
| Communication and dissemination channel | Internal (GRIB outputs and web) | Web platform | Web platform | Web platform | Web platform (internal) |

## 2.1 Czech Republic

CHMI, the national weather service for the Czech Republic, has been publishing biometeorological forecasts for the whole area of Czechia since 1994. In recent years a need had arised to build a new biometeorological forecasting system where the latest advancements in weather prediction and human biometeorology could be integrated. One of these advancements was the definition of the UTCI in 2009. Comparative studies performed against different thermal indices demonstrated the ability of the UTCI to better describe biometeorological conditions across different seasons and locations in the country (Novák 2011a, b, 2013). Following this, the UTCI was considered as the main reference parameter for the new Czech biometeorological forecasting system. The implementation of the UTCI as a forecasting tool required its operational computation from ALADIN, the NWP model currently in use at CHMI (Termonia et al. 2018). This was achieved in two phases. The first phase consisted in setting up an automated procedure for the computation of the mean radiant temperature ($T_{mrt}$). The $T_{mrt}$ is a critical physical quantity for the calculation of the UTCI and represents how human beings experience radiation, specifically its solar (i.e. shortwave) and thermal (i.e. longwave) components. ALADIN provides forecasts of both shortwave and longwave radiation fluxes. Since 2018, those forecast fluxes have been used to routinely calculate $T_{mrt}$ as model output via the methodology described by Fanger (1970). The second phase consisted in including also the operational computation of the UTCI into the model. This was achieved in 2019 alongside other upgrades of ALADIN, such as a higher spatial resolution (from 4.7 to 2.3 km) and a more accurate orography representation. UTCI forecasts are calculated via the operational procedure by Bröde et al. (2012). Their lead time is 54 h

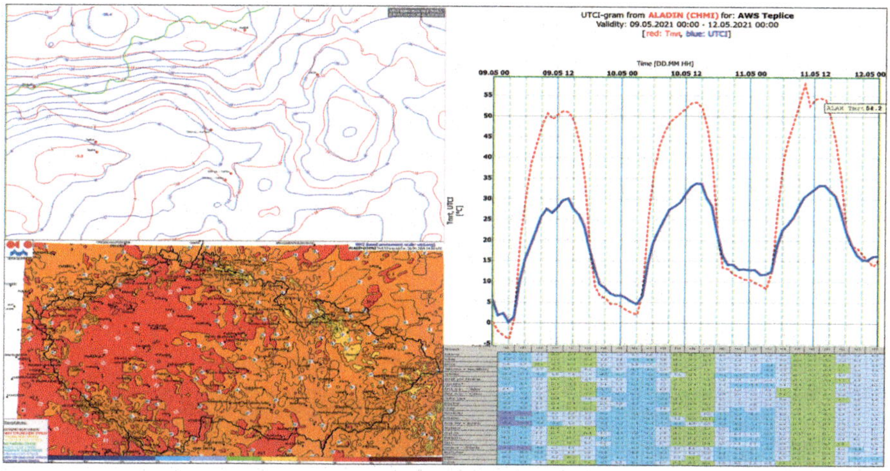


**Fig. 1** Overview of UTCI forecast outputs generated by the Czech Hydrometeorological Institute. Left: fields of maximum (red lines) and minimum (blue lines) UTCI values at 6 h interval, UTCI mapped distribution over the Czech Republic (bottom). Right: meteogram of UTCI and $T_{mrt}$ at Teplice (50.51° N, 13.84° E, 236 m asl, top), table of UTCI values predicted at different Czech locations (72-h lead time, bottom). Outputs are visualised via IBL Visual Weather

(18UTC release) or 72 h (00, 06, 12UTC release). Different types of UTCI outputs are produced, from meteograms to charts and maps (Fig. 1). These are currently available to meteorologists at central and regional CHMI forecasting offices (via SW Visual Weather directly from GRIB outputs or via internal web presentation).

## 2.2 *Italy*

In Italy, national-level forecasts of biometeorology conditions have been limited. Since 2007, however, the Italian National Research Council - Institute of BioEconomy (IBE-CNR), the University of Florence - Centre of Bioclimatology (CIBIC) and the Laboratory of Monitoring and Environmental Modelling for Sustainable Development (LaMMA) have been joining forces to realise an operational chain for the automated production of UTCI forecasts. The chain uses air temperature, humidity, wind speed and radiation data from the Global Forecast System (GFS) refined by a Limited Area Model (LAM). The LAM downscales the four variables to a higher horizontal resolution grid (12 km) than GFS (28 km) via an improved parametrization scheme of soil and surface exchanges. The R package HeatStress has been included in the chain to calculate water vapour pressure and globe temperature (Casanueva 2019). From the latter, $T_{mrt}$ can be computed using the methodology described in Thorsson et al. (2007). In order to reduce computational time, specific additional procedures have been embedded. Location-specific information, for instance, is extracted from available stations and used to calibrate outputs by using a set of Model Output Statistics (MOS) methodologies. From January 2017, hourly UTCI forecast maps of thermal comfort/discomfort for Italy are provided with a 5-day lead time. The maps are available to the general public on a web platform (http://www.lamma.rete.toscana.it/meteo/comfort-termico, Fig. 2) and show predicted UTCI values in full sunlight or shade as well as associated frostbite risk.

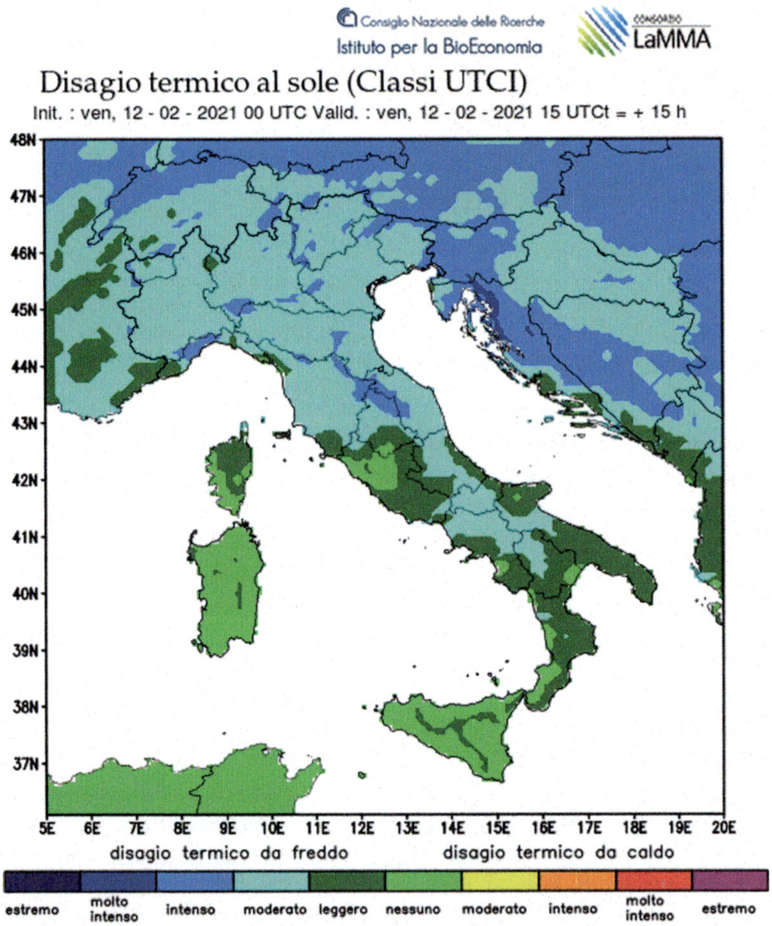


**Fig. 2** Hourly UTCI forecast maps as issued for Italy by a multi-institute consortium composed of the Laboratory of Monitoring and Environmental Modelling for Sustainable Development (LaMMA), the University of Florence Centre of Bioclimatology (CIBIC) and the Italian National Research Council Institute of BioEconomy (IBE-CNR). Maps are made available to the general population via a public web platform (http://www.lamma.rete.toscana.it/meteo/comfort-termico)

## 2.3 Poland

The implementation of the UTCI as a NWS product by the Polish meteorological service began in 2010. The UTCI is computed via application of the operational procedure described by Bröde et al. (2012) with the input parameter $T_{mrt}$ calculated using the procedure provided by the association of German engineers (VDI 3789 1994). Both calculations use as inputs meteorological parameters automatically derived from the GFS 0.25° model. Calculations based on data from the Consortium for Small-scale Modeling (COSMO-7) are also available but are not published to the public.

The first UTCI forecast issued for the general public appeared in June during the 2012 UEFA European Football Championship. Since February 2015, UTCI forecasts are provided daily and made publicly available as the "Heat load in the human body according to the UTCI)" on the web platform of the Polish official biometeorological service (https://biometeo.imgw.pl/%3Fpage%3DUTCI). The service is overseen by the Polish Institute of Meteorology and Water Management – National Research Institute (IMGW-PIB). UTCI forecasts are issued for 21 major cities in Poland with a 48-h lead time ("today" and "tomorrow") and a time resolution of 6 h (Fig. 3). They are initialised each morning and expressed in terms of UTCI categories (−5 to +4). A table describes each thermal stress category, the physiological responses correspondingly induced in the human body and possible protective measures. The choice of using UTCI categories instead of the single numerical values was made in order to avoid potential misunderstandings by end-users, such as interpreting UTCI values as an apparent (feels like) temperature.

The occurrence of strong heat stress conditions (32 °C < UTCI ≤ 38 °C) is also provided as an additional information to the Polish meteorological service. It has been observed in recent years that heatwaves start to affect Poland earlier and earlier in the spring/summer season. This poses a serious threat to local populations as they are not yet adapted to hot conditions just after winter. Specifically, the risk of mortality in the Polish population has been observed to increase more than 25% (mainly as a result of cardiovascular dysfunctions) when the UTCI is above 32 °C, making this a useful threshold for the surveillance of heat stress-related health impacts (KLIMADA 2013).

## 2.4 Portugal

The Portuguese Institute for Sea and Atmosphere (IPMA) has been using the UTCI as a biometeorological tool since 2010, when the first UTCI datasets started being computed from surface observations. Climatological studies were also performed to investigate the applicability of the index in mainland Portugal (Cunha et al. 2011). In recent years, the UTCI has been operationally computed using observations, with a

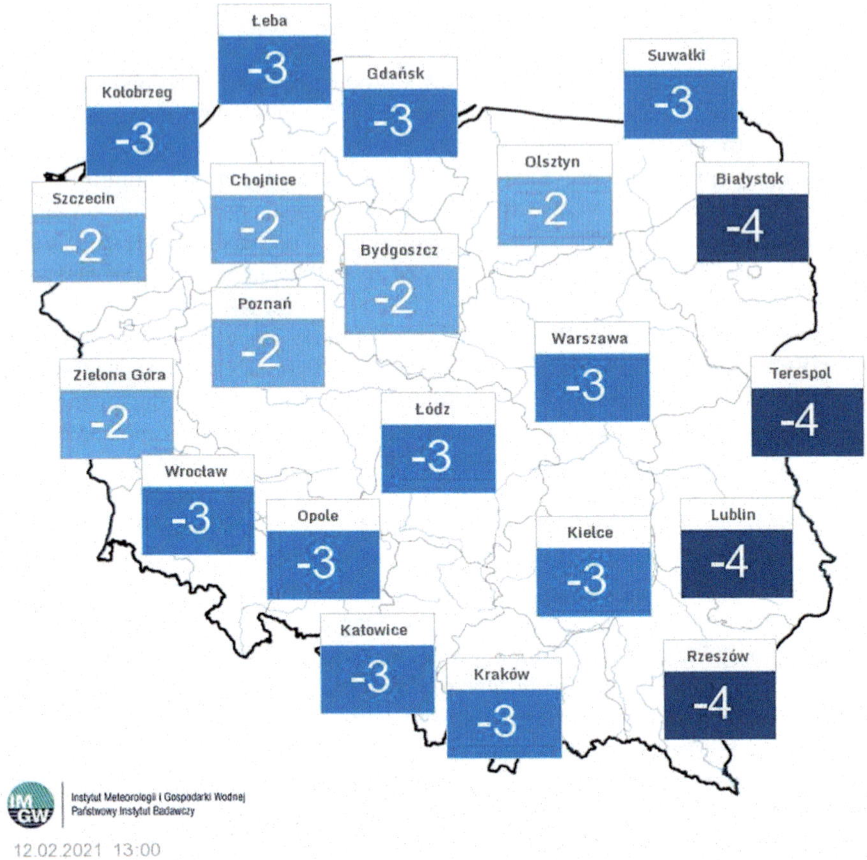


**Fig. 3** City-specific UTCI forecasts as issued for Poland by the Polish Institute of Meteorology and Water Management – National Research Institute (IMGW-PIB). Forecasts are available to the general population on a public web platform (https://biometeo.imgw.pl/%3Fpage%3DUTCI) where protective measures are suggested for predicted UTCI categories

3-h frequency and publicly shown as maps on the IPMA official website for mainland Portugal (http://www.ipma.pt/pt/oclima/biometeo/utci/; Fig. 4a).

Currently only point UTCI forecasts are computed for around 400 locations using statistically post-processed values of 2 m air temperature, relative humidity and 10 m wind speed (Rio et al. 2018). Radiation fluxes and surface temperature are from the ECMWF high resolution forecasting model. Specifically, radiation fluxes are direct model outputs whereas surface temperature is corrected to account for differences between real and model altitudes. Location-specific UTCI forecasts are provided hourly up to 72 h and 3-hourly from day 4 to 6. Whereas UTCI observational maps are available to the public, location-specific UTCI forecasts are exclusively intended for civil protection agencies and public health authorities. To these specific end-users,

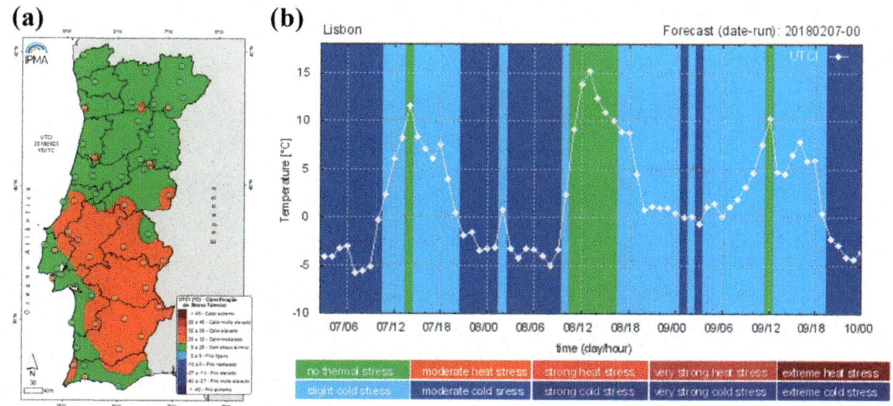


**Fig. 4** Examples of UTCI forecast outputs generated by the Portuguese national meteorological centre (IPMA). Panel **a** Nation-wide map of the UTCI as computed from surface observations for the current day and publicly available on the IPMA website (http://www.ipma.pt/pt/oclima/bio meteo/utci/). Panel **b** Time series of forecast UTCI values and thermal stress categories as provided to civil protection agencies and public health authorities

the biometeorological information carried by local UTCI forecasts is presented as a time series of forecast values with color-coded intervals overlapped to allow a fast assessment of the type and duration of thermal stress periods (Fig. 4b). This information is one of the tools used by local authorities in Lisbon for the activation of emergency plans to safeguard vulnerable communities, such as the homeless.

## *2.5 Europe*

Since 2016, an operational chain has been set up to deliver gridded maps of UTCI forecasts at the pan-European scale. The chain first takes as input the forecast fields of air temperature, humidity, wind and radiation as computed by the latest releases of ECMWF high resolution and ensemble models (00UTC and 12UTC). It then computes $T_{mrt}$ forecasts from solar and thermal radiation fields, and finally UTCI forecasts via the operational procedure (Bröde et al. 2012; Di Napoli et al. 2020a). UTCI forecasts are issued at 6-hourly steps with a 10-day lead time. The spatial resolution is 9 or 18 km according to whether input parameters are from ECMWF high resolution and ensemble model, respectively. Ensemble UTCI forecasts are the mean of the 50 UTCI forecasts that are correspondingly generated using as many ECMWF ensemble forecasts.

Pan-European UTCI forecasts are currently part of a decision-support platform that aims to enhance self-preparedness to single and multiple weather-induced hazard (Horizon2020 ANYWHERE project, EnhANcing emergencY management and response to extreme WeatHER and climate Events, Grant agreement ID: 700099).

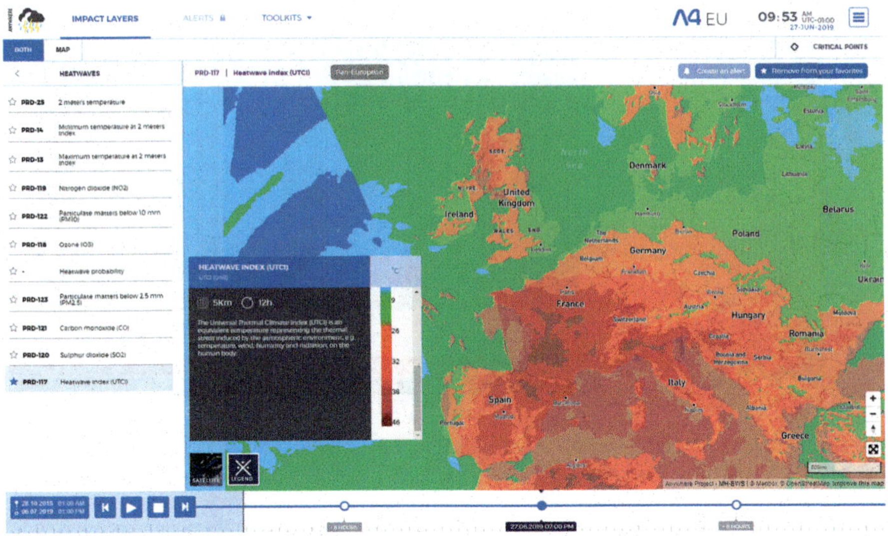


Fig. 5 Pan-European UTCI forecasts predicting the extreme thermal stress conditions achieved during the heatwave that affected the continent in June 2019

The platform is intended for civil protection agencies and local authorities only. Figure 5 shows the UTCI forecasts issued on 27 June 2019 and made available to first responders through the platform. Widespread conditions of very high and extreme heat stress were predicted in Western Europe in agreement with the intense heatwave episode affecting the area in the same period.

## 3 Discussion and Perspectives

The implementation and usage of UTCI forecasts across different European countries answer the call to integrate human thermal models with weather forecasts (Petersson et al. 2019). UTCI forecasts hence represent an opportunity to investigate the capabilities of this index to predict human thermal stress in an operational setting.

As predictors of future biometeorological conditions, the quality of UTCI forecasts needs to be assessed. This can be achieved by comparing the forecasts against observations of what actually occurred. Preliminary verification studies were carried out on UTCI forecasts issued in Poland and Czech Republic. For instance, between May and September 2018 the occurrence of strong heat stress conditions as forecast at 12 UTC in a selection of Polish cities was higher than what was correspondingly observed near IMGW-PIB stations. During a heatwave that occurred between 28 July and 10 August 2018, the difference between UTCI forecast values and UTCI observed values was up to 2 classes (Fig. 6a). In a meteorological station in the Czech

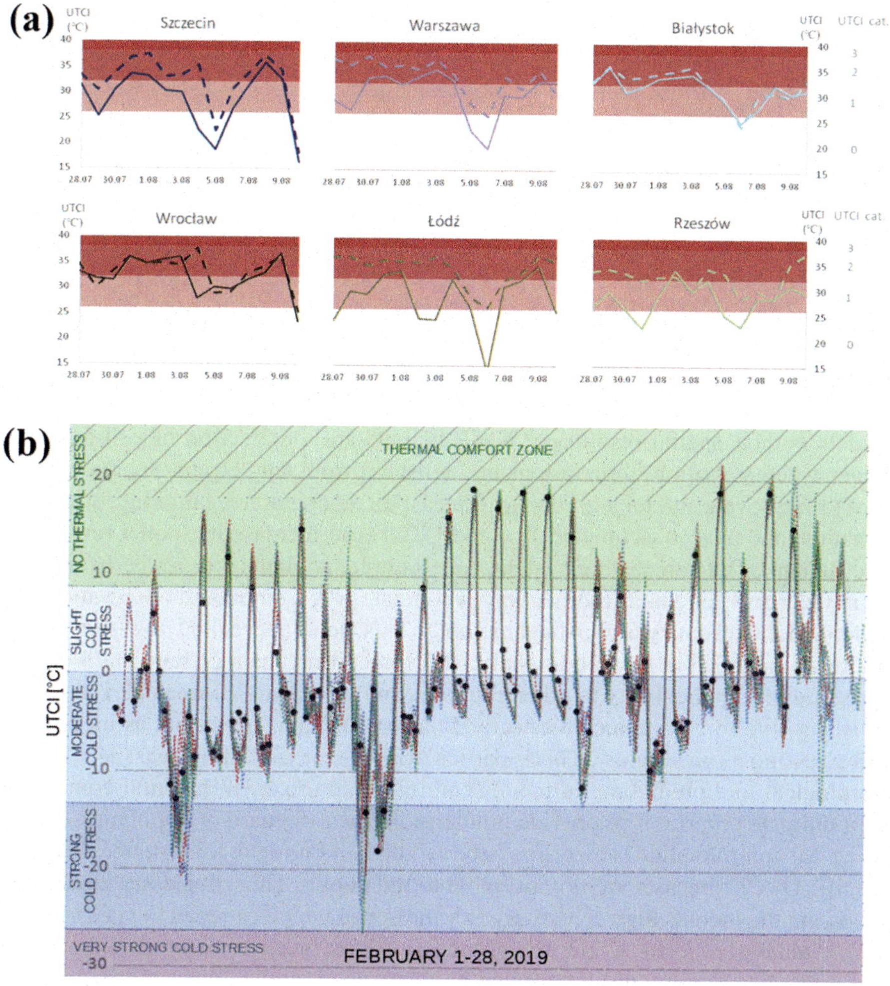


**Fig. 6** Preliminary validation results for UTCI forecasts issued in Poland and Czech Republic. Panel **a** UTCI values at 12UTC as forecast (dashed lines) and observed[1] (solid lines) at indicated Polish cities between 28 July and 10 August 2018. Panel **b** Time series of UTCI forecasts (dashed lines) and objective analysis (black dots) at Teplice, Czech Republic during February 2019

[1] 'Observed UTCI' calculations were based on historical data from telemetry and actinometrical stations of the IMGW-PIB network that are nearest to the city location. If longwave upwelling radiation and longwave downwelling radiation data (necessary to be included in the VDI procedure) were missing, the MRT values were calculated from the simplified formula of prof. Błażejczyk, which is based on measured global solar radiation data.

Republic, UTCI forecast values during February 2019 were instead lower than their observed counterparts (Fig. 6b). Reanalysis data, such as ERA5-HEAT, may also be used as a proxy for observations when assessing the quality of UTCI forecasts (Di Napoli et al. 2020b).

The discrepancy between forecasts and observations can be justified by the accuracy of NWP models, namely their ability to predict atmospheric processes such as cloud formations. Cloud cover and occurrence has been shown to be underestimated at low and middle altitudes (Hogan et al. 2009). This may influence the amount of radiation reaching and leaving the surface, with a potential overestimation of radiation fluxes (Weihs et al. 2012; Di Napoli 2020b). This can be observed, for instance, in the summer season when the formation of convective clouds is characterised by great time and spatial variability. Future model developments are expected to reduce these radiation-related uncertainties. Another factor influencing the accuracy of NWP models is their spatial resolution. NWP outputs are a collection of cell-averaged values whereas station observations are collected from single point locations. One way to improve a model's ability to forecast air temperature, humidity, wind and radiation, and as a consequence $T_{mrt}$ and UTCI is to increase its spatial resolution. The Italian UTCI forecasting system, for example, is planned to refine from 12 to 3 km their grid cell size. Another way is to relate NWP outputs to observational or additional model data via statistical methods. This technique, called *post-processing*, is already included in the operational UTCI forecasting chain of Italy and Portugal suggesting its applicability to other countries and at the pan-European level.

Being able to communicate effectively a forecast is as important as monitoring and assessing its uncertainty. This is particularly true for UTCI forecasts. As a biometeorological tool predicting human discomfort to future environmental conditions, their ultimate purpose is to provide guidance for the safeguard of populations' well-being. Misinterpretation is therefore to be avoided. In Portugal, for instance, location-specific UTCI forecasts are not available to the public. This choice has been made following the index's high sensitivity to wind speed which can lead to very negative UTCI values (e.g., −15 °C in Lisbon) and large differences from 2 m air temperature that might not be correctly interpreted by the general public. Other thermal indices are therefore disseminated on official IPMA forecasts.

In Poland, potential misunderstandings in the interpretation of UTCI forecasts have been avoided by showing a dimensionless number that corresponds to the predicted UTCI category rather than the specific UTCI numerical value expressed in °C. This representation, reinforced by the use of a blue–white–red scale, draws the attention directly on the type of thermal (cold or heat) load local populations will be exposed to. It also recommends via a self-explanatory table which protective measures should be taken against predicted thermal loads. This has the advantage to underline the nature of UTCI forecasts as a biometeorology advisory service. On the other hand, however, this representation relies on category thresholds. Slight differences in the forecasts might be responsible for having UTCI numerical values assigned to one category instead of another. Using post-processing to integrate model forecasts with observations might help reduce the uncertainty in the category assignment. This might become of great importance particularly when UTCI thresholds are deployed as triggers for preparatory actions (Di Napoli et al. 2019; Shartova et al. 2019; Urban et al. 2021).

Another aspect worth noting is the communication platforms used to disseminate UTCI forecasts. These are shown on websites by four out of five operational forecasting systems here described. The widespread distribution of technology, however, suggests that other tools may also represent an effective means for communicating forecasts and associated prevention information, especially to the general public. A smartphone app, for instance, has been recently developed in the Netherlands to display in a user-friendly way city-specific UTCI forecast (Heusinkveld et al. 2017). The Italian and Polish UTCI forecasting systems will disseminate its outputs via an app as well as a web platform in their next implementation.

Using the biometeorological information carried by UTCI forecasts as a decision making tool for thermal environments detrimental to human health has been a potential field of application for the UTCI since its definition. As supported by bioclimate impact research on thermal extremes (see e.g. Basarin et al. 2020), UTCI forecasts would be useful to both the general population and target groups, i.e. outdoor workers and sportsmen as well as the vulnerable groups (e.g. the elderly). They would also help optimise emergency response to thermal extremes when these occur at the same time as other meteorological health hazards, such as wildfires (Vitolo et al. 2019).

Some of the assumptions currently at the basis of the UTCI model—fixed moderate activity level, exposure time of 2 h, typical clothing for urban population—have however prevented the application of the UTCI in specific fields such as occupational health (NIOSH 2016). Improving the index so that it can account for the characteristics of each worker and each working activity would help make it a suitable, applicable tool for the assessment of thermal risk in the workplace too. Differently from the indices currently used in occupational heat health warning systems (e.g. wet bulb globe temperature; Morabito et al. 2019), the UTCI would be able to provide comfort/discomfort information for both heat and cold environments and on the basis of a thermo-physiological model. Given the importance of the topic, research is currently ongoing to extend the application of the UTCI to different activity levels and exposure duration (Bröde et al. 2016). This and further implementations, i.e. the effect of protective clothing in hot environments, will open the path to the UTCI as an application tool also in the occupational field (Gao et al. 2018).

## 4 Conclusions

Since its creation in 2009, the UTCI has been defined to help public weather and health services in the precautionary planning for thermal health hazards, such as heatwaves and cold spells. In recent years, the implementation of UTCI-based forecasting systems in Europe made this a reality. To date, five operational systems produce UTCI forecasts for the prediction of human comfort/discomfort conditions in the Czech Republic, Italy, Poland, Portugal and at the pan-European scale. UTCI forecasts are disseminated to and used by a wide range of end-users, from the general public to civil protection agencies and public health authorities. The application of UTCI forecasts by national weather services suggests that post-processing techniques can help improve the quality of the forecast and biometeorological information has to be communicated in a misinterpretation-free way.

## References


Basarin B, Lukić T, Matzarakis A (2020) Review of biometeorology of heatwaves and warm extremes in Europe. Atmosphere 12:1276. https://doi.org/10.3390/atmos11121276

Błażejczyk A, Błażejczyk K, Baranowski J, Kuchcik M (2018) Heat stress mortality and desired adaptation responses of healthcare system in Poland. Int J Biometeorol 62(3):307–318. https://doi.org/10.1007/s00484-017-1423-0

Bröde P, Fiala D, Błażejczyk K, Holmér I, Jendritzky G, Kampmann B, Tinz B, Havenith G (2012) Deriving the operational procedure for the universal thermal climate index (UTCI). Int J Biometeorol 56(3):481–494. https://doi.org/10.1007/s00484-011-0454-1

Bröde P, Kampmann B, Fiala D (2016) Extending the Universal Thermal Climate Index UTCI towards varying activity levels and exposure times. In: Proceedings of 9th Windsor conference: making comfort relevant, Cumberland Lodge, Windsor, UK, pp 73–79.

Casanueva A (2019) R Package "HeatStress". Zenodo. https://doi.org/10.5281/zenodo.3264930 (Accessed 1 Aug 2019)

Cunha S, Ferreira J, Marques J, Silva A, Mendes M, Numes L (2011) The Universal Thermal Climate Index case studies at Portuguese meteorological institute, EMS Annual Meeting Abstracts, vol 8, EMS2011-533

Di Napoli C, Barnard C, Prudhomme C, Cloke HL, Pappenberger F (2020) ERA5-HEAT: a global gridded historical dataset of human thermal comfort indices from climate reanalysis. Geosci Data J 00:1–9. https://doi.org/10.1002/gdj3.102

Di Napoli C, Pappenberger F, Cloke HL (2018) Assessing heat-related health risk in Europe via the Universal Thermal Climate Index (UTCI). Int J Biometeorol 62(7):1155–1165. https://doi.org/10.1007/s00484-018-1518-2

Di Napoli C, Pappenberger F, Cloke HL (2019) Verification of heat stress thresholds for a health-based heat-wave definition. J Appl Meteorol Climatol 58:1177–1194. https://doi.org/10.1175/JAMC-D-18-0246.1

Di Napoli C, Hogan R, Pappenberger F (2020) Mean radiant temperature from global-scale numerical weather prediction models. Int J Biometeorol 64:1233–1245. https://doi.org/10.1007/s00484-020-01900-5

Fanger PO (1970) Thermal comfort. Analysis and application in environment engineering. Danish Technical Press, Copenhagen, p 244

Gao C, Kuklane K, Ostergren PO, Kjellstrom T (2018) Occupational heat stress assessment and protective strategies in the context of climate change. Int J Biometeorol 62(3):359–371. https://doi.org/10.1007/s00484-017-1352-y

KLIMADA Project (2013) Adaptation to climate change in Poland. Republic of Poland, Ministry of the Environment

Krzyżewska A, Dobek M, Domżał-Drzewicka R, Rząca M (2017) Emergency interventions due to weather-related hypothermia. Weather 72:8–12. https://doi.org/10.1002/wea.2926

Heusinkveld B, Sterenborg G, Steeneveld GJ, Attema JJ, Ronda RJ, Holtslag AAM (2017) Smartphone app brings human thermal comfort forecast in your hands. Bull Am Meteor Soc 98:2533–2541. https://doi.org/10.1175/BAMS-D-16-0082.1

Hogan RJ, O'Connor EJ, Illingworth AJ (2009) Verification of cloud-fraction forecasts. Q J R Meteorol Soc 135:1494–1511. https://doi.org/10.1002/qj.481

Morabito M, Grifoni D, Crisci A, Fibbi L, Orlandini S, Gensini GF, Zipoli G (2014) Might outdoor heat stress be considered a proxy for the unperceivable effect of the ultraviolet-induced risk of erythema in Florence? J Photochem Photobiol 130:338–48. https://doi.org/10.1016/j.jphotobiol.2013.12.009

Morabito M, Messeri A, Noti P, Casanueva A, Crisci A, Kotlarski S, Orlandini S, Schwierz C, Spirig C, Kingma BRM, Flouris AD, Nybo L (2019) An occupational heat-health warning system for Europe: the HEAT-SHIELD platform. Int J Environ Res Public Health 16(16):2890. https://doi.org/10.3390/2Fijerph16162890

Nastos PT, Matzarakis A (2012) The effect of air temperature and human thermal indices on mortality in Athens, Greece. Theor Appl Climatol 108(3–4):591–599. https://doi.org/10.1007/s00704-011-0555-0

NIOSH (2016) NIOSH criteria for a recommended standard: occupational exposure to heat and hot environments. Jacklitsch B, Williams WJ, Musolin K, Coca A, Kim J-H, Turner N. Cincinnati, OH: U.S. Department of Health and Human Services, Centers for Disease Control and Prevention, National Institute for Occupational Safety and Health, DHHS (NIOSH) Publication 2016-106

Novák M (2011a) UTCI—first test in the Czech Republic. Prace i Studia Geograficzne, vol 47. Wydział Geografii i Studiów Regionalnych, Uniwersytet Warszawski, Warszawa, pp 319–325

Novák M (2011b) A comparison of an UTCI to some other selected indices. In: Šiška B, Hauptvogl M, Eliašová M (eds) Bioclimate: source and limit of social development, international scientific conference, 6th–9th September 2011, Topoľčianky, Slovakia. ISBN 978-80-552-0640-0

Novák M (2013) Use of the UTCI in the Czech Republic. Geogr Pol 86(1):21–28. https://doi.org/10.7163/GPol.2013.3

Pappenberger F, Jendritzky G, Staiger H, Dutra E, Di Giuseppe F, Richardson DS, Cloke HL (2015) Global forecasting of thermal health hazards: the skill of probabilistic predictions of the Universal Thermal Climate Index (UTCI). Int J Biometeorol 59(3):311–323. https://doi.org/10.1007/s00484-014-0843-3

Petersson J, Kuklane K, Gao C (2019) Is there a need to integrate human thermal models with weather forecasts to predict thermal stress? Int J Environ Res Public Health 16(22):4586. https://doi.org/10.3390/2Fijerph16224586

Rio J, Silva P, Novo I (2018) Automatic forecasts algorithms. VI Simposio Nacional de Predicción, Memorial Antonio Mestre, Madrid, 17–19 September. https://doi.org/10.31978/639-19-010-0.187

Romaszko J, Cymes I, Dragańska E, Kuchta R, Glińska-Lewczuk K (2017) Mortality among the homeless: causes and meteorological relationships. PLoS One 12(12):e0189938. https://doi.org/10.1371/journal.pone.0189938

Shartova NV, Shaposhnikov DA, Konstantinov PI, Revich BA (2019) Universal thermal climate index (UTCI) applied to determine thresholds for temperature-related mortality. Health Risk Anal 3:83–93. https://doi.org/10.21668/health.risk/2019.3.10.eng

Skutecki R, Jalali R, Dragańska E, Cymes I, Romaszko J, Glińska-Lewczuk K (2019) UTCI as a bio-meteorological tool in the assessment of cold-induced stress as a risk factor for hypertension. Sci Total Environ 688:970–975. https://doi.org/10.1016/j.scitotenv.2019.06.280

Termonia P, Fischer C, Bazile E, Bouyssel F, Brožková R, Bénard P, Bochenek B, Degrauwe D, Derková M, El Khatib R, Hamdi R, Mašek J, Pottier P, Pristov N, Seity Y, Smolíková P, Španiel O, Tudor M, Wang Y, Wittmann C, Joly A (2018) The ALADIN System and its canonical model configurations AROME CY41T1 and ALARO CY40T1. Geosci Model Dev 11:257–281. https://doi.org/10.5194/gmd-11-257-2018

Thorsson S, Lindberg F, Eliasson I, Holmer B (2007) Different methods for estimating the mean radiant temperature in an outdoor setting. Int J Climatol 27(14):1983–1993. https://doi.org/10.1002/joc.1537

Urban A, Kyselý J (2014) Comparison of UTCI with other thermal indices in the assessment of heat and cold effects on cardiovascular mortality in the Czech Republic. Int J Environ Res Public Health 11:952–967. https://doi.org/10.3390/ijerph110100952

Urban A, Di Napoli C, Cloke HL, Kysely J, Pappenberger F, Sera F, Schneider R, Vicedo-Cabrera AM, Acquaotta F, Ragettli MS, Iniquez C, Tobias A, Indermitte E, Orru H, Jaakkola JJK, Ryti NRI, Pascal M, Huber V, Schneider A, de Donato F, Michelozzi P, Gasparrini A. (2021) Evaluation of the ERA5 reanalysis-based Universal Thermal Climate Index on mortality data in Europe. Environ Res 198:111227. https://doi.org/10.1016/j.envres.2021.111227

VDI 3789 (1994) VDI 3789 Part 2:1994-10: Environmental meteorology—interactions between atmosphere and surfaces—calculation of short-wave and long-wave radiation. Beuth, Berlin, p 52

Vitolo C, Di Napoli C, Di Giuseppe F, Cloke HL, Pappenberger F (2019) Mapping combined wildfire and heat stress hazards to improve evidence-based decision making. Environ Int 127:21–34. https://doi.org/10.1016/j.envint.2019.03.008

Weihs P, Staiger H, Tinz B, Batchvarova E, Rieder H, Vuilleumier L, Maturilli M, Jendritzky G (2012) The uncertainty of UTCI due to uncertainties in the determination of radiation fluxes derived from measured and observed meteorological data. Int J Biometeorol 56(3):537–555. https://doi.org/10.1007/s00484-011-0416-7

WMO (2004) Guidelines on biometeorology and air quality forecasts. WMO/TD No. 1184, PWS-No. 10. Kusch W, Fong HY, Jendritzky G, Jacobsen I. In: Kootval H (ed) Geneva, Switzerland

# Chapter 11
# Proposed Framework for Establishing a Global Database for Outdoor Thermal Comfort Research

**Kevin Ka-Lun Lau and Eduardo L. Krüger**

**Abstract** In the last two decades, studies of subjective thermal comfort have been widely conducted in the urban environment, covering different cultures and climatic regions. However, shortcomings can be observed mainly with respect to protocols used in terms of assessment scales of thermal perception, calculated thermal indices and instrumental setup used for micrometeorological measurements. Such data may vary considerably across studies due to constraints at field sites and the availability of instruments, making it difficult for inter-comparisons between studies and climatic regions, calibrations of thermal indices and a true understanding of people's thermal perception in outdoor settings. There is a need for standardisation of methodology and guidance for conducting field surveys in outdoor spaces with implications on climate-sensitive urban design, public health measures and adaptation of humans to a changing climate. The objective of this proposed framework is to develop a standard methodology for outdoor thermal comfort surveys towards the creation of a worldwide outdoor comfort database.

**Keywords** Data repository · Thermal comfort surveys · Thermal monitoring · Outdoor comfort · UTCI

## 1 Introduction

In the last few decades, climatic effects on the comfort of building occupants have been widely studied in terms of mechanisms and most predominant influencing factors, informing building and urban designers for better design for the comfort and health of building occupants (Olgyay 1963; Givoni 1976; Nicol and Raja 1996).

---

K. K.-L. Lau (✉)
Institute of Future Cities, The Chinese University of Hong Kong, Shatin, Hong Kong
e-mail: kevinlau@cuhk.edu.hk

E. L. Krüger
Federal University of Technology of Parana, Curitiba, Brazil
e-mail: ekruger@utfpr.edu.br



E. L. Krüger (ed.), *Applications of the Universal Thermal Climate Index UTCI in Biometeorology*, Biometeorology 4,
https://doi.org/10.1007/978-3-030-76716-7_11

Such advances proved to be important to the health and well-being, as well as productivity and living quality of building occupants (Givoni 1998). The constituents of the physical environment act directly upon human body which tends to achieve biological equilibrium through physical and psychological reactions (Olgyay 1963). As such, the research area dealing with human thermal comfort is focused at the energy exchanges between human body and the climatic environment of the surroundings, and aims to achieve the "comfort" conditions as defined by integrating climatic variables such as temperature, humidity, air movement and radiation.

The shorter time people spent (e.g. in the range of minutes) in the outdoor environment than the indoors also influences the thermal exposure. Höppe (2002) suggested that the steady-state assumption of indoor thermal comfort does not provide realistic assessments for outdoor settings. His previous study based on the Instationary Munich Energy-Balance Model (Höppe 1989) showed that thermo-physiological parameters such as skin and core temperatures take at least one hour in outdoors to achieve the steady-state level. The complex outdoor environment also creates large variations in thermal conditions that the outdoor space users are exposed to. Lau et al. (2019a) showed that subjective thermal sensation changes considerably when pedestrians travel outdoors and suggested that the environmental conditions exposed have a lag effect on thermal perception of pedestrians. Therefore, the assessment of human thermal comfort in outdoor environment requires a different methodological framework and analytical approach in order to address the distinctive relationship between subjective thermal sensation and environmental conditions experienced by outdoor space users. Such short-term acclimatization effects on reported thermal sensation in the context of UTCI applications are discussed in a previous chapter of this book (Chap. 5: Long and short-term acclimatization effects on outdoor thermal perception versus UTCI).

## 1.1 Subjective Assessment of the Thermal Environment

Conventionally, questionnaire surveys are used to study human thermal comfort to obtain a subjective assessment of the thermal conditions that the respondents are exposed to. There are a wide range of subjective assessment scales for the thermal environment, including perceptual or affective, global or localised, instantaneous or covering certain period of time (ISO 10551 2019). Subjective judgement also varies from the surrounding environment to the person of assessment, from general thermal conditions to specific components such as temperature and air movement, from permanent to temporary situation. The ISO 10,551 provides five subjective judgement scales to describe the thermal state of a person, including thermal perception, thermal comfort, thermal preference, personal acceptability and personal tolerance. The ASHRAE Standard 55 (2017) also provides a scale for thermal perception (commonly known as the ASHRAE 7-point scale) and thermal acceptability (Table 1). However, these standards were not designated for outdoor conditions and their applications in previous studies vary considerably across local contexts.

**Table 1** Protocols for subjective perception of thermal environment (Johansson et al. 2014)

| Parameter | Standard | Interview question and measurement scale |
|---|---|---|
| Thermal sensation or perception | ISO10551<br>ASHRAE | 'How are you feeling now?'<br>7 Point scale: cold (−3), cool (−2), slightly cool (−1), neutral (0), slightly warm (+1), warm (+2) and hot (+3)<br>9-point scale: above plus 'Very cold' (−4) and 'Very hot' (+4) (mainly for use in extreme environments)<br>'What is your general thermal sensation?'<br>7-Point symmetrical thermal perception scale (equal in wording to the ISO 10,551) |
| Thermal comfort (affective evaluation) | ISO10551 | 'Do you find this environment…?'<br>4-Point: comfortable (0) as the point of origin followed by slightly uncomfortable (1), uncomfortable (2), very uncomfortable (3) |
| Thermal preference | ISO10551 | 'Please state how you would prefer it to be now'<br>7-Point: much cooler (−3), cooler (−2), slightly cooler (−1), neither warmer nor cooler (0), a little warmer (+1), warmer (+2) and much warmer (+3) |
| Personal acceptability | ISO10551 | 'On a personal level, this environment is for me…'<br>Two-category statement: acceptable rather than unacceptable (0) and unacceptable rather than acceptable (1)<br>Continuous scale: clearly acceptable, just acceptable, just unacceptable and clearly unacceptable |
| Personal tolerance | ISO10551 | 'Is it…?'<br>5-Point: perfectly tolerable (0), slightly difficult to tolerate (1), fairly difficult to tolerate (2), very difficult to tolerate (3) and intolerable (4) |

There is a lack of standard guidelines or procedures for subjective assessment of the outdoor thermal environment while there are considerable discrepancies in the use of questions and assessment scales. The ASHRAE 7-point scale was commonly used (Krüger and Rossi 2011; Lau et al. 2019b) while 5-point (Nikolopoulou and Lykoudis 2006; Metje et al. 2008) and 9-point scales (Kántor et al. 2012) were also used in some studies for specific purposes. There is usually a middle point in the assessment scale, but the terms used to describe this middle point include "neutral", "comfortable", "neither cool nor warm" and "acceptable". The assessment of other meteorological components, such as solar radiation, air movement and humidity, was also used in some studies (Stathopoulos et al. 2004; Villadiego and Velay-Dabat 2014; Lau et al. 2019b). Moreover, the personal state of thermal comfort (affective evaluation) and thermal preference were sometimes included in the thermal assessment (Oliveira and

Andrade 2007; Ng and Cheng 2012). The inconsistencies in subjective scales and semantics used lead to possible errors in comparisons between results from different studies.

Personal factors such as biological sex, body weight, and skin colour were also found to be associated with the subjective assessment of the thermal environment (Krüger and Drach 2017). It is suggested that variations in thermo-physiological promote changes in adaptation to the thermal environment. Previous studies also suggested that human behaviour is another determinant of thermal perception (Knez and Thorsson 2006), while reasons for visit and cultural background were widely regarded as psychological mechanisms of thermal adaptation in outdoor environments (Nikolopoulou et al. 2001). However, these factors were not addressed by several studies and the methods of assessment need to be standardised to produce more accurate and reliable results.

## *1.2 Thermal Comfort Indices for the Outdoor Environment*

There have been more than 100 different thermal indices developed to describe the heat exchange between the human body and its surrounding environment (Błażejczyk et al. 2012). Energy balance models of human body were developed and widely used in the 1970s–1980s, with a number of biometeorological indices developed for the assessment of thermal stress and strain (Höppe 1997). One of the commonly used indices is the Predicted Mean Vote (PMV) which provides a practical and easily programmable heat balance model of human body (Fanger 1970). It has since been a widely adopted biometeorological index to describe the predicted mean thermal perception under indoor conditions. Pickup and de Dear (2000) developed a physiologically valid outdoor comfort index (OUT_SET*) by adapting the indoor comfort index SET* to outdoor settings. This involves an estimation of the amount of shortwave and longwave radiation absorbed by the human body and hence determines an outdoor mean radiant temperature.

The Munich Energy-balance Model for Individuals (MEMI) was later developed to incorporate individual heat fluxes, body temperatures, sweating rates and skin wettedness into the assessment of the thermal conditions of the human body in a physiologically relevant way (Höppe 1984). It also forms the basis of the Physiological Equivalent Temperature (PET) which is defined as "the air temperature at which, in a typical indoor setting (without wind and solar radiation), the heat budget of the human body is balanced with the same core and skin temperature as under the complex outdoor conditions to be assessed" (Höppe 1999, p. 71). PET has been used in studies of outdoor thermal comfort in different climates and urban settings (Lin 2009; Ng and Cheng 2012; Krüger 2017).

Another commonly used thermal index which has been widely employed in the last decade is the Universal Thermal Climate Index (UTCI), whose applications are highlighted in this book. The UTCI is defined as "the air temperature which would produce under reference conditions the same thermal strain as in the actual thermal environment" (Błażejczyk et al. 2010). It is therefore a one-dimensional quantity which represents the human physiological reaction to the actual thermal conditions defined by multiple dimensions. It was developed based on the UTCI-Fiala model, which was adapted to predict human responses to outdoor climate conditions. The model also considers behavioural adjustments of the clothing insulation according to outdoor air temperature as well as the effect of air movement, walking speed and clothing's thermal and evaporative resistances (Havenith et al. 2012). The UTCI has been widely used in the assessment of outdoor thermal environments (Bröde et al. 2012; Krüger et al. 2017; Oh et al. 2019—for a more detailed description of such application, refer to Chap. 2 of this book: Literature Review).

## *1.3 Micrometeorological Measurements*

The measurement of micrometeorological conditions is an integral part of outdoor thermal comfort studies since it provides background data for comparisons to subjective thermal perception. Oke (2006) presents a set of guidelines for meteorological observations in urban areas while ISO 7726 (1998) and ASHRAE Handbook of Fundamentals (ASHRAE 2017) also provide a description of instruments that suit thermal comfort measurements for indoors. However, additional considerations are necessary for the adequate exposure of instruments, the measurement of wind speed, and the estimation of the mean radiant temperature $T_{mrt}$ (Johansson et al. 2014).

Temperature and humidity sensors may be affected by radiation sources like solar radiation and heated urban surfaces, leading to overestimation of the air temperature. As such, shielding and ventilation of sensors are required to minimise the radiative exchange between the instrument and its surroundings and avoid the accumulation of warm air around the probe. Cheng et al. (2012) argued that the radiation shield may not be sufficient to prevent overestimation of air temperature so correction to the results may be required. Wind speed is also an important variable in the assessment of thermal comfort and the type of sensors may affect the accuracy of measurements. Two-dimensional anemometers are commonly used but the turbulence outdoors may result in an underestimation of actual wind speed.

$T_{mrt}$ is a critical variable in the assessment of thermal comfort, particularly during warm and sunny weather conditions (Mayer and Höppe 1987) since it represents the aggregated short- and long-wave radiation fluxes in the surroundings that a human body is exposed to (Johansson et al. 2014). It can be determined by two common approaches, namely integral radiation measurements with the inclusion of angular factors and global thermometer combined with measurements of air temperature and wind speed (Thorsson et al. 2007). The large variations in the use of instruments cause inconsistencies and issues in comparisons between studies.

### *1.4 Objectives of the Study*

The present study aims to: (1) prioritise the elements of outdoor thermal comfort studies such as subjective thermal sensation, affective evaluation of thermal comfort, thermal preference for better understanding of human thermal comfort at international level; (2) develop an internationally recognised standard methodology for conducting field studies of outdoor thermal comfort; and (3) establish a database of outdoor thermal comfort surveys by collating existing data from studies conducted in different climates. The methodological framework of the research is described, and the potential applications are discussed. This forms the basis for the broader objectives of the development of the global database of outdoor thermal comfort studies. In connection with the aims of this book, advantages of the proposed repository toward an extensive usage of UTCI are also introduced in this chapter.

## 2 Data Acquisition

### *2.1 Identification and Acquisition of Relevant Data Sources*

Figure 1 shows the methodological framework of the proposed study. At the preliminary stage, a desktop study was conducted to identify relevant data sources for the inclusion in the database. Articles indexed in journal databases such as PubMed, Web of Science, Scopus and SpringerLink were retrieved and shortlisted for relevance, using relevant keywords including (but not limited to) "outdoor thermal comfort", "human thermal comfort", "thermal perception", "thermal assessment", "outdoor environment" and "questionnaire survey". The authors of relevant studies were contacted for their interest in contributing to the database. At the same time, a call for contributions to a pilot study was made in the newsletter of the International Association for Urban Climate in mid-2019 (http://www.urban-climate.org/wp-content/uploads/IAUC072.pdf). In the pilot study, there were 20 responses from researchers worldwide covering a range of climatic regions (Fig. 2).

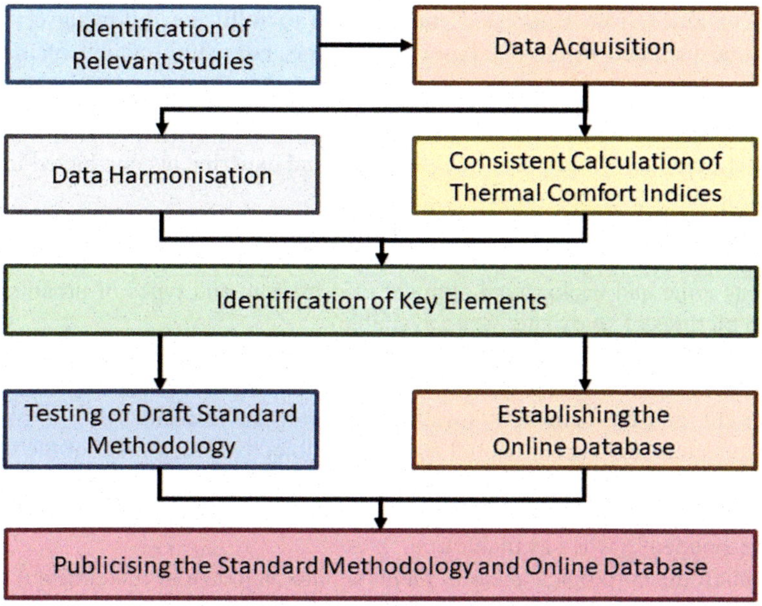


**Fig. 1** Methodological framework of the development of the data repository

**Fig. 2** Locations of the field data from different countries included in the pilot study

Data obtained from shortlisted studies have to fulfil the following criteria in order to be included in the database. A template, consisting of unit of measurements, code names, and coding conventions, was provided for data contributors. Recommendations were as follows:

- Data should be collected from field surveys and experiments conducted in semi-outdoor or outdoor environments;
- Metadata of the study are required, including (but not limited to) dates of questionnaire survey and relevant micrometeorological measurements, number of samples, climate zone and background climatic information, and types of urban settings (with pictures of study site, when available);
- Both subjective (questionnaire survey) and instrumental (micrometeorological measurements) data are required and they should be simultaneously collected to represent the right-here-right-now response from the respondents;
- The questionnaire survey should consist of subjective assessment of the thermal environment, estimated metabolic rate and clothing level of the respondents, immediate thermal history (if any), biometric information, and age and sex of the respondents (when available);
- The micrometeorological measurements should include four fundamental parameters for calculating thermal comfort indices, namely air temperature, humidity, air movement, globe temperature (or three-dimensional measurements of radiation fluxes for calculating mean radiant temperature, $T_{mrt}$). Technical specifications of instruments/sensors and detailed instrumental settings will also be required;
- Raw data are required, i.e. not from processed or published data. However, data must have been published in peer-reviewed journals or conference papers. Therefore, publication metadata should be provided. Coding of the data should also be clearly defined by data contributors.

## *2.2 Data Harmonisation*

Subjective thermal perception is often compared to objective micrometeorological measurements in order to understand the subjective–objective relationship of thermal assessment. As different assessment scales (e.g. number of points on the scales) were adopted by previous studies, there is a need to harmonise the datasets obtained from different studies. Table 2 shows the different assessment scales adopted by questionnaire surveys of previous studies. In this study, the ASHRAE 7-point scale will be used as the standard assessment scale for thermal sensation such that studies using 5-point or 9-point scale will be converted to the ASHRAE 7-point scale. The rescaled data were evaluated against the original data based on the mean, variance, kurtosis and skewness values. To further assess whether the rescaled data retain the original structure, data of selected studies were compared to other studies in similar climatic regions.

**Table 2** Different assessment scales adopted by thermal comfort studies

| | | | | | | | | |
|---|---|---|---|---|---|---|---|---|
| *Kántor et al.* (2012)—*9-point scale* | | | | | | | | |
| Very cold (−4) | Cold (−3) | Cool (−2) | Slightly cool (−1) | Neutral (0) | Slightly warm (+1) | Warm (+2) | Hot (+3) | Very hot (+4) |
| *Lau et al.* (2019b)—*7-point scale* | | | | | | | | |
| Cold (−3) | Cool (−2) | | Slightly cool (−1) | Neutral (0) | Slightly warm (+1) | Warm (+2) | | Hot (+3) |
| *Aljawabra and Nikolopoulou* (2018)—*5-point scale* | | | | | | | | |
| Cold (−2) | | Cool (−1) | | Neutral (0) | Warm (+1) | | Hot (+2) | |

## 2.3 *Applicability of UTCI in Regional Comparisons*

Thermal comfort indices are important to provide an objective assessment of the thermal environment. As pointed out by de Dear (1998), there are potential sources of "noise" in the thermal comfort indices since different versions of computer algorithms may have been used to calculate such indices. In order to avoid these inconsistencies, the UTCI is selected and calculated from the raw data of micrometeorological measurements acquired from data contributors. The UTCI has the advantage to represent the assessment of the thermophysiological effects of the atmospheric environment, which is one of the key issues in human biometeorology. The software BioKlima (Błażejczyk 2011) was used to calculate the thermal comfort indices. It provides easy calculations of more than 60 various biometeorological and thermophysiological indices. The mandatory inputs of meteorological variables include air temperature, relative humidity, mean radiant temperature, wind speed, metabolic rate and clothing level (thermal insulation). The resulting UTCI data can then be integrated into the datasets for subsequent data analysis.

As several thermal indices have been used in outdoor studies for quantifying the thermal conditions experienced by individuals, the data repository used the UTCI as the principal thermal index in order to reduce the inconsistencies across different studies. The UTCI was found to be well correlated with other thermal indices, especially those based on human heat balance, e.g. Standard Effective Temperature for outdoor conditions (SET*), Perceived Temperature (PT) and Physiological Equivalent Temperature (PET) (Błażejczyk et al. 2012). This indicates that the UTCI is capable of representing the thermal conditions exposed to human body and the physiological responses to such thermal conditions. The universal assessment scale of the UTCI also allows comparison across different climatic regions and local contexts. This is particularly important for thermal comfort studies as individual adaptations to the thermal conditions vary considerably across local contexts (see Chap. 5: Regional adaptation of the UTCI: comparisons between different datasets in Brazil). By including data acquired from different climatic regions worldwide, the data repository facilitates further studies of how individuals are acclimatised, and their adaptation mechanisms, in terms of physiological, psychological, and cultural perspectives. At a later stage, improvements in UTCI predictions for various climates can be useful for urban designers to provide thermally favourable outdoor spaces in those locations.

One of the potential UTCI applications is in weather forecasting based on synoptic data. Błażejczyk et al. (2012) suggested that the UTCI shows good correlations with synoptic weather variables, by comparing long-term records of air temperature, wind speed, and $T_{mrt}$ acquired from meteorological stations in Freiburg, Germany. It implies the applicability of human-heat-balance-based indices in weather forecasting and provides information of human thermo-physiological conditions for general public and health practitioners to better prepare for extreme weather (hot or cold) conditions. Such application is presented in more detail in Chap. 9: The Universal Thermal Climate Index as an operational forecasting tool of human biometeorological conditions in Europe). As shown in Chap. 2 of this book (Literature review), the applicability of the UTCI was also studied at micro-scale climate by examining field data measured in cold, hot and dry, and hot and humid climates. It was found that the UTCI is generally consistent with most of the thermal indices and it is able to capture the variation of wind speed in cold climate while other thermal indices fail to show such variations (Błażejczyk et al. 2012). Cities and urban built-up areas have specific climatic conditions due to the complex urban geometry (e.g. orientation and disposition of building blocks). Important features of urban climate such as the urban heat island phenomenon, reductions in ventilation, complex spatial variations of exposure to solar radiation at pedestrian level can be well-represented by field measurements and numerical simulation and mapping of the UTCI (cf. Chap. 7: Application of UTCI in high-resolution urban climate modeling techniques). There is thus a great potential of the data repository in terms of facilitating such analyses and feeding them with field measurements carried out in different climatic regions, thereby estimating outdoor thermal conditions based on the UTCI.

## 3 Analytical Procedures

The following subsections outline the necessary steps that shall be taken for establishing the global database for outdoor thermal comfort research.

### *3.1 Identification of Key Elements in Outdoor Thermal Comfort Surveys*

Subjective assessment of the thermal environment includes thermal perception, thermal comfort (affective evaluation), thermal preference, personal acceptability and tolerance (ISO 10551 2019). Data before and after harmonisation must be tested among subjective assessment and objective measurements for the sensitivity in thermal assessment. The non-parametric Spearman's rank correlation coefficient will be used to determine the correlations among subjective assessment and objective measurements, which indicates the significant elements in thermal assessment with respect to micrometeorological conditions that the respondents were exposed to.

Linear regression analysis will be then conducted to investigate the relationship between both original (raw data) and binned values of meteorological variables against diverse thermal comfort indices including the UTCI. At that stage, linear models will be developed to examine how well subjective thermal assessment can be predicted by selected indices from observed meteorological measurements.

The models will be validated using two approaches. Firstly, studies from similar climatic regions or similar urban settings will be divided into training and validation datasets. This ensures the applicability of the models in relatively consistent climatic and environmental conditions. Secondly, the entire datasets will be randomly divided into training (80%) and validation (20%) datasets in order to evaluate the overall predictability of the linear models. The parameters identified will be included in the draft standard methodology which will be further tested by selected research teams in different climatic regions.

## 3.2 Testing of the Draft Standard Methodology

The key elements identified in data harmonisation will be included in the draft standard methodology which will consist of subjective thermal assessment and micrometeorological measurements. The testing of the draft standard methodology will be conducted in selected countries by corresponding research teams in order to test its feasibility and identify issues or difficulties found by the research teams.

The questionnaire surveys will include the assessment scales and elements determined by the linear models to form the subjective part of data collection. At the same time, the instrumental settings will also be provided for testing the draft standard methodology. The testing will be conducted in different seasons in order to examine the applicability of the draft methodology in both extreme and transitional conditions. The target sample size is 100 responses in each of the four seasons (or relatively different climatic conditions) in order to maintain sufficient samples to compare between different studies and refine the methodology if necessary.

## 3.3 Establishment of the Online Database

Based on the findings obtained in previous stages, the key elements of human thermal comfort in outdoor environments will be identified and used for establishing the online database. Browser-based applications will be used for readily available and easy-to-use visualisation and user interface. Open-source JavaScript libraries will be used to visualise the data based on the data analysis previously conducted. The primary focus of the database is to provide useful information about the conditions that are perceived as comfortable so that the users such as urban planners and designers can take into account these conditions in their practices.

Four types of information will be included in the database. First, subjective assessment of thermal comfort will be provided to indicate how people perceive their thermal comfort under specific conditions. Second, the corresponding meteorological conditions and derived comfort index results will be provided in order to allow users to understand what conditions are required to achieve thermal comfort. Drawing such relationships will be facilitated not only for a given meteorological variable, which can be analysed individually, but for post-processed comfort indices such as the UTCI, which integrate diverse variables. Third, age and sex of the respondents will be included for any specific use or design of outdoor spaces. Finally, the urban settings where the data were collected will be specified.

During the process of development, a website with online forum will be established to provide an online platform for communication between researchers and contributors. The questions or issues encountered during the process will be shared and data contributors can answer or raise any questions of their own concern. The online platform can also engage potential users during the development process in order to maximise the applicability of the online database. A communication platform will be provided in order to facilitate users' feedback.

## 4 Way Forward

The primary objective of developing this global database for outdoor thermal comfort survey is to provide the empirical basis for establishing outdoor thermal comfort models by understanding the influential elements of human thermal comfort in the outdoor environment. However, the content of the database has a large potential beyond this due to the large amount of high-quality field data that can be used to explore the issues regarding human thermal comfort in the outdoor environment. The followings are some examples of potential applications of this global database.

The database provides numerous possibilities for developing empirical relationships between different assessment scales of subjective thermal perception. Human thermal comfort research has been using a wide range of subjective assessment scales, for example, the seven-point ASHRAE thermal sensation scale, thermal acceptability and preference assessment. The database therefore provides a platform for evaluating the assumptions behind different assessments and their applicability in outdoor settings. By adopting the UTCI as one of the principal thermal indices, regional comparisons may contribute to a better understanding of the contextual effects on subjective thermal perception and the corresponding thermal conditions exposed by individuals in the outdoor environment. It also accommodates the need for understanding the effect of individual characteristics such as biometric parameters and human behaviours on thermal perception.

The contextual effects were studied in some previous work but there have been no comprehensive understandings of how these effects influence subjective thermal perception in different climates. Therefore, there are opportunities for researchers to investigate the characteristics of the outdoor environment and their relationship with human thermal comfort of pedestrians and users of outdoor spaces. Urban planning and design professionals can be informed thereby improving the design of outdoor spaces in order to encourage their usage, which in turn has implications on human health and well-being, as well as energy consumption of buildings.

Since the data provided by researchers have been previously published in peer-reviewed academic journals and undergone the process of quality, they are reliable and ready to use for scientific and design work. The database also allows professional practitioners to extract relevant information for their design. For example, design professionals can acquire the understanding of thermal comfort requirements for specific urban contexts and climatic regions without conducting the field work themselves.

The long-term goal of the database is to establish a standard methodology for conducting outdoor thermal comfort research. The draft version of the standard methodology provided in the later stages of the development of the database allows robust testing of the methodology. It also facilitates comparison of results between different climatic regions and urban settings in order to enhance the understanding of outdoor thermal comfort. This potentially contributes to the discussion of the difference between indoor and outdoor studies, which has been widely discussed in the last two decades.

**Acknowledgements** This work was supported by the Vice-Chancellor's Discretionary Fund of the Chinese University of Hong Kong, Hong Kong.

## References

Aljawabra F, Nikolopoulou M (2018) Thermal comfort in urban spaces: a cross-cultural study in the hot arid climate. Int J Biometeorol 62:1901–1909

ASHRAE (2017) ASHRAE Standard 55—thermal environmental conditions for human occupancy. American Society of Heating, Refrigerating and Air-Conditioning Engineers, the United States.

Błażejczyk K (2011) BioKlima—Universal tool for bioclimatic and thermophysiological studies. https://www.igipz.pan.pl/Bioklima-zgik.html

Błażejczyk K, Broede P, Fiala D, Havenith G, Holmér I, Jendritzky G, Kunert A (2010) Principles of the new Universal Thermal Climate Index (UTCI) and its application to bioclimatic research in European scale. Miscellanea Geogr 14(1):91–102

Błażejczyk K, Epstein Y, Jendritzky G, Staiger H, Tinz B (2012) Comparison of UTCI to selected thermal indices. Int J Biometeorol 56(3):515–535

Bröde P, Fiala D, Błażejczyk K, Holmér I, Jendritzky G, Kampmann B, Havenith G (2012) Deriving the operational procedure for the Universal Thermal Climate Index (UTCI). Int J Biometeorol 56(3):481–494

Cheng V, Ng E, Chan C, Givoni B (2012) Outdoor thermal comfort study in a sub-tropical climate: a longitudinal study based in Hong Kong. Int J Biometeorol 56(1):43–56

de Dear RJ (1998) A global database of thermal comfort field experiments. ASHRAE Trans 104:1141

Fanger PO (1970) Thermal comfort. McGraw Hill, New York

Givoni B (1976) Man, climate and architecture, 2nd edn. Applied Science Publishers, London

Givoni B (1998) Climate considerations in building and urban design. John Wiley & Sons, Canada

Havenith G, Fiala D, Błazejczyk K, Richards M, Bröde P, Holmér I, Jendritzky G (2012) The UTCI-clothing model. Int J Biometeorol 56(3):461–470

Höppe P (1984) Die Energiebilanz des Menschen (The Energy Balance in Human). Wissenschaft Mitteilung Meteorological Institute University Munchen, Munich, Germany

Höppe P (1989) Application of a dynamical energy balance model for the prediction of thermal sensation and comfort. In: Proceedings of the 11th ISB-congress. West Lafayette, USA, pp 267–272

Höppe P (1997) Aspects of human biometerology in past, present and future. Int J Biometeorol 40(1):19–23

Höppe P (1999) The physiological equivalent temperature—a universal index for the biometeorological assessment of the thermal environment. Int J Biometeorol 43(2):71–75

Höppe P (2002) Different aspects of assessing indoor and outdoor thermal comfort. Energy Build 34(6):661–665

ISO 7726 (1998) Ergonomics of the thermal environment—instruments for measuring physical quantities. International Standard Organisation, Geneva

ISO 10551 (2019) Ergonomics of the physical environment—subjective judgement scales for assessing physical environments. International Standard Organisation, Geneva

Johansson E, Thorsson S, Emmanuel R, Krüger E (2014) Instruments and methods in outdoor thermal comfort studies—the need for standardization. Urban Climate 10:346–366

Kántor N, Égerházi L, Unger J (2012) Subjective estimation of thermal environment in recreational urban spaces—part 1: investigations in Szeged. Hungary. Int J Biometeorol 56(6):1075–1088

Knez I, Thorsson S (2006) Influences of culture and environmental attitude on thermal, emotional and perceptual evaluations of a public square. Int J Biometeorol 50(5):258–268

Krüger E (2017) Impact of site-specific morphology on outdoor thermal perception: a case-study in a subtropical location. Urban Clim 21:123–135

Krüger EL, Rossi FA (2011) Effect of personal and microclimatic variables on observed thermal sensation from a field study in southern Brazil. Build Environ 46(3):690–697

Krüger EL, Drach P (2017) Identifying potential effects from anthropometric variables on outdoor thermal comfort. Build Environ 117:230–237

Krüger EL, Tamura CA, Bröde P, Schweiker M, Wagner A (2017) Short-and long-term acclimatization in outdoor spaces: exposure time, seasonal and heatwave adaptation effects. Build Environ 116:17–29

Lau KKL, Shi Y, Ng EYY (2019a) Dynamic response of pedestrian thermal comfort under outdoor transient conditions. Int J Biometeorol 63(7):979–989

Lau KKL, Chung SC, Ren C (2019b) Outdoor thermal comfort in different urban settings of subtropical high-density cities: an approach of adopting local climate zone (LCZ) classification. Build Environ 154:227–238

Lin TP (2009) Thermal perception, adaptation and attendance in a public square in hot and humid regions. Build Environ 44(10):2017–2026

Mayer H, Höppe P (1987) Thermal comfort of man in different urban environments. Theoret Appl Climatol 38(1):43–49

Metje N, Sterling M, Baker CJ (2008) Pedestrian comfort using clothing values and body temperatures. J Wind Eng Ind Aerodyn 96(4):412–435

Ng E, Cheng V (2012) Urban human thermal comfort in hot and humid Hong Kong. Energy and Buildings 55:51–65

Nicol F, Raja I (1996) Thermal comfort, time and posture: exploratory studies in the nature of adaptive thermal comfort. Oxford Brookes University, School of Architecture

Nikolopoulou M, Baker N, Steemers K (2001) Thermal comfort in outdoor urban spaces: understanding the human parameter. Sol Energy 70(3):227–235

Nikolopoulou M, Lykoudis S (2006) Thermal comfort in outdoor urban spaces: analysis across different European countries. Build Environ 41(11):1455–1470

Oh W, Ooka R, Nakano J, Kikumoto H, Ogawa O (2019) Environmental index for evaluating thermal sensations in a mist spraying environment. Build Environ 161, 106219

Oke TR (2006) Initial guidance to obtain representative meteorological observations at urban sites. Instrum Obs Methods Report no. 81. WMO/TD-No. 1250

Olgyay V (1963) Design with climate: bioclimatic approach to architectural regionalism. Princeton University Press, the United States

Oliveira S, Andrade H (2007) An initial assessment of the bioclimatic comfort in an outdoor public space in Lisbon. Int J Biometeorol 52(1):69–84

Stathopoulos T, Wu H, Zacharias J (2004) Outdoor human comfort in an urban climate. Build Environ 39(3):297–305

Thorsson S, Lindberg F, Eliasson I, Holmer B (2007) Different methods for estimating the mean radiant temperature in an outdoor urban setting. Int J Climatol A J Roy Meteorol Soc 27(14):1983–1993

Villadiego K, Velay-Dabat MA (2014) Outdoor thermal comfort in a hot and humid climate of Colombia: a field study in Barranquilla. Build Environ 75:142–152

# Afterword

The Biometeorology Series initiated by Prof. Glenn McGregor in 2009 had as one of its aims "to demonstrate how a biometeorological approach can provide insights to the understanding and possible solution of cross-cutting environmental issues". This book attempted to fulfill this goal with a modest contribution to these issues, presenting an overview of current applications of the UTCI, ranging from outdoor thermal comfort (OTC) studies to mapping techniques at different levels.

Important initiatives are shown here, aiming at a wider range of UTCI application such as Claudia Di Napoli's work with UTCI-based forecasting systems involving a multi-team effort from diverse research institutions, and Kevin Lau's ambitious purpose of setting up a worldwide repository of OTC data.

Parting from an introductory part describing the background behind the development of this book and of the UTCI (Chap. 1), we present the current ways to deploy the index in Chap. 2 and devote the literature review (Chap. 3) to applications of the UTCI. That chapter showed that papers on the UTCI cover different topics, but with a prevalence of OTC research and studies related to microclimate and urban planning. Applications in other essential areas such as meteorology, climate change and tourism or health are still less explored with the UTCI. These areas will potentially gain in importance in the years to come with the progress of global warming and weather extremes and in the aftermath of the pandemic. In this book, coincidentally, some chapters have focused on OTC and others dealt with bioclimatic spatiotemporal analyses at different levels.

In Chap. 4, the accuracy of UTCI predictions of outdoor thermal comfort was evaluated from gathered thermal perception data in two Brazilian urban areas. Overall, a good accuracy was found for the UTCI and the evaluated personal and urban morphology aspects hardly affected UTCI's predictive power.

Chapter 5 used data from a longitudinal study and from a cross-sectional study to analyze long- and short-term acclimatization effects on reported thermal comfort perception using dissimilar approaches. In both studies, the UTCI and the associated Dynamic Thermal Sensation (DTS) index were relevant parameters for the performed



E. L. Krüger (ed.), *Applications of the Universal Thermal Climate Index UTCI in Biometeorology*, Biometeorology 4,
https://doi.org/10.1007/978-3-030-76716-7

analyses. In the longitudinal study, long-term or seasonal acclimatization was found to have a relevant influence on subjective thermal assessment, whereas in the short-term negligible and statistically irrelevant relationships have been identified. In the cross-sectional study, part of the sample (Melbourne) did not reveal clear patterns of relationship regarding long-term acclimatization, but effects of short-term thermal history were observed. Moreover, field data from mainland China pointed to observable impacts of long-term acclimatization and short-term thermal history on subjective thermal perception. In the longitudinal study, remarkable relationships between precedent visual comfort and reported thermal sensation were additionally observed, reinforcing the need for further investigation on such interactions. Future OTC studies could further benefit from the consideration of such multisensory interactions.

From the various field datasets presented and analyzed in Chap. 6, it is suggested that obtained neutral temperatures, expressed as UTCI units, were well aligned with the mean UTCI value for the monitoring campaigns and that thermal sensitivity was inversely correlated with the range of conditions surveyed. A regional adaptation effect was verified, which closely corresponds to measured climatic data during the surveys, despite not necessarily resembling prevailing or most commonly found climatic conditions at those locations. Such results stress the importance of the repository of OTC data presented in Chap. 11. The primary objective of developing such worldwide database is to provide a platform for establishing and testing outdoor thermal comfort models such as the UTCI, including regional adaptation studies and also allowing further analyses on the effect of individual characteristics such as biometric and site-related parameters, such as discussed in Chap. 4.

Chapter 7 looked at health implications in summer pointing to the potential of the UTCI to become a credible risk indicator for heat-related health effects towards its usage in guidelines for the health sector. Using field datasets, it suggested that the manifestation of mild heat-related symptoms in pedestrians in Athens, Greece, was associated with UTCI and that the category of moderate heat stress could be considered a threshold for experiencing these symptoms . The study presented in that chapter is part of a larger research initiative, the UBiPlan project (Urban BIometeorology and PLANning), aimed at improving design guidelines towards optimized thermal conditions in outdoor urban spaces with reduced health risks.

Chapter 8 presented an overview of mapping approaches of the UTCI developed for different spatial scales, from global to local. Both, global and regional mapping of the UTCI provide a great opportunity to analyze spatial patterns and peculiarities of thermal stress variability. Parting from this overview, the subsequent chapters presented maps at different scales, modeling at city-scale and at a regional scale in the pan-European realm.

In Chap. 9, we presented a feasible fine-scale modeling approach for spatiotemporal analyses of the UTCI in urban planning. The modeling domain was built for the city of Prague, Czech Republic. The study was focused on the analysis of the spatiotemporal variability of the UTCI. Results representation in the form of maps with a spatial UTCI distribution represents an advantage for the necessary dialogue between urban climate experts and municipality stakeholders, making existing relationships between morphology and site-specific attributes comprehensible to urban planners and to the layman in modeling research. Moreover, the level of detail of the model allowed us to test adaptation and mitigation strategies, such as urban greenery, type of vegetation and surface albedo.

Chapter 10 presented a relevant and useful way of representing UTCI spatiotemporal distribution over different parts of Europe as an aid to precautionary planning for thermal health hazards. The forecasting tool presented in the chapter for predictions of human comfort/discomfort conditions in terms of UTCI data can be used by a wide range of end-users, from the general public to civil protection agencies and public health authorities.

## Concluding Remarks

Could this book entail other areas of application of the UTCI? Most certainly! The literature review showed us a wide spectrum of areas of application covered by the index, which could have also been part of this account. Yet, this would surpass the desired length of the book and I am convinced that the most prominent areas of application up to now have been duly addressed. Further developments are, however, expected in the near future. Newly, the UTCI is going to be considered as a feasible index for evaluating human biometeorology in one of the guidelines of the Association of German Engineers (Verein Deutscher Ingenieure 'VDI'). VDI 3787 (Blatt 2) is aimed at providing "evaluation methods of human biometeorology as a standard for taking into account climate and air quality (bioclimate) in relation to man in overall physical planning". The UTCI is included as the major index (with mentions of other indices like PET, PT and SET* as well) in the draft document for a revised version of that guideline, which was scheduled for publication in 2020. Unfortunately, the procedure is delayed up to now due to the COVID-19 pandemic, which prohibits regular committee meetings. With the inclusion of the UTCI, it might be expected that applications in planning will increase.

Currently, the City of London applies the UTCI in their recently published Thermal Comfort Guidelines (https://www.cityoflondon.gov.uk/assets/Services-Environment/thermal-comfort-guidelines-for-developments-in-the-city-of-london.pdf). The proposed planning guidelines, part of the British planning system, take into consideration microclimatic requirements in order to "assess the impact of new developments on the microclimate of the City's streets, parks, public roof gardens and terraces and other public spaces". Their guidelines place OTC as the most important factor in the quality of a public space and apply the UTCI as the metric to be used for assessing OTC conditions in the City.

Other areas of application will naturally arise with the need for a permanent monitoring and reliable predictions of human thermal stress, mostly heat stress, resulting from global warming. In short, there is a bright future in sight for the UTCI that greatly justifies the efforts made during its elaboration.

Editor
Eduardo Krüger